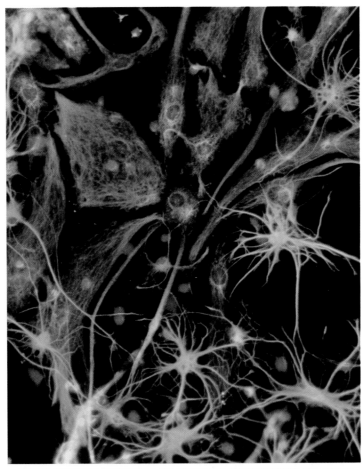

STARLIKE ASTROCYTES AND OTHER SO-CALLED GLIAL CELLS SERVE AS SCAFFOLDING FOR THE
BILLIONS OF NEURONS THAT MAKE POSSIBLE MEMORY AND THE HUMAN MIND.

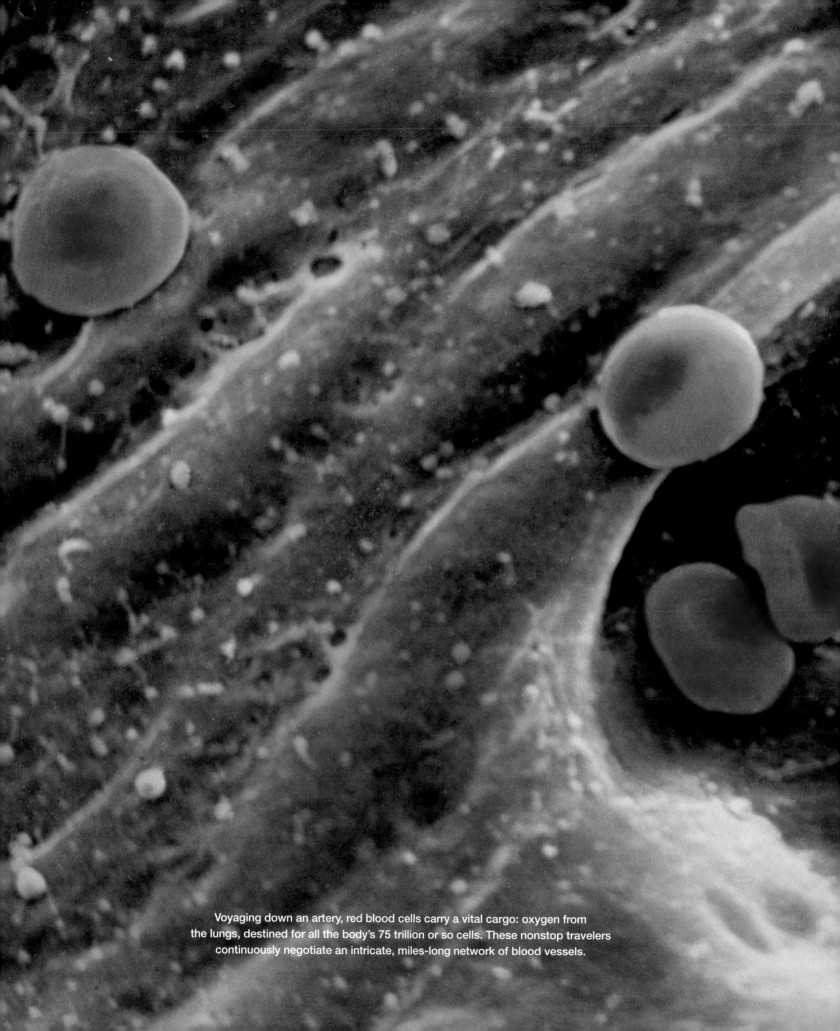

Voyaging down an artery, red blood cells carry a vital cargo: oxygen from the lungs, destined for all the body's 75 trillion or so cells. These nonstop travelers continuously negotiate an intricate, miles-long network of blood vessels.

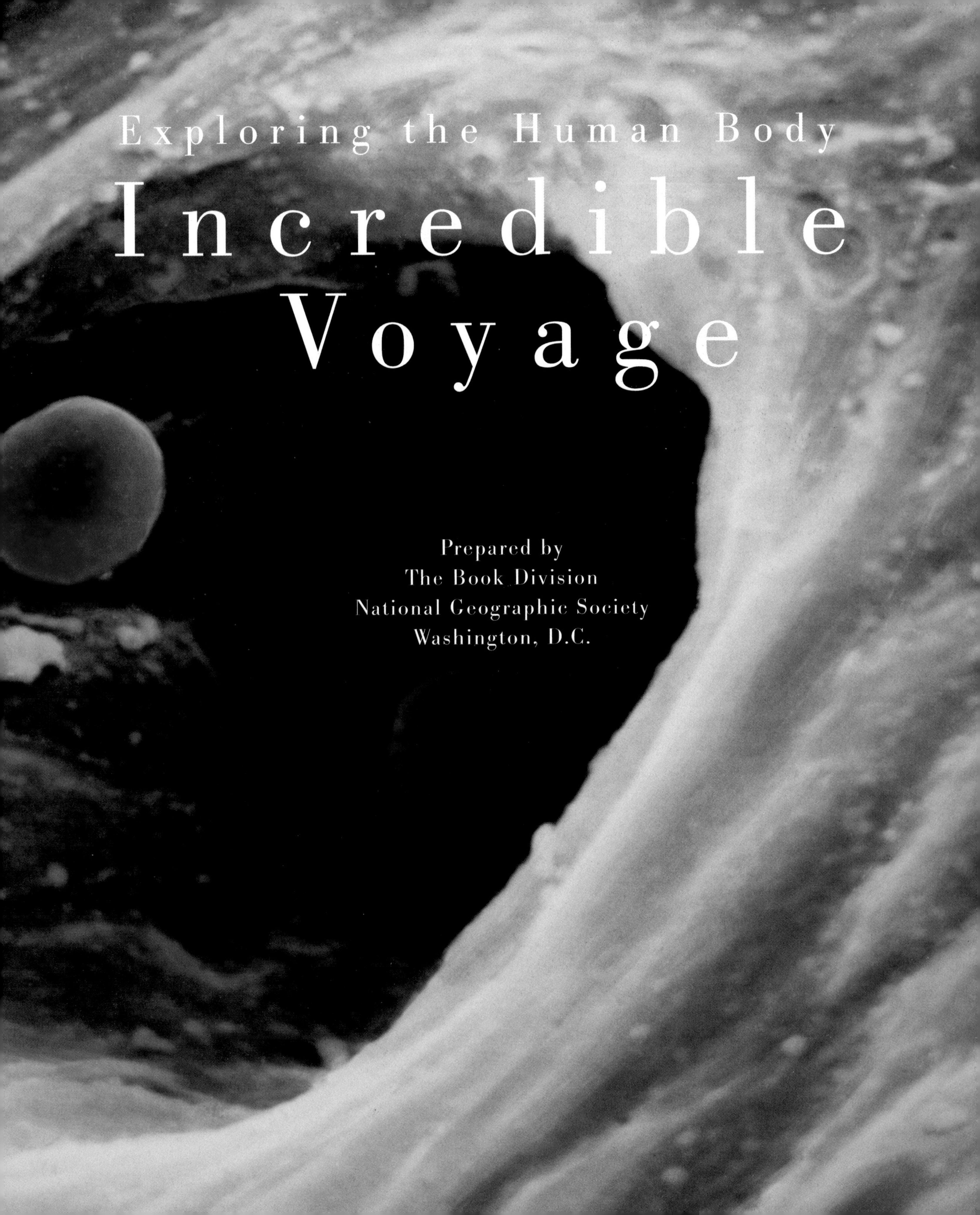

Exploring the Human Body

Incredible
Voyage

Prepared by
The Book Division
National Geographic Society
Washington, D.C.

CONTENTS

Symphony of strength, balance, exquisite grace, and more, the human body consists of
myriad cells and different systems that—usually—work in concert.

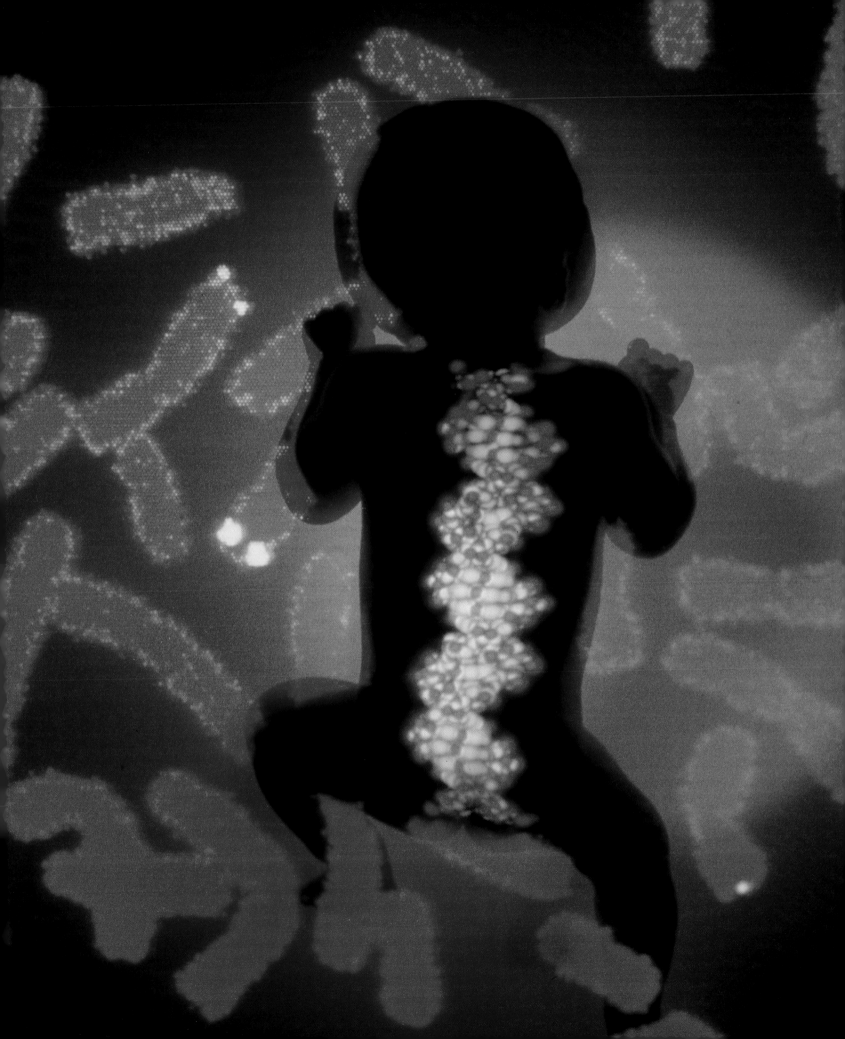

Foreword

SHERWIN B. NULAND

To summarize—in a single word—the difference between modern scientific medicine and every form of traditional, folk, or alternative treatment that either has preceded it or is practiced today is an easy task. The word is "seeing."

All schemes but the scientific consider each living thing as a whole, defined more by its relation to the environment and the cosmos than to minute physical and chemical activities going on within its structure: In such systems an organ is more a concept than a distinctly understood object. The biomedical search for understanding the human body, however, is more dependent on discovering what our inner structure actually looks like. Current investigations go far beyond our organs and tissues.

Since the discovery in the mid-19th century that the basic unit of all life is the cell, scientists have been attempting, with ever increasing success, to find ways to see inside these tiny building blocks. In recent decades they have been able not only to identify the internal components themselves, but also to scrutinize molecules. The most striking example of this identification is the biological revolution that began in 1953 with the elucidation of the structure of DNA. To think of that molecule now is to see the double helix in the mind's eye.

In his introduction to the National Geographic Society's magnificent 1986 publication *The Incredible Machine*, Lewis Thomas proclaimed that human beings "are indeed a splendid invention." He traced the derivation of the title's "Machine" back to its Indo-European root and then forward again to its cognate, "magic." "Incredible magic," he wrote. "That's what we are."

In introducing this paean to the magic within us, I find myself, like Dr. Thomas, drifting toward words that illuminate not only the title, but also every page. *Incredible Voyage* is a great journey with writers and photographers who approach their assignments with the wonder shared by all of us who have devoted our careers to the study and contemplation of the extraordinary edifice of cells that is the human body. The skills of the contributors enable us to see—in the great tradition of Western biomedical science. Displayed before us are our very selves. We can look at our own smallest parts and share what scientists have seen as they have discovered one after another of nature's best-hidden secrets.

In tribute to our need to see, the text of this book calls forth images that add entire dimensions to the illustrations that accompany it. This voyage through our deepest and smallest parts is high adventure, and nothing less than expert guides take us through it. We are in good hands.

The images are not only of the here and now. As this book wends its fascinating way through the past in order to bring us to the marvels we know today, it takes a long look forward to what we may reasonably expect the future to hold. And here we come to a word that shares a syllable with "voyage," the French *voyant*, meaning a visionary or seer. In its progressive nature the Western approach to medicine is alone among the systems of earlier or present days. All other medicines present an overarching formulation meant to explain all of the microcosm and macrocosm. Such a scheme is changeless and without the inherent capability of

The machinery of heredity: Spine-like model of DNA adorns a baby on a backdrop of highly magnified human chromosomes, which contain DNA. The infinite variety of these tightly wound, helical molecules enables life's diversity.

improving, as new information is discovered or new interpretations are made. The dynamism and excitement of progress evident on every page of *Incredible Voyage* match the dynamism and excitement within our own bodies and within the biomedicine that studies them.

Dynamism, in fact, is a notion that is inescapable when considering just what it is that makes up life. Leo Tolstoy was not trained in science, but he understood the essential nature of humankind as few scientists have. He seems to have intuited something about biology as well. "Our body is a machine for living," he wrote. "It is organized for that, it is its nature."

We are composed of some 75 trillion cells, divided into approximately 200 separate types. Every type performs a particular job, be it absorption of nutrients, secretion of hormones, detoxification of poisons, or any of the other many kinds of work that must be done to maintain the life of the entire organism. The very word "organism," in fact, derives from the fact that the constituent parts of multicelled living things are organized for a common effort—the maintenance of life and health. Specifically, the parts are organized into organs.

And so, the human organism is an organization of organs. The three words are so alike that they will correctly be thought to derive from a similar linguistic source. They do, in fact, share the single Indo-European root *uerg*. And herein lies the revelation of that dynamism of our cells and our selves. For uerg means "work," which is, of course, the secret behind the entire enterprise. From our smallest cell to the totality of our beings, we are organized to do the work of living. It is not our structure per se that makes us what we are, but the way in which every part of that structure functions as a uniquely contributing part of the whole. Each of those 75 trillion cells finds the meaning of its individual existence through participation in the miraculous project that is a person. A human being is, indeed, a machine for living.

When Andreas Vesalius wrote the first textbook of anatomy based wholly on the dissection of human corpses—the first accurate depiction because it was based on the actual seeing of what is within—he chose to call it *De Humani Corporis Fabrica*, using "fabrica" as it was used in 1543, the year of publication. Then the word signified "workings," much as does the French word for "factory," which is *fabrique*. The proper translation for Vesalius's book is thus *On the Workings of the Human Body*, which signifies his appreciation that what really counts is not so much the structure but the way it functions—its dynamism.

It was not just for its functional perspective that the great volume of Vesalius came to be considered the most important single book in the 2,400-year-old history of Western medicine. The factor that made it truly unique for its time was the way in which its profusion of strikingly beautiful illustrations not only clarified the text but also explained structure and workings at a single glance. So faithful were the woodcuts to Vesalius's dissections that the images stood alone as instructional guides. They also shone as pieces of art. The *Fabrica*, as the book is called by historians, was without doubt medicine's first example of the maxim that one picture is worth a thousand words.

The talent displayed in Vesalius's dissections is matched by the talent of the illustrator, and no wonder. The pictures were made by one of Titian's ablest pupils, Jan Stephan van Calcar. The only previous drawings that had in any way approached them for accuracy and functional visualization were those scattered in the works of Leonardo da Vinci, another artist who sought to depict anatomy in action.

Incredible Voyage follows in the grand tradition of books in which illustrations are not only singular in their instructiveness, but also strikingly beautiful works of art. Combined with the lively and informative text, these images do justice to their theme, the wondrous machinery of the human body and our fascination with seeing it.

To search out the nature of that machinery has been the life's work of hundreds of thousands of investigators over a period of almost two and a half millennia. Until barely two centuries ago, researchers of the body had only their five senses to guide them. The era of useful instrumentation did not begin until 1816, when a tubercular Frenchman named René Laënnec invented the stethoscope. Shortly afterward, in 1832, the hazy images seen through old microscope lenses were improved, and attention could be turned to the efficient study of the minute structure of tissues. Entire new vistas of visualization opened up with Wilhelm Konrad Röntgen's discovery of x-rays in 1895. Within a few years, researchers were experimenting with liquids, which, when swallowed, allowed a physician to follow the passage of material through the entire length of the digestive tract. With the addition of fluoroscopy, scientists could see through the living body; they found themselves gazing at the undulations of the gut and gaping in astonishment at the powerful thrusts of a throbbing heart.

Meantime, progress was being made toward the goal of penetrating the body's orifices by means of tubes fitted with lenses and sources of light. By the end of the 19th century, a wide variety of such 'scopes existed—their names fitted out with prefixes like cysto, broncho, gastro, and sigmoido—by which the urinary, respiratory, and digestive tracts might be inspected. It soon became an easy matter to visualize the lining of these structures, and actually to snip out samples of them, using new biopsy forceps. With continuing improvements in photographic techniques, it became possible to take pictures through the tubular devices that were each decade probing more effectively into organs and cavities. In recent years the invention of fiber optics has provided scoping with a flexibility that brings remote places closer and makes accessible a great deal that was previously beyond reach.

With gradual acceptance of the germ theory following on the work of chemist Louis Pasteur and surgeon Joseph Lister, the field of surgery burgeoned, and each body cavity in time became fair ground into which an expeditionary force might be sent. A formal profession of medical illustrators came into being shortly after the turn of the 20th century. Master craftsmen accompanied doctors into their operating suites and autopsy rooms, emerging with accurate and often artistic representations not only of the tissues and organs themselves but also of their relationships. Still photography and then cinematic techniques increased the usefulness of film. The opportunity to follow the movement and flow of tissues and fluids added immeasurably to the strictly instructional value of the new methods—and to their usefulness as research tools.

By the middle of the current century, the notion of taking a voyage through the body—and recording it for others to see—had become a reality. The techniques of photography were by then being applied to the microscopic examination of tissues. With the aid of sophisticated endoscopes, improved computerized scanning equipment, and invasive devices of various sorts, life came into view as it is actually being lived within the deepest recesses of the body. The camera was revealing the terra incognita that, a century earlier, Louis Pasteur had called "the world of the infinitely small." *The Incredible Machine* and now *Incredible Voyage* bring the spectacular sightings made by the most advanced of biomedical researchers to all of us.

Humankind, in all its diversities and similarities, is revealed in this book. Each of us is unique, and yet each of us is genetically so similar to all others of our species that the differences would seem at first minuscule. We are closer than we think, too, to other mammals. Even the distinction from our nearest primate relative, the chimpanzee, exists in only one percent of the sequences of nucleotides that make up our DNA. We have evolved by small steps over the course of the 3.5 billion years since the first living things appeared on our planet. There is good reason to believe Charles Darwin's statement: "And as natural selection works solely by and for the good of each being, all corporeal and mental endowments will tend to progress towards perfection." That perfection has not yet been reached. Far from it. But still, as Lewis Thomas tells us, "From our point of view, the human being is the highest achievement of the natural world, the best thing on the face of the Earth."

Earlier I recalled Dr. Thomas's derivation of the word "machine," and we might usefully now turn again to the term "voyage." Back we go to our Indo-European predecessors, who used the root *uegh* to mean "to and fro." That ancient source inspired a host of words over the millennia, among them "voyage" and "vehicle" and even "weighty" and "wee." It is precisely the weighty and the wee that are the substance of this volume. *Incredible Voyage* is weighty in size as well as in subject matter. It is about whole people—about you and me and the way we live our lives. The theme of the essential uniqueness of each individual runs through the narrative. We are found here in all sorts of activities—exciting and humdrum. And while the entire heft of our bodies is involved in the work, in the play, and in the love that bring value to what would otherwise be mere existence, each wee cell inside us is doing its part to make those things happen.

Together with people pictured in action, we see in this volume their deepest tissues at work. Even enthralling descriptive text and the miracles of modern imaging will never approach in magnitude the miracle of actual life, but such things are subjects of awe in themselves. All humanity—with its magic large and small—is in this book.

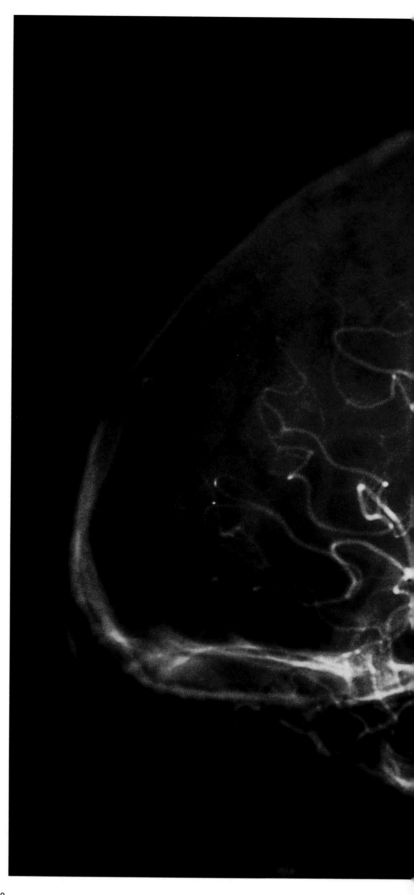

. . . .

Highways of the mind reveal themselves in a cerebral angiogram—a vital diagnostic x-ray imaging technique that helps detect blockages in a patient's circulatory system by injecting blood vessels with a radiopaque substance. This particular image shows actual flow patterns through the right carotid artery and its branches, which serve the brain.

. . . .

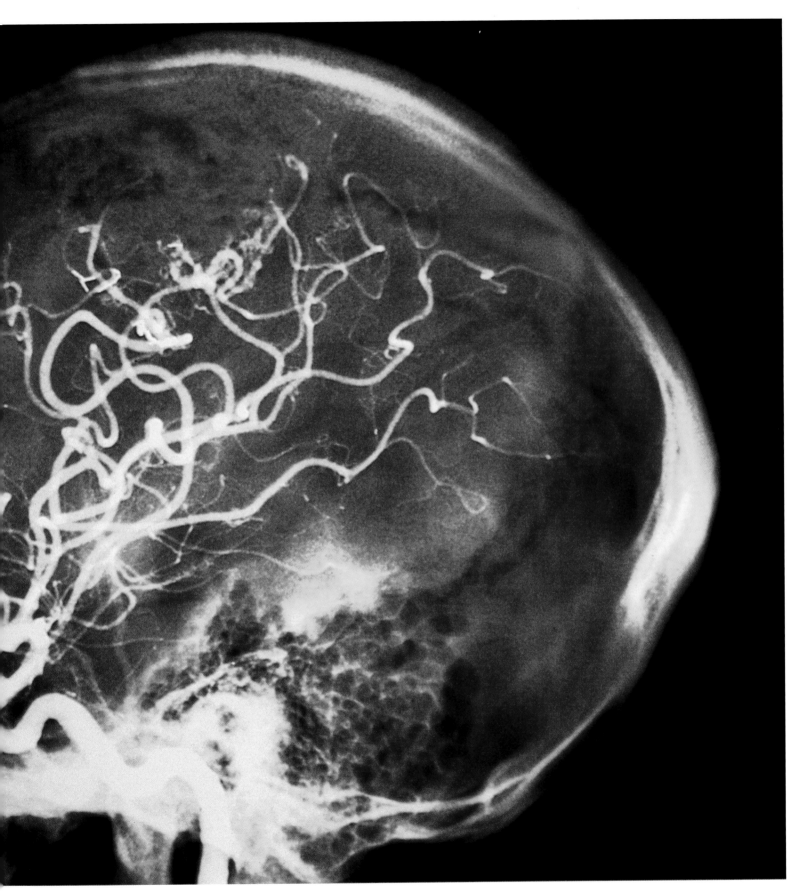

Wide-eyed and wonderful, a new human life culminates a remarkable nine-month-long developmental voyage: a single fertilized egg becomes an independent organism consisting of trillions of cells and an astonishing array of interactions.

Origins

BOYCE RENSBERGER

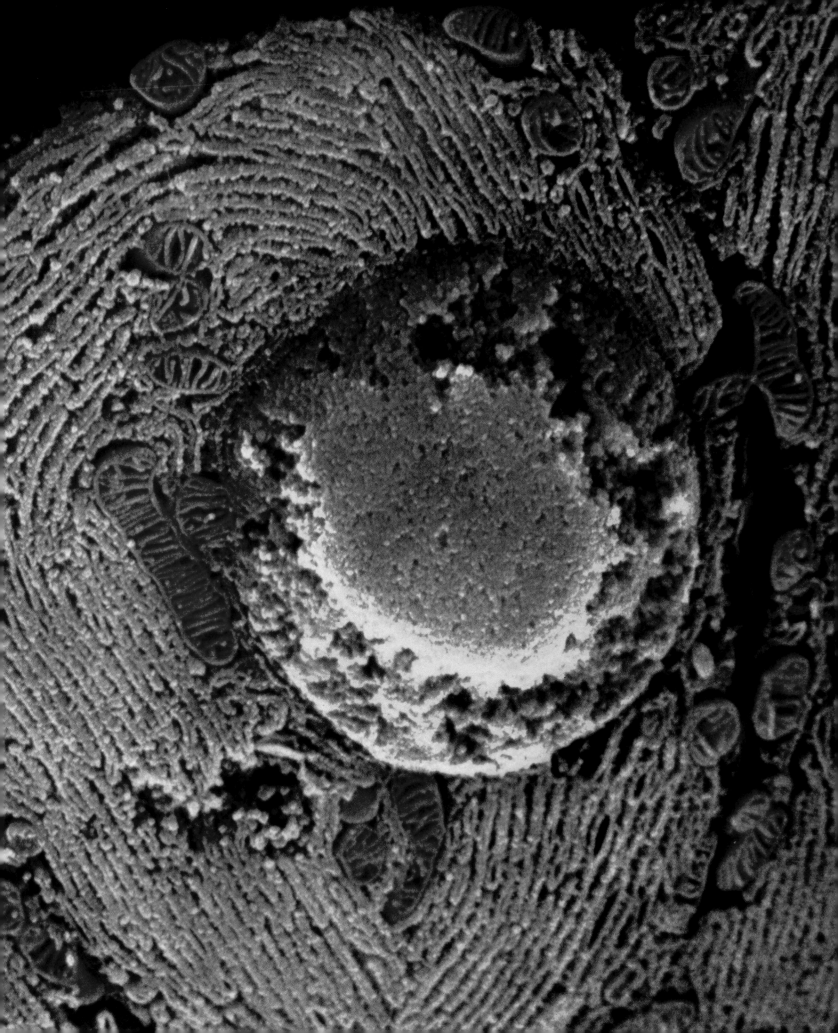

Your hands grasp this book. They are alive. Your eyes scan these words. They, too, possess life. As these sentences flow into your living brain, scores of other tissues and organs throughout your body quietly perform hundreds of internal housekeeping processes that keep you alive and well.

Be they large organs such as the kidney and the stomach, or small, specialized structures such as the glandular cells that produce insulin or the white blood cells that scan your body for microbial invaders, all these structures are alive. There is life in the layer of cells in your lungs that delivers inhaled oxygen to your bloodstream. Life is in the muscle cells that contract rhythmically in your heart to pump that blood. Even inside solid bone there exist units of life—isolated cells waiting in microscopic chambers in case a fracture signals them into bone-repairing action.

In other words, although most of us think of life as something possessed only by the body as a whole, each of the parts of our bodies is alive, right down to every one of the 75 trillion or so cells that make up an average-size human being. Any of those cells can be removed from the body and placed in a dish of nutrient broth, and it will go right on living—and growing. Many such cells even revert to the amoebalike behavior of their primordial ancestors and crawl around the bottom of their dish.

In the body, cells work together in marvelous coordination to form tissues which, in turn, cooperate to form organs. Each of our bodies, in other words, is an extraordinary society of cells founded at conception, grown through nine months of gestation plus another 15 or 20 years, and maintained for several more decades. Of course, things do go wrong. Diseases and accidents happen. And, eventually, the body ages and, ultimately, succumbs to death.

Life is cellular. All these events are the results of processes that go on inside cells and among cells. Scientists, as they probe the innermost workings of these microscopic units of life, are beginning to grasp how life works—and to understand one of the most profound mysteries of the universe.

Today's understanding of the innermost workings of the human body was made possible by the slow overthrow of an ancient belief that the phenomenon of life was fundamentally mysterious, even mystical, and certainly unknowable.

ITSELF: IT CAN BE UNDERSTOOD

The way the human body developed from conception and the way it could move under its own power was attributed to something that early scholars called the vital force. That force, according to some vitalist philosophers, was the breath of life that God gave Adam after molding him from clay. Many scholars taught that the physical body could not move under its own power and that the ability to do so—called animation—came from the soul. In other words, a supernatural force was responsible for the ability of living things to move.

As the centuries passed, biologists came to reject the teachings of vitalism. They gradually discovered that the phenomena once thought mystical were entirely natural; they could be explained by the known laws of chemistry and physics. In the early 1800s vitalism faded.

Its most prominent enemies then were two Germans, botanist Matthias Schleiden and zoologist Theodor Schwann. Independent of one another, they promoted the cell theory. Radical then, today it is recognized as one of the most fundamental concepts in biology. Schleiden and Schwann argued that the fundamental units of life were not whole organisms. The basic units of life, they maintained, were microscopic cells, each of which was a living entity. Schleiden even suggested that if the cell is the unit of life, then any aggregation of cells into a larger organism is actually a colony and not an individual.

The idea is not so far-fetched. Scientists have discovered several species of a peculiar organism called a slime mold. Slime molds live part of their life cycle as one-celled amoebas, crawling around moist areas of the forest floor. If food grows scarce or the environment dries out, thousands of amoebas congregate to form a

Slice of life: Seen in cross section and artificially stained, a typical human cell reveals its own tightly packed anatomy. The yellow nucleus holds massed tangles of DNA. Scarlet "sausages" are mitochondria, the cell's energy processors. The purplish endoplasmic reticulum, a protein-processing facility, reveals onionlike layering.

multicellular organism that looks rather like a slug. After a while this sluglike mass of slime mold changes into a form that looks and acts like a fungus. A long shaft grows upward from the slug and swells at the top, producing a bulb. The cells on the outside of the bulb dry, forming a hard shell around it. The shell eventually cracks open, and the cells inside disperse with the wind as dustlike spores. Spores that land on moist surfaces split open, and the cycle is complete.

Cell biologists today study the slime mold's remarkable transformation for clues to how the human embryo develops, its own cells changing form and function just as the slime mold's do. The fact that such lowly creatures can be laboratory stand-ins for humans illustrates one of the most profound discoveries of modern biology—cells of all species are fundamentally alike. In fact, a large proportion of the genes that govern human cells perform the same jobs for the cells of slime molds and all other organisms.

For all their insight, Schleiden and Schwann failed to grasp one of the key attributes of cells—where they come from. The two speculated that cells materialized in some mysterious fashion from body fluids. In the 1850s, Rudolf Virchow, another German, found the correct explanation. All cells arise from the splitting of older cells. With this crucial modification of cell theory, biologists could envision how organisms develop, one original cell dividing to produce two, two dividing to make four, and so on. Virchow also was the first to understand that most diseases are the result of processes gone awry within or among cells. Diseases, he proposed, have a cellular base, and life is a natural process that the sciences of physics and chemistry can explain. Virchow's views encouraged other researchers to study the cell and cell division. Through the late 1800s, scientists developed better microscopes and new ways of staining cells so that their internal components could be seen more easily and followed in detail as they divided.

By the turn of the century cell theory had become fully formed and offered an explanation of how a complex organism develops. The long-lost genetics experiments of the Austrian monk Gregor Mendel had been rediscovered, and biologists understood that the father's sperm and the mother's egg each carried a packet of hereditary factors. It was known that the fertilized egg was the founding cell of each new individual, that the combined genetic endowments were copied every time a cell divided, and that one complete copy was transmitted to each new cell.

STRIATED MUSCLE CELL, SHOWING PROTEIN FILAMENTS

HEMOGLOBIN-RICH RED BLOOD CELLS, WHICH TRANSPORT OXYGEN

SLICE OF SKIN, EACH BULL'S-EYE A HAIR FOLLICLE

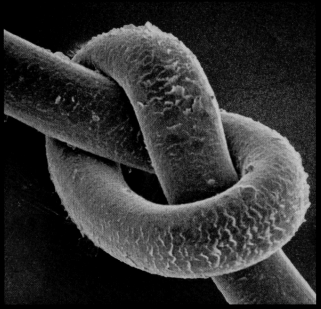

STRAND OF HUMAN HAIR, LAYERED CELLS PARTING AT BENDS

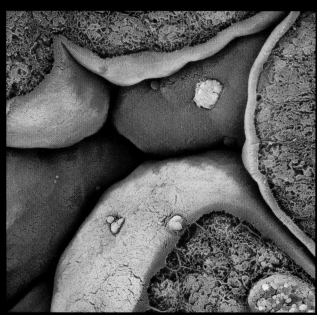

INTESTINAL CELLS, CARPETED WITH ABSORPTIVE FILAMENTS

Studies in form and function, the cells of the human body vary startlingly, from highly regular, bagel-shaped red blood cells to seemingly amorphous blobs. All, however, contain essentially the same organelles and use the same biochemical processes to carry out the basic functions of life. The body comprises some 200 cell types, in addition to substances those cells produce—such as bone—or that, like hair, are made of bound cells.

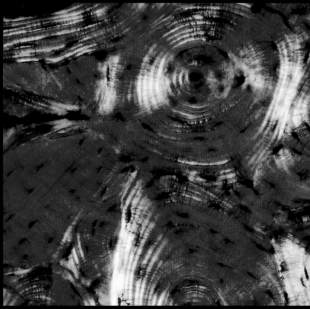

SECTION OF ADULT BONE, MADE BY CELLS AND PIERCED WITH CANALS

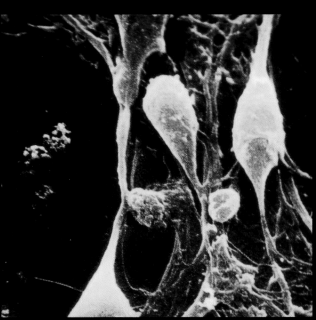

LINKED NERVE CELLS IN THE BRAIN

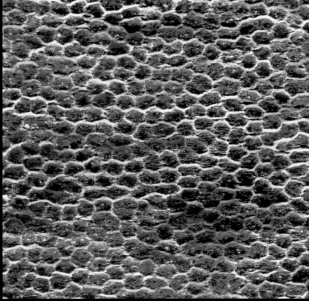

TRANSPARENT CELLS OF THE EYE'S CORNEA

In 1912 biology's transition from life-as-miracle to life-as-machine received a major boost from German-born biologist Jacques Loeb, who had emigrated to the United States. That year he published his landmark book entitled *The Mechanistic Conception of Life*, which described his experiments on sea urchin eggs. Loeb discovered that if he removed the eggs from a female urchin, he could make them start developing as embryos without the aid of sperm. A small dollop of certain chemicals would launch the development of an organism from a single cell.

To Loeb and most other biologists, the discovery confirmed the mechanistic view. Loeb, like others who had studied Mendel's experiments with sweet peas, realized that chemical substances in the chromosomes are responsible for heredity.

Loeb's "mechanistic view of life" is today the dominant view of the two disciplines on the cutting edge of biomedical research—molecular biology and cell biology.

....
LIVING MOTION:
LIFE ON THE GO
....

The now discredited vitalist philosophers were right about one thing. The ability of living organisms to move is extraordinary, a phenomenon we often take for granted, but one so remarkable that it demands attention. Other than the things that possess life—be they large animals or single cells—nothing else in the known universe can pick itself up and travel in a purposeful way. Rocks roll down hills, pulled by gravity. Clouds are blown by wind. Winds are driven by the rising of sun-heated air. All these forms of motion are powered by outside forces; none is driven by internal mechanisms.

Life is radically different. The ability of a human—in fact any animal—to move comes from something inside us. More specifically, it comes from tiny motors and machines inside each cell. These are not mere metaphors. There really are microscopic motors and machines inside every one of our cells, and they are absolutely essential not only to the internal lives of each cell but also to the whole body.

The machinery is most abundant in muscle cells. Look at the inside of your wrist. Now make a fist and then release your hand. You should see the skin of your wrist change contour, as if objects were moving about underneath it. With your other hand, feel the flesh close to the elbow while clenching and unclenching your fist. The motions you feel are the machines that operate your fingers. The muscles that work your fingers are in your arm, linked by long cables—called tendons—to the levers that we call fingers. Each muscle is like a winch hauling in a cable, and the tendon is the cable connecting the pulling force to the load.

The work of pulling on tendons is done by muscle cells, which are very thin and very long. Whether they be the muscles of your arm, the ones that focus your eyes, the ones that pump your blood, or any of the thousands of other muscles throughout the body, all work in the same basic way. Inside each muscle cell, or fiber, is a bundle of countless parallel filaments of two different kinds, called actin and myosin. These two are mixed together, each myosin filament surrounded by six actin filaments. In cross section, the arrangement is as precise as the hexagonal pattern of a honeycomb.

Think of the actins as so many parallel ropes. Each myosin filament has many knobs that protrude in all directions; think of them as hands. When a muscle works, each myosin "hand" reaches out, grabs one of the actin "ropes," and pulls. Randomly, the hands let go, reach ahead, grab, and pull again. While some hands keep their grip, others let go and reach forward. Each myosin "hand" is one microscopic myosin molecule, but as countless millions cooperate, they deliver enormous force. The two

....
Containing thousands of genes, a single chromosome reveals a long chain of DNA coiled and supercoiled
like abused phone cord (above). Stretched out and magnified some two million times, DNA's helical molecule
becomes discernible (opposite).
....

kinds of filaments slide past one another like interlaced fingers folding, pulling the ends of the muscle together.

Though too small to be seen with a conventional microscope, a myosin "hand" is a marvel of biological engineering. For one thing, it is nothing like the popular notion of molecules as rigid ball-and-stick models. Myosin is a protein, one of the most common types of substances in cells, and like all proteins, it is made of many smaller subunits. In addition to having a lumpy outer surface, many proteins are flexible because the subunits are not rigidly linked. Myosin is one of the most flexible proteins. It is a hand with a wrist and with "fingers" that can grasp actin. And, just as any motorized machine of metal does, myosin consumes a unit of chemical energy with each cycle of reaching, grabbing, and pulling. That energy comes from calories in the food we eat.

Marvelous though myosin may seem, cell biologists found in the 1980s that it is not so special. Since then, dozens of similar "motor molecules" have been found. All proteins, they are able to grasp objects, flex, and even carry them along tracks. They are hard at work inside cells, using chemical energy to produce several forms of living motion unknown until recently. Some transport containerized cargo from the cell's protein factories to its export dock. Others haul imported goods in the opposite direction. One kind of motor molecule is active when cells divide, pulling duplicated components apart to furnish the two new cells.

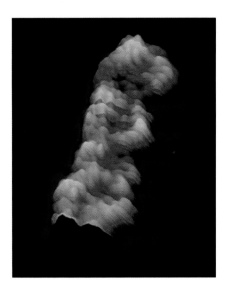

If it sounds as though the inside of a cell resembles a bustling industrial city, that's perfectly appropriate. The old concept of a cell as a lump of jellylike protoplasm is obsolete. Modern cell biology has learned that every kind of cell is a beehive of activity. The insides of cells are packed with thousands of objects—from those containers of cargo to about a dozen different kinds of internal organs or organelles, a term that means "little organ."

Modern science now knows that living motion and all the other wonderful abilities of organisms are not the products of a random biological broth inside cells or of some mystical vital force. The phenomenon of life arises because of special properties that keep the biochemical

processes in every cell under rigid control. One such property is that the insides of cells are divided into several kinds of chambers with different sets of chemicals in each.

Only certain chemical reactions can occur within a given organelle. Once the reaction is complete, the organelle then hands off the newly made molecules to another kind of organelle. It does this because of the way its enclosing membrane works. The membranes of organelles are thin, flexible sheets of fatty molecules that automatically curl themselves to form sealed chambers; they allow passage in or out only to specific molecules. Entry and exit are controlled by other specialized molecules embedded in the membrane that act as gatekeepers. Cell biologists call these gatekeeper molecules receptors or, sometimes, docking proteins.

Receptors are lumpy protein molecules, each kind with a docking site so irregularly shaped that only one kind of molecule can fit into that site. When that type of molecule, the received, fits together with the receptor, the receptor changes its own shape. The encounter has different effects. The received molecule may be taken into the cell as a whole, or it may be taken inside the organelle, or the shape change of the receptor may trigger a change inside the cell or organelle.

The role of receptors has proved to be central to many of life's essential processes. Biomedical researchers have discovered more and more processes in life that are the result of receptor molecules binding only to certain other molecules and then acting with them or upon them. One metaphor likens receptors to molecular locks that can be opened only with the right molecular keys. Often a complex sequence of chemical reactions that occurs in a cell happens in the right order only because the combined structure of receptor and received then becomes the "molecule" that is received by the next receptor.

Once the chemical reactions are complete in an organelle, it may dispatch the newly made substances in a form "intended" to travel to some other organelle in the cell. Often this shipping is containerized—the molecules held inside tiny bubbles of membrane, called vesicles. Vesicles are studded with receptors, and when a receptor on a vesicle meets a receptor on an organelle or on another vesicle with a complementary shape, the two lock, causing the two membranes to fuse. The vesicle's cargo is released automatically into the other container.

Place a living cell removed from the body under a microscope, and you can easily see the activity of vesicles being shuttled about. Thousands of tiny dots, each a vesicle holding cargo, glide this way and that or jiggle and lurch from place to place.

Another property essential to life is an elaborate transportation network with the molecular equivalent of highways or rail lines linking organelles. The existence of this network came to light as recently as the 1980s. Long filaments lace the cell in many directions like a complex, three-dimensional subway system. And, as befits a transportation network, that system includes what cell biologists call molecular motors to haul the cargo. Each of the motors is chemically similar to the myosin "hand" in muscle cells. However, instead of being attached to the myosin filament as it reaches, grabs, and flexes, these motor molecules attach themselves to vesicles.

Thus a bubble of intracellular cargo may be covered with motor molecules. Instead of powering muscles, those molecules act on the cell's system of tracks, hauling vesicles along them. Every second in every single cell, thousands of molecular motors are shuttling thousands of vesicles around.

The cargo may be newly manufactured molecules needed in distant regions of the cell. Or a vesicle may contain molecules that the cell took in at its surface, such as hormones or nutrients from the bloodstream, that must be transported to the cell's interior. Three-foot-long filaments inside the single-celled nerve fibers that link an adult's spinal cord to his or her toes are the longest tracks. Molecular freight liners make this journey at a rate of about one foot per day. Inside more typically sized cells, the entire trip takes only seconds.

INSIDE CELLS: LOOKING AT THE SMALL PICTURE

Living motion—more amazing and more wonderful than the ancient vitalists could have possibly imagined— lies at the essential heart of life, within every single cell

in the human body. Typical human cells are small, very small. Each of the ridges of skin that make up your fingerprints is about 20 cells wide. It would take about 250 cells to cover the period at the end of this sentence. It would take about 2,500 lined up single file to span an inch. Not all cells are so tiny. An ostrich egg, which like all eggs is a single cell, is bigger than a grapefruit. Some human cells are good size, too. Muscle cells can be more than an inch long, though they are as thin as typical cells. Nerve cells that reach from the spinal cord to our toes or fingers are even longer.

Most cells in the human body, though, are so small that it is difficult to visualize what goes on inside them. This problem is made worse by the type of diagram often used in biology textbooks. The typical cutaway of a cell shows just a few scattered organelles with lots of open space around each. Real cells are crammed with structures such as the organelles and the transportation trackways mentioned earlier. Some of the photographs in this book can give you an idea of what it's really like inside a cell. The interior is more crowded with hardware than a computer or a car. Cells have thousands of internal structures, most of them in motion.

Most organelles can be seen under a microscope, going about their business. Hundreds of tiny balls and blobs, each a vessel full of chemicals, tumble about. Many vesicles creep smoothly in and out of view, some squirm in place or jerk along. Watch the spectacle long enough through a microscope and you may see dark, sausage-shaped objects appear suddenly then slide out of sight. The microscope reveals an amazing scene, showing that motion is a property of life to its core.

What do organelles do? Each is a specialist. Some handle the cell's import business, some carry out recycling programs, others are involved in manufacturing, construction, or export. All the organelles are essential to the maintenance of the cell's life and, therefore, of the whole body's continued well-being. Our exploration starts just outside the cell, at the membrane that encloses it like a skin.

The cell membrane, also called the plasma membrane, is the outer boundary of every cell. The membrane is one of the most active—and most important—organelles. Thousands of receptors and other structures that control passage in or out of various kinds of molecules stud its surface. Intestinal cells, for example, have membrane receptors that take in various products of digestion. These cells relay those molecules to the bloodstream. The outer membranes of heart muscle cells have receptors for the stimulant hormone called adrenalin. This is the "fight-or-flight" hormone that a body produces in extra amounts when under stress. One of its effects is to speed the heartbeat. Sometimes this readiness to take in objects goes tragically awry. For example, certain cells of the immune system take in the human immunodeficiency virus (HIV) because it carries molecules on its outer surface—receptors, again—that resemble those that the blood cells are supposed to take in. Mistakenly, the human cell grants entry to this virus that causes AIDS.

The cell's outer membrane is made exactly like the membranes that enclose most of the organelles inside the cell. It consists of two layers of molecules. Each molecule has a head and two dangling legs. The head end attracts water, the other repels it. This difference gives these molecules a remarkable property. In water they spontaneously link to form two-layered sheets—heads to the outside where the water is, legs to the inside where there is no water. In other words, the outer surfaces of both layers are made of the heads, which have a natural attraction for one another, and the water-repellent legs are between the two layers of heads. The sheets also form spheres filled with water. The membrane is remarkably strong and not easily ruptured. So strong is the cell membrane that when scientists inject cells, pushing ultra-sharp tips of glass needles against the cell membrane, the membrane dimples deep into the cell before breaking through. When the membrane does give way, the cell does not burst. The membrane holds together, and as the needle is pulled out, seals instantly. This type of two-layered membrane encloses every cell of each living organism—as well as almost every organelle in those cells.

Cell membranes control the entry and exit of substances not only with receptors but also with simpler kinds of gatekeepers. Some of these are little more than holes, called channels, that allow passage to small molecules, including water. Channels are lined with protein molecules arrayed like a grommet to keep the membrane from squeezing the hole shut. Some gatekeepers are more complicated, using the cell's energy—derived ultimately from food—to take in or put out electrically charged atoms, or ions.

In several kinds of cells the membrane also functions as the organ of ingestion. One cell type in the human body is particularly adept at this—the macrophage. The main role of the macrophage in the body is eating. It is one of the cell types that has no fixed address but roams about, slithering through virtually all parts of the body like an amoeba over a pond bottom. Macrophages cruise the blood vessels; they slither about in intercellular spaces; there are even special macrophages that creep over surfaces in the lungs, gobbling up inhaled dust.

When a macrophage encounters a food particle—say a bacterium that has invaded the body or a dead human cell—it engulfs its food just as an amoeba does, pushing out lobes of its plasma membrane until they surround the food on all sides and fuse. It is as if your fingers closed around a coin in your palm and fused, sealing it inside. In the case of the cell, the enclosed volume now is within a bubble of membrane—a vesicle that drifts toward the macrophage's interior. The food-laden vesicle travels to one of the macrophage's stomachs, or digestive organelles, called lysosomes. A typical cell may have about a hundred or more lysosomes, each enclosed by a membrane. Lysosomes harbor many kinds of digestive enzymes, each able to break down a specific kind of molecule. The enzyme cocktail is so powerful that if a lysosome ruptured, its corrosive contents would quickly digest the cell to death. When a food-bearing vesicle meets a lysosome, the two organelles fuse—as two drops of water merge into one. The meal is delivered to the lysosome. Bacteria are not on the diets of most human cells, which feed instead on partially digested food carried from the gut by the bloodstream. Lysosomes break these down into smaller molecules.

For many of life's molecules the process of digestion is rather like unstringing a string of pearls. The molecules are, in fact, chains, or polymers. A polymer is a long molecule made up of many identical or highly similar links strung together. A protein molecule is a polymer composed of subunits called amino acids. DNA is a polymer made of smaller molecules called nucleotides. Starches are polymers of sugar. For each kind of molecule, lysosomes have special enzymes that chew away at these long-chain molecules, reducing them to separate links, most of which the cell will use to make new, larger molecules tailored to its own specifications. In most cells virtually all organelles are broken down and rebuilt anew every few weeks. DNA, the repository of the genes, is a major exception.

Most cells also take in fully digested nutrients by way of other kinds of gatekeeper molecules in the cell membrane. These are called ports. One, the glucose port, imports molecules of glucose sugar. Amino acid ports bring in various amino acids. These molecules are sources of energy as well as building blocks for the cell's growth. Glucose, the sugar that plants make by photosynthesis, contains the stored solar energy that all cells need to power their own internal processes. Cells cannot use glucose energy directly, however, so enzymes floating free in the cell's liquid change the sugar (glucose) into another substance, called pyruvate, which still holds the stored energy. Pyruvate is taken into other organelles floating in the cell—long, sausage-shaped objects that are called mitochondria.

These are the cell's energy suppliers. Much as a power plant takes various forms of energy and converts them into electricity used to run different machines, each mitochondrion converts the energy from digested food—pyruvate molecules—into a form that all parts of the cell can use. Cells that must meet high energy demands have more mitochondria than typical cells, which contain only a few hundred. Muscle cells, for example, contain thousands of mitochondria, and they occupy half the volume of heart-muscle cells.

The chemical reactions that occur inside a mitochondrion are highly complex. In a nutshell, pyruvate is processed to release its energy in the form of high-energy electrons, the same form of energy that is electricity. In cells, by contrast, the released electrons are recaptured to weld together atoms that make up a molecule called ATP, short for adenosine triphosphate. ATP is the universal power source of cells. Its energy is what drives molecular motors as they carry vesicles, and it drives muscles and gives nerve cells the power to make electricity to transmit signals. Even the body heat of warm-blooded animals comes from ATP.

ATP's energy is needed because not all the chemical reactions hosted by cells occur spontaneously. Some do, because the molecules possess inherent chemical properties leading them to bind in specific ways. Other molecules, however, are like portable electronic devices that will not operate unless batteries are included. Mitochondria manufacture ATP batteries that are transported throughout the cell. And, like a battery, ATP can run down and require recharging. Once the energy in a molecule of ATP is used, the "battery" must be recharged.

ATP works on a simple principle. Each unit of ATP is made of a molecule of adenosine bound to three phosphates linked end to end. Each bond linking a phosphate to the other parts of the molecule contains energy. ATP molecules drift about in the cell, and if one encounters molecules that would react but for a lack of energy, ATP sheds one phosphate, releasing the energy that held it. This energy may allow two molecules to combine, or it may cause an enzyme to split one molecule into two smaller ones. The ATP "battery," now lacking one of its original phosphates, returns to a mitochondrion and is "recharged" by having a new phosphate attached to it.

GENES: LIFE'S MASTER CONTROLLERS

One of the most important cellular activities that uses ATP is the manufacturing business, which is overseen by the most widely publicized component of the cell—the gene. Genes are the focus of one of the most important scientific activities being carried on today—the federally funded Human Genome Project. This is a massive, global effort to identify all the 60,000 to 80,000 genes that exist inside each human cell. Scientists predict that, when this feat is accomplished, it should be possible by studying a person's genes to diagnose predispositions to many diseases and begin preventive treatment before a disease can arise. It should even be possible to correct many genetic defects by giving patients new, properly functioning genes to replace their old, damaged ones.

To appreciate the full potential of today's genetic research, it is necessary first to understand something about what genes are and how they work. What, really, is a gene? It is far more than something that determines hair color, nose shape, height, and other distinguishing features. To be sure, genes do these things, but their role is vastly more important.

Genes dictate that people have two arms and one nose. They ensure that we have skeletons, that the pancreas makes insulin, and that the heart pumps blood through our bodies. Genes contain the instructions for building a large brain and a fingered hand with an opposable thumb. And they rule chemical events such as the digestion of food, the kidney's ability to cleanse blood, and how the immune system fights infection. In sum, genes govern everything from the ways cells work to the design of the entire body.

For all their seeming omnipotence, genes are essentially passive. They simply exist inside the nucleus much as computer programs exist on a hard drive. Genes are, in fact, the software of life, and like computer programs their function is to let their codes be read. A gene's code tells the cell how to assemble different kinds of amino acids to make different kinds of protein molecules. There are 20 kinds of amino acids, and they can be linked in innumerable sequences just as the 10 Arabic numerals can be arranged in an infinite variety of ways. Proteins serve both as structural component—the bricks and mortar of life—and as chemists—enzymes that perform many of the chemical reactions needed for life.

Most cells of the body contain a full set of genes. But only a small percentage of those genes—estimated at between 5,000 and 10,000 or between 8 percent and 12 percent—are operational in any given cell. Most are so-called housekeeping genes; they contain the instructions that tell the cell how to make the enzymes and structural molecules that are common to the growth or maintenance of all cells. In a single cell perhaps only a few hundred of the remaining genes are also active, fulfilling that cell's specialized responsibilities within the body. In other words, every cell contains the gene for hemoglobin, the protein that carries oxygen in blood, but that gene stays dormant throughout the body except in certain cells that give rise to red blood cells. All cells contain the genes for insulin, the substance the body needs to metabolize blood sugar, but those insulin genes

are turned off everywhere but in the cells destined to become pancreas cells.

Nearly all a cell's genes exist as segments of larger structures called chromosomes. The only exceptions are a few genes that reside in the mitochondria. In every human cell, with the exception of two cell types, there are 46 chromosomes. (Mature red blood cells have lost their original chromosomes, and sperm and egg cells have just 23.)

Think of chromosomes as 46 very thin but startlingly long filaments, each made up of a single, tightly-coiled molecule no wider than a typical molecule but millions of times longer.

Unwind the shortest chromosome and it will measure about three-quarters of an inch long; the longest is about three inches. If all 46 chromosomes were unwound and linked end to end, they would stretch about six feet, an extraordinary fact when you consider that the cell into which these six feet of filaments are coiled and packed is only 1/2,500th of an inch across.

Chromosomes usually exist as long filaments running in all directions inside a nucleus. If that nucleus were the size of a Volkswagen Beetle, the chromosomes would be 340 miles long but as thin as sewing thread. This is the form in which chromosomes do their work of guiding the manufacture of protein molecules. Just before cell division, however, the long filaments coil. Those coils then coil upon themselves repeatedly, winding the long filaments into short, stubby packages that look like fat X's. It is in this form that chromosomes become visible under a microscope.

The threads are made of one of the longest molecules known to science—DNA, which stands for deoxyribonucleic acid. As we learned earlier, DNA is a polymer, a linked chain of smaller molecules called nucleotides. There are four kinds of nucleotides, and they can be linked in any sequence. A gene is one segment of this thread, typically containing between a few hundred and a few thousand nucleotides. In the metaphor that scientists use most commonly, each nucleotide is like a letter of the alphabet; a gene is like a long sentence. The two genes needed to make hemoglobin, the protein that carries oxygen in the blood, contain a total of 861 nucleotides. The sequence of letters—the sentence—instructs cells how to make the protein molecule. A chromosome, then, is a series of genes, or genetic sentences linked end to end, rather like a book chapter.

....

Arrayed in pairs for analysis, the 46 chromosomes from one human cell embody that person's genetic heritage. Nearly all body cells contain identical chromosomes, yet only specific genes are "turned on" in certain cells. The pair at lower right consists of two X chromosomes, indicating a female. FOLLOWING PAGES: A mother embodies her grandmother's portrait, affirming the persistence of genetic patterns.

....

The entire collection of genes that govern each human cell, called a genome, numbers between 60,000 and 80,000. Though commonly referred to as "blueprints of life," the genome is more aptly thought of as the "recipe for life." Blueprints show the builder what the building should look like. Genes do no such thing. They don't look anything like the finished product. Instead, they tell the cell what ingredients to put together to make various kinds of protein molecules. Thus, the human genome contains recipes for making between 60,000 and 80,000 different kinds of protein (the exact number has not been determined yet), and when these proteins are manufactured in the right combinations and proper quantities, the result is a living cell.

Each cell contains two copies of most genes—one inherited from the mother and one from the father. A few genes are more numerous, half the complement from each parent, and a few are unique, only one copy per cell. These occur on the sex-determining chromosomes, called X or Y.

The two most common combinations of sex chromosomes are XY (an embryo inheriting this combination is male) and XX (female). Genetic diseases usually arise only if both copies of a gene are defective. In many cases one good gene carrying out its job properly is enough to keep a person healthy. The defective gene either leads to an abnormal protein that cannot function or to no protein at all.

Genes, despite their powerful reputation, are not autocrats. Most function only when triggered into action by chemical signals from other parts of the body or from the environment outside the body. This fact underlies an assertion that biologists have had great difficulty in getting across to the public: A human being—indeed, any organism—is the product not of "genes or environment" or "nature or nurture." Rather, it is the product of subtle interaction between the two.

The environment determines which genes are turned on and when. Chemical signals contribute. Hormones travel through the bloodstream to distant cells, where they enter the nucleus and bind directly to the DNA, stimulating it to action or blocking it from acting. Outside influences can also turn genes on or off. Stress, for example, can cause some glands to make hormones that alter the activity of genes. Intellectual activity can stimulate brain cells to initiate processes that make new connections with other brain cells.

Genes lack power in another way. Even when triggered by a molecular signal, the only thing they can do is sit still and allow special gene-reading molecules to move down their length and read their code. The gene-readers then make a working copy of the gene, which is carried out of the nucleus to the cell's protein factories. There, other machines read the copy and follow its detailed instructions to assemble the specified protein from a stockpile of raw materials called amino acids.

At any moment, scores or even hundreds of regulatory signals arrive at any given cell. They play the genome as a pianist plays a piano—triggering various genes to sound off in various combinations. Simultaneously, many cells are sending out chemical signals that act on other cells. The growth and health of the body depend on a complex interplay of hundreds of chemical signals shuttling about as if the body were some vast telephone switching system.

The signals do not act directly on the gene. For each gene in a chromosome there is a nearby stretch of genetic

Characteristic features of Down's syndrome result from a rare event in which duplicates of chromosome 21 do not separate before fertilization, resulting in a fetus that possesses three copies instead of the usual two.

coding that is not a gene. It is the gene's regulator, a length of DNA subunits—nucleotides—in a sequence that the signal molecule can recognize or, more literally, a sequence with a shape that the signal molecule can fit onto. Much as a receptor is shaped to fit only certain other molecules, a signal molecule is fashioned to recognize a particular sequence of nucleotides. The signal molecule enters the nucleus, but does not roam aimlessly. It clasps a DNA strand and crawls along it, somewhat like an inchworm looping along a twig. The signal molecule walks along a DNA strand until it happens onto the right regulatory sequence.

When the signal finds the regulatory sequence, it snuggles up close to it and changes its own shape slightly. This makes the combination of signal and DNA suitable for the binding of one of the most amazing and complex enzymes in all of biology—a molecule called RNA (ribonucleic acid) polymerase. This is the gene-reading molecule that will make the working copy of the gene, another long-chain molecule called "messenger RNA." In the computer analogy, messenger RNA is the version of the program that is held temporarily in the computer's memory bank for active use.

GENETIC MESSAGES

So, how does DNA encode the gene, and how is that gene transcribed into messenger RNA? First, let's look at the structure of DNA. It is a long chain of smaller units linked end to end like boxcars in a train. DNA's boxcars are nucleotides and, like real boxcars, they can be lined up in any sequence. Most of the time, DNA exists as two parallel trains, twisted around one another like two wires. Despite the fame of the shape, called a double helix, it is immaterial to DNA's basic workings. What really matters is the sequence of nucleotides and the way each nucleotide on one chain is linked to a nucleotide on the other chain. Straighten out the famous double helix, and the whole array looks something like a ladder, with each rung made up of two halves, one projecting out from each chain.

When the half-rungs of parallel DNA strands are locked together, the code cannot be read. Saw the ladder in half down the middle of the rungs, and DNA is ready for action. This sawing process is just what happens in preparation for the two great events in the life of any DNA strand—the replication of the chromosomes for cell division and the transcribing of a gene so that it can exercise its command.

There are only four kinds of half-rungs, or nucleotides, in the DNA ladder, each acting as one letter in the genetic alphabet. The four letters are A, T, G, and C, which stand for the chemical names adenine, thymine, guanine, and cytosine. A gene is the genetic sentence spelled out by the nucleotides on one chain and reads something like this: TACGCGAATTTT, continuing with hundreds or thousands more "letters." On the other strand, the nucleotides are in a complementary sequence that follows two simple rules: An A can link only to a T and a G can link only to a C. In other words, the rungs are always made of A plus T or G plus C.

Complementarity is one of the fundamental concepts of molecular biology. A is complementary with T, and similarly G binds only with C. This phenomenon is critical both when a gene is transcribed into messenger RNA and, as we will see later, when DNA is duplicated for cell division. When the gene reader, RNA polymerase, goes to work, it first saws apart the two strands of DNA by cutting down through the middle of the rungs. In another metaphor, RNA polymerase "unzips" the double helix.

This process exposes the gene for reading. It also exposes the opposite strand, but in most cases, that strand is not a gene but merely a "keeper" of the coding strand. The job of the gene reader is to make a working copy of the gene, called messenger RNA. The assembly is guided by complementarity because RNA follows the same rules as DNA with one exception, instead of T, RNA has a U, for uracil. Once the gene reader has transcribed the gene into a length of messenger RNA, this working copy is transported, probably by motor molecules, out of the nucleus to meet another of the cell's amazing molecular machines, the ribosome.

Each cell has thousands of ribosomes. They read the messenger RNA's code and assemble the specified protein molecule. Proteins, you may recall, are chains of smaller units called amino acids. There are 20 kinds of amino acids, and by linking them in specified sequences, ribosomes can make any of the tens of thousands of different kinds of proteins encoded in the human genome.

All sentences of the genetic code are written in three-letter words called triplet codons. Each word is the code

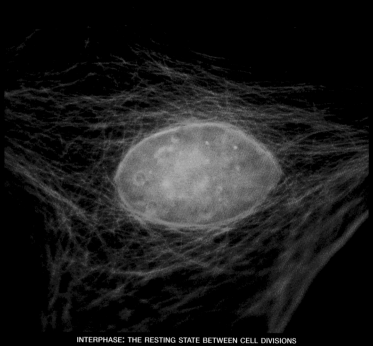

INTERPHASE: THE RESTING STATE BETWEEN CELL DIVISIONS

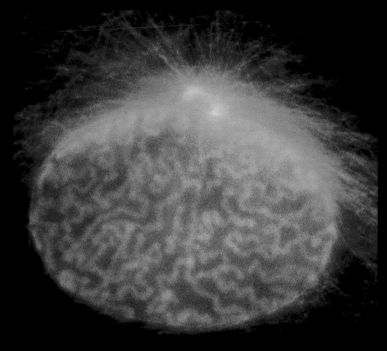

PROPHASE: DNA DUPLICATES ITSELF THROUGHOUT THE NUCLEUS

PROMETAPHASE: CHROMOSOMES BEGIN TO SEGREGATE AS FILAMENTS FORM

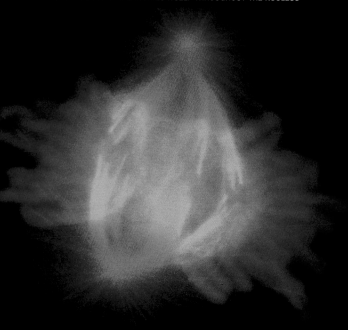

METAPHASE: CHROMOSOMES ALIGN AT THE EQUATOR

Magic of mitosis: Life reinvents itself wherever cells divide. Here, green-stained filaments called microtubules enfold electric-blue DNA. During the interphase between cell divisions, DNA is dispersed through the oval nucleus, guiding the cell's activities. As division begins, DNA duplicates itself and begins to resolve into fat individual chromosomes. Filaments then reach in from opposite sides of the cell, link with the chromosomes, and help separate duplicated pairs. Gradually, identical sets of all chromosomes cluster at what will become separate nuclei. Finally (opposite) the cell itself sunders, forming two totally independent daughter cells that resume normal cell activity.

DIVISION ACCOMPLISHED: DAUGHTER CELLS SEPARATE

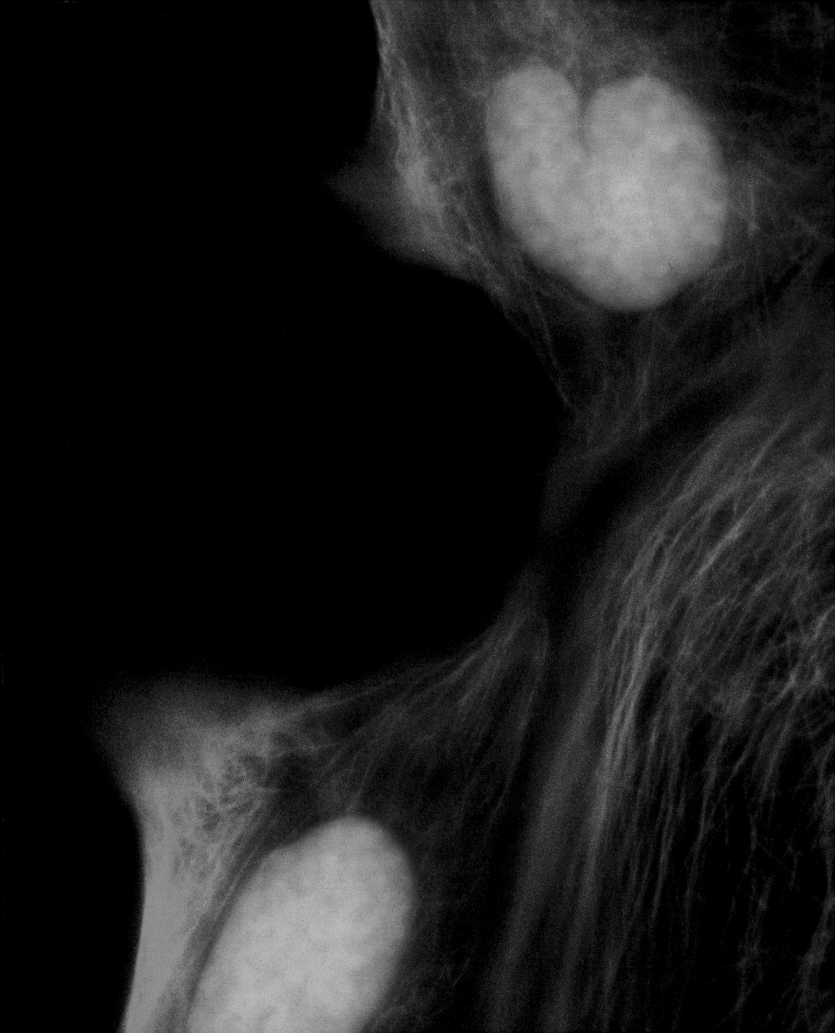

for one of the twenty amino acids. If, for example, the ribosome encounters the genetic codon AUG, the ribosome finds the amino acid called methionine. If the next word is AAG, the ribosome finds a different amino acid called lysine, and links it to the previous amino acid. One by one, the ribosome creeps along the strand of messenger RNA, finding the required amino acid and binding it to the previous one.

This is made easy because the liquid that bathes the ribosomes is a rich soup of amino acids. Step by step, codon by codon, the ribosome slithers along the messenger RNA, flexing, swiveling, and extending like some wonderful machine. After linking the requisite amino acids, the ribosome reaches the end of the messenger RNA and separates from it.

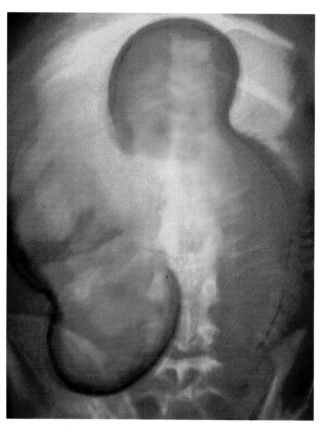

The result, a long strand of amino acids, is not yet ready to function as a protein. It must fold, coil, and wrinkle into a lumpy form. These actions occur according to a precise pattern because amino acids have intrinsic chemical abilities to repel or attract one another. For example, certain amino acids bind strongly to one another, forcing the part of the strand between them into a permanent loop. It is the final shape of the wadded-up string of amino acids, plus the chemical proclivities of those that wind up on the outer surface, that give the finished protein its special properties.

Protein synthesis is a bewilderingly complex process, but it is incredibly fast. In a typical cell thousands of ribosomes produce an estimated 2,000 new protein molecules every second. One reason for such high productivity is that several ribosomes can work their way along the same messenger RNA simultaneously, each cranking out an identical protein. Some of the new molecules are used right away in their native cell. Others are modified by other organelles and packaged for shipment to the outer membrane, where they will be released. This is the fate of such proteins as hormones that enter the bloodstream or of digestive enzymes needed in the stomach.

Genes play fundamental roles in virtually every phenomenon of life. It is no surprise, therefore, that many, if not most, of the ills that afflict human beings are either caused by or made more likely by defective genes. Errors in the genetic code, either inherited from parents or acquired as a new mutation early in life, cause more than 4,000 known diseases. Often the errors seem minor, a single substituted or missing letter out of thousands.

The best studied genetic disease is sickle-cell anemia, the most common genetic disease among African Americans. It is caused by the substitution of just one "letter" out of the 861 needed to make hemoglobin. The defective gene directs the manufacture of a hemoglobin molecule with a slightly odd shape, one that gives it a chemical propensity to polymerize. Under certain circumstances, the thousands of hemoglobins in a red blood cell will lock together, forming long rods that warp the cell's normally round shape into a crescent, or sickle. As a result, the cells jam in tiny capillaries and block blood flow, causing pain and, often, death. Victims of another blood disease, hemophilia, lack a proper gene for one of the enzymes needed for blood to clot in a wound.

Mirror images even before birth, twins nestle in the womb in a sonogram (above) that also shows the mother's spine. Identical twins—which arise from a single fertilized ovum—display shared genetic traits, including physical features (opposite).

Scientists have identified the defective gene that causes muscular dystrophy, another relatively common genetic disease. That defective gene is one of the biggest known, consisting of more than two million nucleotides. Its role appears to be to link the working part of muscle cells—the actin and myosin—and structures outside muscle cells. Without it, victim's muscles are effectively paralyzed and wither away.

Cystic fibrosis, the most common fatal hereditary disorder affecting white Americans, is caused by a flaw in a gene whose product is supposed to function as a chloride ion channel. With a defective ion channel, cells cannot process chloride ions properly, and a thick mucus forms inside the lungs, impairing breathing and encouraging infections. The same defect causes several other kinds of problems in other parts of the body.

....

ULTIMATE MEDICINE: DECODING THE GENOME

....

The Human Genome Project was established to locate and decipher all human genes and make the information available to scientists, who will find better treatments, diagnostic methods, and preventives for diseases. The means to that end are to identify all human genes, learn where they are positioned within chromosomes, and determine their precise nucleotide sequences, all by the year 2005.

Almost 5,800 genes had been mapped to specific chromosomes and another 1,400 were known but not yet mapped as of late 1997. That is only about 2 percent of the entire nucleotide sequence of the genome, but researchers say they have focused most of their efforts on developing new, automated methods of rapid DNA sequencing and that they expect to dramatically accelerate the pace of research in the near future.

Among the earliest practical benefits of genetic research are improved methods of diagnosing disease. Some genetic diseases do not produce symptoms until long after birth, when it may be too late to undertake preventive measures. Once the gene and its defect are known, doctors can easily check a blood sample to see whether the patient's cells harbor the defect long before the disease arises. In some cases, it may be possible to test an embryo. Couples at risk of a disease may even opt for in vitro fertilization, producing a "test tube" embryo that can be tested to see whether it is suitable for implantation.

The prospect of being able to screen a person for many genetic diseases has raised concerns that insurance companies and employers, among others, might obtain the information and use it to discriminate against people. The leaders of the Human Genome Project say they are fighting this possibility and have set aside as much as 5 percent of their budgets to study ethical, legal, and social implications of their research and to develop methods and practices that will keep genetic information private.

Perhaps the most daring consequence of genetic research is "gene therapy." If a person lacks a properly functioning gene, doctors may be able to implant good, new genes in the patient's cells. A few gene therapy experiments have been tried with only modest success, but hope remains high for dramatic cures as now-primitive methods are improved.

One difficult problem is delivering new genes to cells where they are needed. Cells growing in a laboratory dish can easily be given new genes, and such experiments show that they immediately adopt the genes and use them just as they do their original genes. Cells inside the body, however, are not so accessible. One technique that has already been tried is to remove some of a patient's blood-making cells from the bone marrow, place them in a laboratory dish and give them the new gene, and then inject the cells back into the patient's bloodstream. The cells take up residence in the bone marrow and make the substance encoded by the new gene. The bloodstream distributes it throughout the body.

Unfortunately, this approach works only for a limited number of diseases. Often the gene product is needed within specific cells and, therefore, must be manufactured there. One technique involves installing the new gene inside a harmless virus that is engineered to enter cells without causing disease. Viruses generally replicate by dumping their own genes into infected cells, which automatically read and duplicate the viral RNA as if it were their own. Thus, correctly designed viruses can be used to deliver a corrective gene.

The full promise of the Human Genome Project and gene therapy remains many years away, but it is likely to bring about one of the most dramatic advances in medicine ever achieved.

LIFE RENEWS ITSELF: CELL DIVISION

Perhaps the most dramatic event in an individual cell's life comes when it divides, carrying out the process by which one cell turns itself into two cells. Cell division occurs at least four million times every second in an average human body. Most of the body's cell types are continually growing old, dying, and being replaced by new cells. Cells may be damaged and die. Cells even commit suicide for various reasons. For example, when a gene is damaged beyond repair, special molecules in the nucleus detect this and trigger a genetic program that causes

exposed to the harshest environments. For example, cells lining the intestinal tract, where powerful digestive juices can attack them, live for only one or two days. Skin cells live a matter of weeks, dead ones sloughing off the outside of the body as new ones are formed in the so-called basal layer deeper inside. Muscle cells, by contrast, do not divide, but muscles do harbor their own kind of progenitor cells that are stimulated by injury to multiply and rebuild damaged muscles.

Scientists once thought that neurons, the thinking cells of the brain and the sensors and wires of the nervous system, never divided once the full complement was achieved in childhood. New research, however, indicates that some neurons can and do divide in special cases.

the cell to break into tiny pieces which then are "eaten" by neighboring cells.

Normally, these events are no problem. In most tissues, progenitor cells wait for such moments. These special cells divide, producing one cell that takes on the specialized characteristics of the lost cell and another cell to persist as an unspecialized progenitor cell.

The life span of cells varies greatly throughout the body. Those with the shortest lifetimes are some of those

Cell types that divide more frequently are more likely to turn cancerous. That's because every round of cell division requires the cell to copy its chromosomes, and errors can arise in the process. Geneticists call those errors mutations. Skin cancers and various tumors of the digestive tract are the most likely mutations to arise. Sarcoma, a muscle cancer, is relatively rare; cancer involving nerve cells is almost nonexistent. Brain tumors usually arise not among neurons themselves but among

Bursts of color and form result when various bodily substances are isolated, crystallized, and photographed under polarized light. At left, the female hormone progesterone. At right, testosterone, chief hormone for maleness.

a different type of cell in the brain, known as glial cells, which feed and care for neurons.

Central to cell division is mitosis, the process in which cells make a duplicate suite of their chromosomes and then apportion one set to each of the two new cells. The process of copying the chromosomes is something like that of making the working copy of a gene, but on a much larger scale.

Chromosome replication is effected by a special protein molecule, called DNA polymerase, that creeps along an existing chromosome, splitting the double helix into separate strands and using each exposed series of nucleotides as a template, or mold, for the assembly of a new, complementary DNA strand. Each new strand is paired with the old strand that served as the template. The two strands wind around one another, reproducing the double helix.

position. The process must occur flawlessly. Any error could produce a mutation that might kill the cell or do the opposite, turn it into an essentially immortal cancer cell. One way cells ensure accuracy is by having quality-control proteins; their job is to examine newly duplicated DNA strands and search for defects. They do this by checking the two strands for noncomplementary pairs of nucleotides. For example, if the protein finds a nucleotide other than an A, it snips out the erroneous nucleotide and splices into its place a nucleotide that is the proper complement of the original strand.

The process of copying the entire genome of some three billion nucleotide pairs goes fast but takes a long time. One DNA polymerase enzyme can replicate about 50 nucleotides per second, but even though many such enzymes work simultaneously, it takes seven to ten hours to complete the job. If each nucleotide were one letter

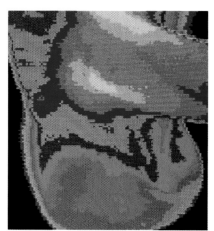

THERMOGRAM OF MALE GENITALIA AND SCROTUM

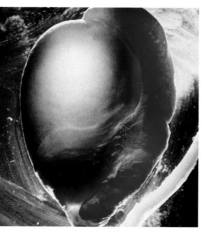

THE SCROTUM CONTAINS THE TESTICLES

COILED, SPERM-MAKING TUBULES IN THE TESTICLES

As with the making of messenger RNA, the key to the process is complementarity—the fact that each exposed A, T, G, or C acts as a template for the assembly of the only nucleotides that will fit with them.

Thus, wherever there is an A on one strand, a T must fit opposite it on the new strand—and vice versa. Like-wise, with C and G. For example, if one DNA strand bears the sequence TCAG, the sequence on the opposite strand must be AGTC. Viewed as two parallel strands, the sequence of pairs would read T-A as one rung of the ladder, C-G as the next, then A-T and G-C.

The new strand is assembled from a stockpile of loose nucleotides drifting nearby. As one enzyme "unzips" the double helix, others bring new nucleotides into position and link them to make the new strand, which is an exact copy of the old strand that formerly occupied the same

printed in a book, one complete DNA replication would be comparable to reading nearly five thousand 500-page books. When the replication process is complete, the original set of chromosomes will have been duplicated. Forty-six strands of DNA will have become two identical sets of 46 strands.

As the DNA-copying process nears its conclusion, the new double-stranded DNA begins shriveling, coiling, and super-coiling into its compact form. Again, a motor molecule does the work, grabbing free loops of DNA and pulling them into tighter form, rather like a person coiling a rope. At this stage the newly twinned chromosomes look like pairs of tiny worms, each attached to its mate at one point along its length, forming the familiar X shape. In some chromosome pairs, the attachment is so close to one end that the X looks like a V or Y.

At this point the dividing cell faces a problem. The 92 chromosomes—46 pairs of identical, linked chromosomes—are entangled. They must be sorted so that one member of each pair goes into each of the two nuclei that form. The cell constructs a sorting machine that encloses the 92 chromosomes and sprouts long fibers that grow from two opposite sides toward the middle. The fibers grab the chromosome pairs at special regions near the middle of each. The chromosomes do the work, grabbing the fibers and hauling themselves toward opposite ends of the cell.

Once again the job is entrusted to proteins resembling motor molecules, which are attached to opposite sides of each chromosome pair. Because the motors are on opposite sides of each chromosome pair, their pulling physically separates the twins, allowing one twin from each linked pair to break its link and move toward opposite sides of the cell. Once the chromosomes reach their destination, a new nuclear membrane forms around each.

As this is happening, the shriveled chromosomes begin unwinding, reverting to their sprawled out, filamentous form so that the genes can be available for reading. Simultaneously, the cell begins pinching in two. This task is carried out by a ring of filaments that works just like a tiny, circular muscle. A ring of actin, the same protein filament that is in muscle, forms around the cell's equator. A ring of myosin, the same motorlike molecule with grasping hands that is in muscle, forms with the actin. Once the two new nuclei are safely enclosed in new membranes, the myosin

starts pulling, squeezing the mother cell's membrane like a drawstring until it pinches the cell in two. Each daughter cell is half the volume of its mother, but as the genes swing back into action, the cells will grow to full size and will develop the specialized features that allow them to perform their proper role in the body.

CREATION: A LIFE BEGINS

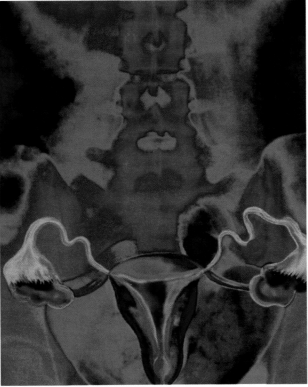

THERMOGRAM SHOWS FALLOPIAN TUBES WINDING TO JOIN UTERUS

Scientifically, there is no such thing as "the moment of conception." There is no one biological event or moment that can be considered the start of a new human embryo. Rather, conception is a process of many steps over several days. In fact, if one defines the embryo as the structure destined to become the baby, its earliest rudiments do not form until about two weeks after the sperm meets the egg.

Until that time, the mass of cells that develop from the fertilized egg, called a conceptus, is busy with the preliminary task of constructing a primitive placenta. Only after that life-support system is established can construction of the baby begin. The embryo does not reach the stage traditionally called "fetus" until eight weeks have passed.

Incidentally, none of these facts addresses religious beliefs that might influence one's position in the abor-

Makings of the next generation arise from specialized cells in the sex organs of females (above) and males (opposite). Millions of sperm are manufactured daily inside long tubules coiled inside each testicle. Eggs, however, are released usually only once a month from the ovaries (above, red lumps near lower corners). The eggs travel via fallopian tubes (orange) toward the uterus (bluish triangle).

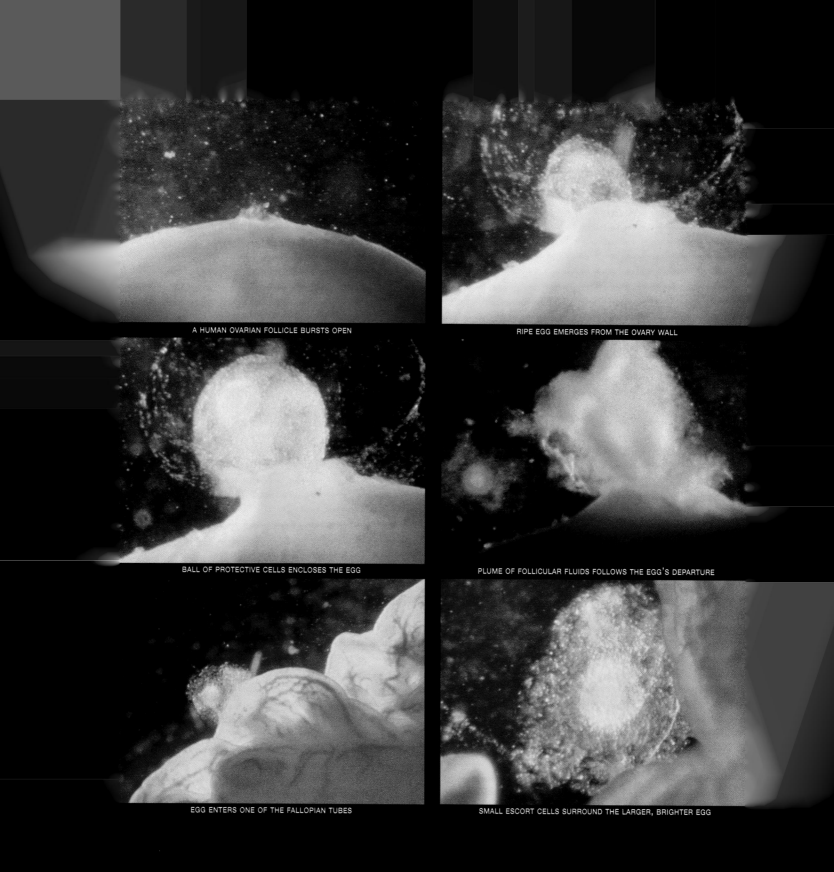

A HUMAN OVARIAN FOLLICLE BURSTS OPEN

RIPE EGG EMERGES FROM THE OVARY WALL

BALL OF PROTECTIVE CELLS ENCLOSES THE EGG

PLUME OF FOLLICULAR FLUIDS FOLLOWS THE EGG'S DEPARTURE

EGG ENTERS ONE OF THE FALLOPIAN TUBES

SMALL ESCORT CELLS SURROUND THE LARGER, BRIGHTER EGG

ncredible voyage: A follicle in the outer wall of a woman's ovary ruptures, releasing a mature egg that sets out on a convoluted trip o the uterus. A ball of smaller protective cells surrounds the egg— about the size of the period at the end of this sentence—as it emerges. Following this release, a cloud of hormones spills forth, signaling the woman's body to prepare. Fleshy fingers of a fallopian tube (opposite), richly supplied with blood vessels, sweep the traveler inside and direct it on toward the uterus.

tion debate or on legal declarations as to when the embryo acquires rights or what rights they might be. Those are matters that, while they may be based on scientific knowledge, must be judged outside the realm of science.

The creation of a new life begins at a key point in the woman's menstrual cycle—ovulation. Ever since the woman was a fetus herself, her two ovaries have held dormant eggs. At puberty her reproductive system began going through, on average, 28-day cycles, the ovaries taking turns at releasing one egg in each cycle for possible fertilization. Once released from the ovary's surface, the egg waits. Nearby is the fallopian tube, fringed with octopuslike tentacles that sweep the ovary, gathering the egg into the tube that leads to the uterus, or womb.

At this point the egg carries 46 chromosomes in pairs of 23, the full complement possessed by all human cells. Indeed, the human egg is a fully functioning cell, equipped with all necessary organelles and metabolic enzymes.

The egg is about a thousand times bigger than most human cells and is just visible to the naked eye. About the size of the dot on this *i*, it looks even bigger because it is surrounded by a protective corona of thousands of smaller cells.

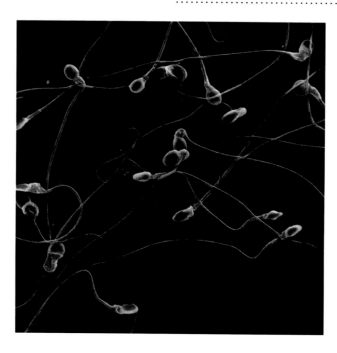

Once inside the fallopian tube, either the egg or its escorts exude special chemicals that attract sperm.

Preceded by the chemicals that act like a perfume to attract sperm, the egg glides through the fallopian tube. Lining that tube are cells with hairlike cilia that beat back and forth in coordinated waves, wafting the attractant chemical toward the uterus as muscles in the tube wall contract rhythmically to push the larger, heavier egg along. The egg, nestled in its protective retinue of escort cells, surfs toward the uterus. The trip from ovary to uterus takes about four days.

Fertilization, however, occurs well before the egg reaches the uterus. If sperm have been deposited in the woman's vagina, typically 200 million to 600 million in

an ejaculate, they detect the attractant signal and swim toward the egg, lashing their long tails furiously. Aside from its tail, the sperm is among the smallest of cells.

Its only goal is to swim to an egg to deliver a parcel of genes. Unlike the egg it performs little metabolism. Because its genes are dormant, the sperm lacks the metabolic machinery for reading the genetic code and executing its instructions.

EGG AND SPERM: THE JOURNEY

The head of a sperm contains little more than a nucleus. The tail, by contrast, is richly endowed with mitochondria, the organelles that supply ATP, the energy-carrying molecule. The flagellum, or long, whiplike tail, of a human sperm cell is a vivid reminder of its evolutionary link to many kinds of free-swimming, one-celled organisms. In addition, the tail of a sperm cell has the same design and contains the same internal molecules as the flagella of all other organisms.

A sperm needs huge amounts of energy because it must swim a distance several thousand times its own length; the mitochondria are wrapped about the base of its tail, where they can deliver their energy to a tail-whipping apparatus powered by motor molecules that work something like those of muscles.

Attracted by the egg's potent chemical perfume, millions of sperm swim from the vagina into the uterus, and out the other side. They enter the fallopian tube as the egg is still making its way slowly from the opposite end. During their journey, most of the sperm are killed by a chemical environment that, ironically, is hostile to them.

At the time they emerge from the man, sperm cells are covered with a coating of linked sugars and proteins that must be stripped away. Unless this happens, a sperm cannot fertilize an egg. Removing that coat is one function of the female's chemical barrage.

Of the hundreds of millions of sperm that started the journey, only a few hundred reach the fertilization site somewhere in the tube. This high death rate explains a common cause of male infertility. If a man produces only a few million sperm, too few will survive the precarious passage to the egg. Hundreds of millions are needed to boost the odds of success. This is because more than one sperm must meet the egg. The egg's surrounding corona of small cells must be removed, and it takes many sperm to do the job. In addition to carrying the father's genes, each sperm also carries a packet of enzymes—a kind of chemical "warhead"—that can digest away the gluelike substance that holds corona cells together.

Adrift in folds of a fallopian tube, an unfertilized human egg (above) courses toward the uterus surrounded by hundreds of escort cells, visible as tiny bumps. Fertilization, if it occurs, takes place in the tube, where sperm (opposite) are thought to be drawn by the egg's potent perfume.

When a sperm reaches the corona, it releases its enzymes, loosening the escort cells. As more sperm deliver their enzymes, they eventually dislodge the escort cells and expose the egg.

....
EGG AND SPERM: THE ENCOUNTER
....

At this point the sperm touches the egg and changes the rhythm of its tail from a steady, wavelike motion to stronger lurches, as if to push its way into the egg. But fertilization cannot occur yet. The egg is shielded by another barrier, a gelatinous coating called the zona pellucida. In this thick wall wait thousands of special gatekeeper molecules. Like locks, they can be opened only with the proper keys. Unless they bind to a correctly shaped molecule on a sperm, they somehow keep the zona in a tightly defensive form.

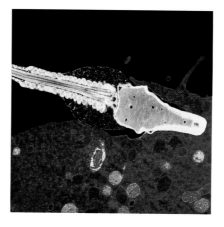

The purpose of these so-called recognition factors—preventing sperm of the wrong species from fertilizing the egg—was more obvi- ous in the evolutionary past when fertilization took place in the open sea, as it still does with many marine organisms. Under those circumstances, an egg might meet sperm from another species. If it allowed the sperm to enter, the egg would have been wasted.

If the human egg's receptor accepts the sperm's complementary protein, fertilization proceeds. Yet another receptor mechanism is believed to play a role, causing the outer membranes of sperm and egg to fuse. As with the vesicles inside cells that fuse with organelles to deliver their contents, the membranes of sperm and egg become one and the sperm's cargo of chromosomes can enter the egg. At this point, the sperm stops wriggling, its head and tail still a bump on the spherical surface of the egg. Now the egg takes the initiative. Projections on its surface grasp the sperm somehow and pull it in, tail and all. This is the event that many scientists consider the moment of fertilization. Yet

another 24 hours must pass before the father's genes meet the mother's to begin the process of repeated cell divisions that will turn one cell into two, two into four, four into eight, and so on.

But first, the egg has other work to do. It must block entry to any more sperm. If another sperm were to enter the egg, its extra set of genes would prove lethal after only a few rounds of cell division. Eggs have two methods of closing the door—a quick one that lasts only a few seconds, and a slower one that uses the few seconds to erect a permanent barrier. The fast method takes advantage of the fact that many atoms in the fluid surrounding the egg carry innate electrical charges. Normally the egg's surface has a negative charge, which attracts the sperm. But as soon as one sperm has entered, the egg activates special channels in its membrane that pull calcium ions into the cell. These are positively charged, and their entry immedi- ately reverses the charge on the egg's membrane.

The sudden change in charge has a more far-reaching effect; it sets in motion a chain of events that will cause the cell to divide in two and then into four, eight and so on. As Jacques Loeb, the pioneer of mechanistic biology, discovered in the early 1900s, this sequence can be triggered without sperm. The chemicals he added to sea urchin eggs to trigger development caused exactly the same reversal of electrical charge. A small dose of potassium chloride causes sea urchin eggs to behave as if they had been fertilized by sea urchin sperm. This trick works because the first few cycles of cell division in a newly fertilized egg—human or urchin—make no use of the sperm's genes. All the enzymes and machinery for carrying out cell division are present in the egg before the sperm reaches it. They merely await an electrical signal.

The egg's new positive charge will last only a few seconds before being neutralized. But that is time enough for the egg to create a long-lasting barrier. The egg releases a substance that acts on the jellylike zona rather like an epoxy hardener, causing its separate molecules to cross-link, forming an impenetrable shell around the egg. The shell will encase the egg and sperm for several days, long enough for all the remaining sperm to die.

Even at this point, the father's and the mother's genes remain separated in their own nuclei, typically resting on opposite sides of what is sometimes called a zygote. They will not join for more than a day. They cannot merge at this stage because the egg still has the full complement of 23 pairs of chromosomes. The sperm, by contrast, has only 23 single chromosomes, one member of each original pair. The sperm waits while the egg randomly selects one member of each pair, packages those 23 lone chromosomes into a new nucleus and discards the other 23, bundling them into a miniature cell that is pinched off from the egg and allowed to disintegrate. While this is going on, the sperm's compact nucleus, which had carried the father's chromosomes, swells many times to reach its normal size, and the tail disintegrates.

The nuclei do not join at this point, as scientists once thought they did. Instead, while the two nuclei remain separate, each now duplicates its 23 single chromosomes to make 23 pairs. That done, the nuclei move toward one another, apparently pulled by motor molecules. The nuclei meet, their membranes disintegrate, and all 92 chromosomes—two sets of 46 pairs—mingle. From this point, the fertilized egg behaves like an ordinary cell undergoing cell division. It constructs the same kind of apparatus that other cells use to separate duplicated chromosomes and pulls one member of each pair into a

Enter the sperm (opposite), its head packed with genes, its tail a powerhouse of mitochondria. Once inside, the sperm's compressed filaments of DNA begin to unwind, and the tail disintegrates (above).

Bristling with sperm, a fertilized egg (below) bars entry to all late arrivals, shielding itself within a protective shell known as the zona pellucida. The egg begins to divide, forming two identical cells (opposite, upper left) that become sixteen by day three (opposite, upper right). The cells, stained blue here, continue to share a single, red-tinted zona. Seen in more lifelike tones, a two-celled conceptus moves through the fallopian tube (opposite, center). By about day four, the conceptus, now a hollow, multicellular ball called a blastocyst, "hatches" from the zona (opposite, bottom) and prepares to implant in the uterine wall.

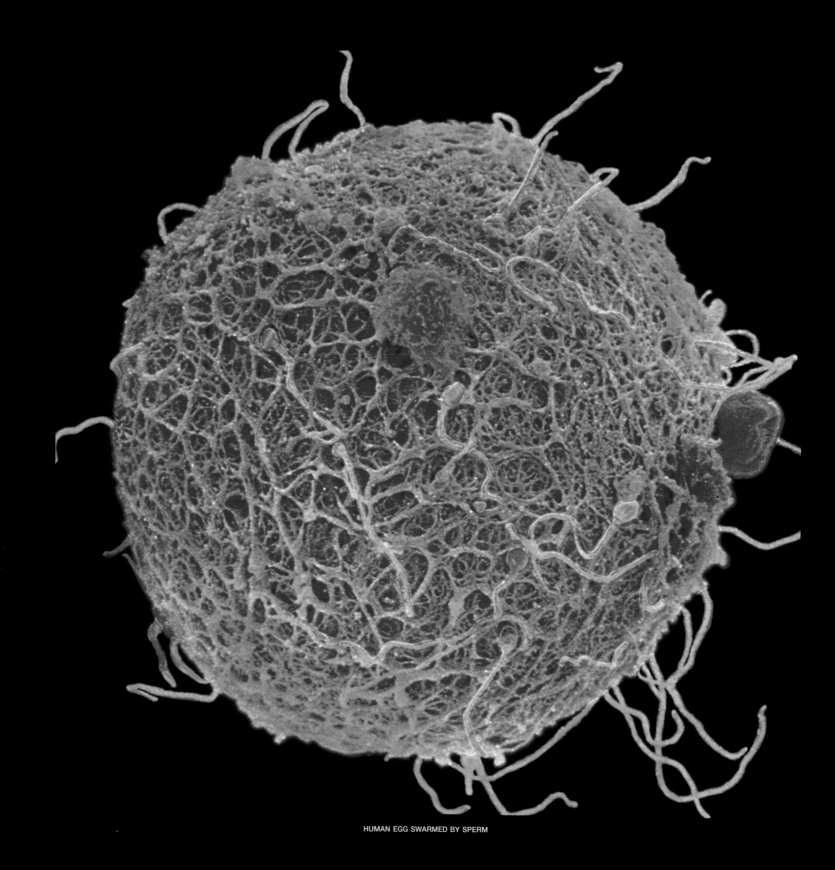

HUMAN EGG SWARMED BY SPERM

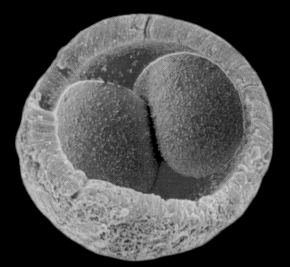

TWO-CELLED CONCEPTUS WITHIN CUTAWAY OF ZONA PELLUCIDA

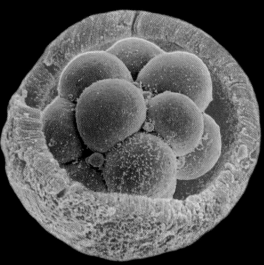

SIXTEEN-CELL STAGE, ONE DAY LATER

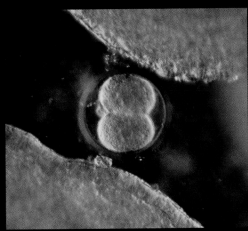

TWO-CELLED CONCEPTUS TRAVELS THE FALLOPIAN TUBE

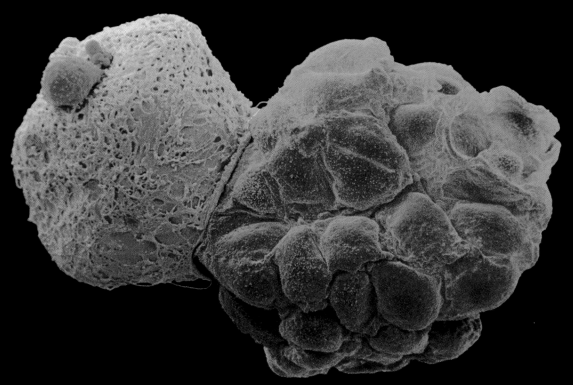

BLASTOCYST "HATCHES" FROM THE ZONA

new nucleus. The single cell pinches in two between the nuclei and—30 hours after the sperm entered the egg—a new genetic potential has been established.

Yet, even at this stage, work on the structures that will become the baby will not begin for another 13 days. The conceptus is still in the fallopian tube, being carried slowly toward the womb.

If not fertilized, the egg dies about 10 to 15 hours after ovulation. Sperm have a somewhat longer life span, about 48 hours after entering the vagina.

Thus, for a woman to become pregnant, sperm must be deposited at some time during a period of about two and a half days in each menstrual cycle—from 48 hours before ovulation to 10 to 15 hours after. Eggs that remain unfertilized are disposed of by macrophages, the body's ubiquitous scavenger cells, which also eat dead sperm.

Some time during the second day after fertilization, the two cells of the conceptus undergo mitosis, duplicating their chromosomes, and divide to produce four cells. Also on Day 2, the four become eight, the cells halving in size with each cell division.

Early researchers who used microscopes to examine the conceptus at this stage thought its clump of eight ball-shaped cells made it look like a mulberry. So they applied the Latin name of the fruit, *morula*, to describe the eight-celled conceptus. The term has stuck.

The four-cell and eight-cell stages of the conceptus are commonly those at which a "test tube baby" is implanted in a prospective mother's uterus. Fertilization of an egg in a dish or test tube, a procedure called in vitro fertilization, comes from the Latin, meaning "in glass." Thus far more than 50,000 babies have been born through the application of this technique.

The morula also is the stage at which the cells first exercise their genes. Until the eight-cell stage, all the cells' genes have been dormant. All cell divisions and other metabolic events have been controlled by enzymes already present in the unfertilized egg and by messenger RNA that was transcribed from the mother's genes before the egg left the ovary.

Now, on Day 3, the new combination of genes—the new genetic individuality—first expresses itself. Developmental biologists estimate that between this point and implantation as many as half of all conceptuses die. Most of these are carrying severe genetic defects that were hidden until the genes began to act. A healthy young couple trying to reproduce naturally has about one chance in four of accomplishing their goal in a year of trying.

Also during the third day after conception, the morula changes its shape so that it no longer resembles eight clustered balls. The newly activated genes direct the cells to make receptors and other surface structures. The new receptors act on each other, a protein on one cell grasping the same protein on the surface of another. As the number of these links spreads over the cells, the cells squeeze together, producing a more compact structure. This process is known as compaction. Along with receptors, the cells make other kinds of junctions, some of which serve as channels through which small molecules may pass, allowing the cells a simple kind of communication.

Other junctions function like spot welds, holding cells firmly together. These different junctions provide much of the strength that binds cells of one type together in a tissue. A typical cell forms many thousands of junctions to its neighboring cells.

As development proceeds, many links between cells will form and vanish as cells take on different characteristics and create various kinds of specialized structures. The appearance of channels and junctions among cells during compaction is evidence that they are beginning to cooperate to produce a larger scale of biological organization, the tissue. Tissues are aggregations of similarly specialized cells that perform the same function.

During the morula stage, the conceptus that seemed destined to become one baby can break apart. One ball of cells can fragment into two balls of cells. If this happens, each may go on to develop perfectly normally, yielding identical twins. But if the morula breaks up only part way, the result may be more tragic. Some cells may develop as if they were in separate embryos, while others may continue as part of just one embryo. Depending on how or where the morula comes apart, the result may be conjoined, or so-called Siamese, twins—two babies who share one or more parts of the same body.

As Day 4 nears, it brings another epochal event—the beginning of differentiation, the process in which cells become different from one another, some taking on one shape and function while others pursue a different fate.

Without differentiation, the proliferating cells of the conceptus would form nothing more than a growing, amorphous lump of homogeneous tissue. Actually, the lump could not become very big because it would have no way of transporting oxygen and nutrients to its interior. Most human cells must lie within a few cell thicknesses of an oxygen source or die.

DIFFERENTIATION: CELLS DECIDE

How do the cells of the early conceptus—all alike as they are—become different? How do they determine what differences to effect? Why don't lung cells slip off to consume invading bacteria, and why don't brain cells produce digestive juices? Scientists are still searching for answers to these and other questions. Differentiation is one of the most puzzling phenomena in all of biology. The mystery lies in the fact that every cell of the body contains the same set of 46 chromosomes bearing the same suite of genes. The cells of the eye hold the instructions on how to make toenails. Skin cells have all the genes needed to contract rhythmically like a heart muscle cell. The cells that make bone contain the code for making digestive juices. So, how do the homogeneous cells of the early conceptus "know" which genes to use and which to leave dormant? All the body's cells use the same "housekeeping" genes to stay alive, but somehow each has selected

A sculpture that continually remodels itself, the human embryo undergoes dramatic changes early. Three weeks after fertilization it has grown as big as a sesame seed; its neural tube (opposite), destined to become brain and spinal cord, begins to close, starting at the middle. One week later a primitive head appears (above), as do branchial arches, a curving tail, and limb buds.

Development proceeds unevenly, some structures moving toward final form faster than others. At five or six weeks, upper limb buds change into stubby arms, and each paddle begins to separate into fingers. Lower limb buds, however, still have feet with no discernible toes. Within the next two to three weeks, the fingers separate, thanks to the programmed death of intervening cells, determined by DNA; eyes begin to acquire lids. At eight weeks the external ears are little more than holes, but by four months they are nearly complete. Approximately midway through gestation, the fetus becomes unmistakably human (opposite), though bodily proportions will change for many more months. This fetus, linked by a purplish umbilical cord to the placenta at the lower left, lies inside the amniotic sac.

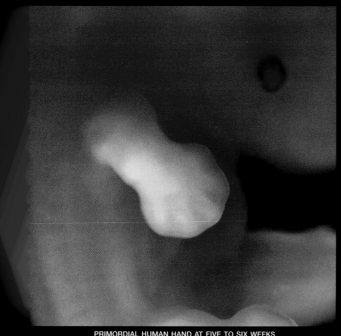

PRIMORDIAL HUMAN HAND AT FIVE TO SIX WEEKS

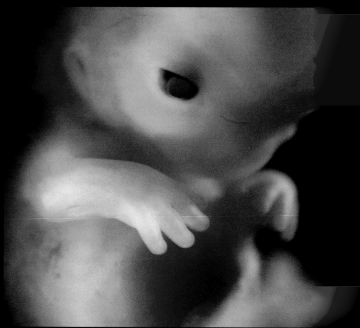

FINGERS SEPARATE BY ELEVEN WEEKS

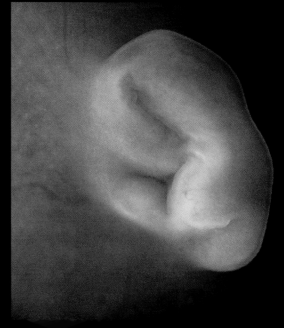

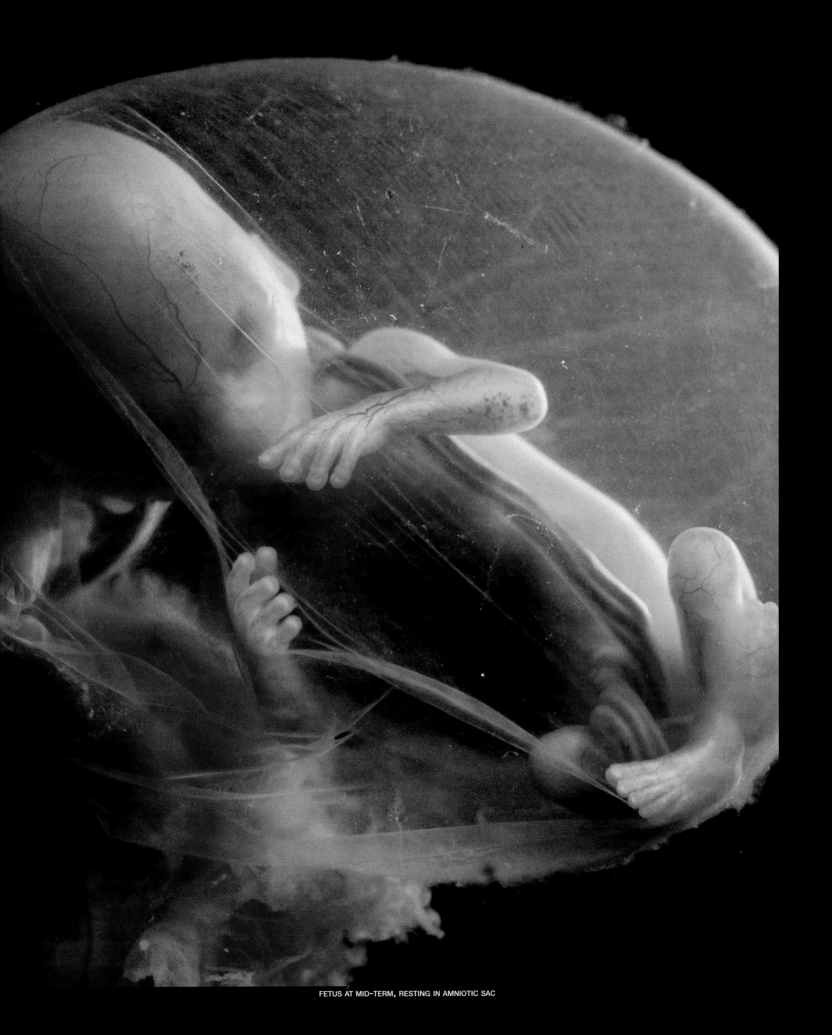

FETUS AT MID-TERM, RESTING IN AMNIOTIC SAC

to keep just a few of the remaining genes in operation. How each cell adopts a particular role and stops using the rest of its potential is one of the central problems of modern biology—one scientists are just beginning to solve.

A clue to this process has begun to emerge from studies of what happens during the fourth doubling of cells, from 8 to 16 by Day 3. After the eight cells squeeze together during compaction, the surfaces of the cells just inside the zona pellucida (it still encases the cells) sprout numerous fingerlike projections. At the same time, most of the organelles inside each cell move toward the outer surface, positioning themselves just under the projections. The opposite side of the cell contains very few organelles and develops no special features. Each cell somehow seems to be concentrating its resources against its outer surface.

Simultaneously, the eight cells divide, each splitting so as to apportion the organelles very unequally. One daughter cell is blessed with much of the enriched outer surface, while the other gets short shrift—little or none of it. Typically, an average of 11 of the 16 daughter cells acquire some of the outer surface and the enriched complement of organelles. Five cells get the short end of the stick; they are in the center of the ball, each possessing a nucleus and just enough organelles to stay alive.

These five seemingly impoverished cells will not stay poor for long. It is from them that the baby will develop, but for now the conceptus has more urgent business. It sets aside a few undifferentiated cells with a minimal complement of organelles to bide their time, while the rest of the cells carry out the vital task of constructing an apparatus that can invade the wall of the mother's uterus and build a placenta. Unless and until that can be accomplished, the five poor cells in the middle of the ball—which embryologists term simply the "inner cell mass," will have no future.

With this unequal division, the conceptus carries out the first of hundreds of cellular differentiations that must occur in a precisely choreographed pattern to construct both a placenta and a baby.

Now, about four days after fertilization, when the conceptus is reaching the end of the fallopian tube where it enters the uterus, the pre-embryo must "hatch" from the hard shell of the zona. The outer cells make and release an enzyme that dissolves the zona, and the conceptus emerges.

Also at about this time, it begins the major task of commandeering the woman's body, essentially telling it that a pregnancy is about to begin. Embryonic cells of the blastocyst produce a hormone called chorionic gonadotropin (CG). It acts on the woman's body, stopping the menstrual cycle that otherwise would break down the uterine wall that has become engorged with blood in preparation for pregnancy. Without this signal, the uterus would shed its lining as it normally does in every menstrual cycle.

When the conceptus finally enters the womb, it does not immediately attach itself to the woman's body; thus, as doctors figure it, a pregnancy has not begun. The conceptus does, however, absorb oxygen and nutrients from uterine fluids. While the conceptus drifts independently for a day or so, the cells double and redouble in number to 64. And the little cluster of cells undergoes a second major change in architecture. It transforms itself from a solid clump of cells to a hollow ball lined with cells. This happens when the outermost cells open special portals that allow fluid to flow into the internal spaces between cells inside the clump. The outer cells flatten out, forming a thin layer that encloses a growing volume of fluid. Inside the ball the clump of undifferentiated cells—destined to become the baby—remains unchanged and waits, stuck against the inside wall of the ball. The conceptus is now called a blastocyst.

BECOMING BABY: THE BLASTOCYST

Now about six days old, the conceptus prepares to implant itself in the uterine lining. Still no bigger than the unfertilized egg but now comprising 120 cells, it burrows into the wall of the womb, using enzymes that literally digest away some of the mother's tissues.

At this point female blastocysts—only female ones—undergo a surprising change. One of the two sex-determining chromosomes in each cell is deliberately disabled. Each female cell has two X chromosomes, while males have one X and one Y chromosome. Both the male's sex chromosomes function throughout life. Not so with the female. In each cell of the blastocyst, one of the two X's is put out of commission permanently. Special molecules bind to it and prevent it from using its genes. This so-called X inactivation makes women genetic mosaics. Because the choice of X chromosome is random

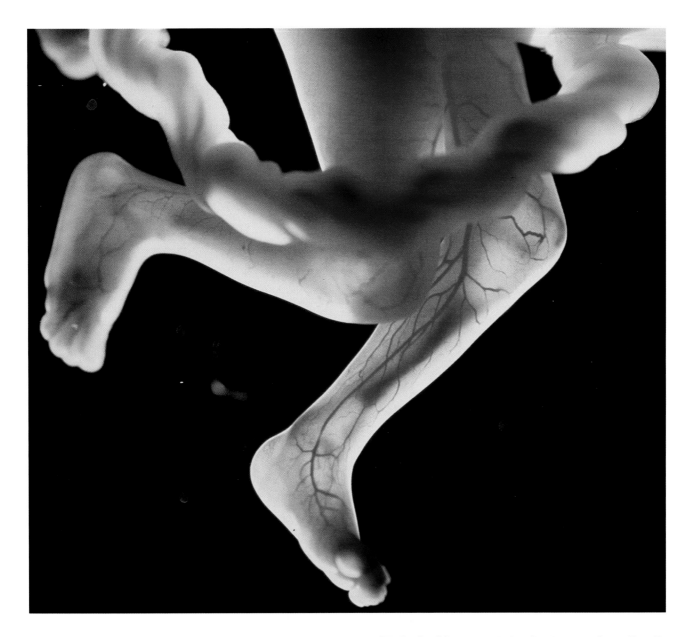

in each cell, it will be the X from the father that is knocked out in some cells and the X from the mother in other cells. As those cells continue to divide, their descendants will inherit the same pattern of inactivation. Thus, certain patches of cells in certain parts of a woman's body will be under the influence of the father's X and others under the sway of the mother's. There is little evidence that this process matters, but it may explain such mosaic-like patterns as skin being oilier over one part of the body than another.

While the blastocyst is implanting, at about Day 7, some of the outer cells begin the construction of a primitive placenta, the embryonic tissue that will extract oxygen and food from the mother's bloodstream and release embryonic waste products into it. These cells begin to divide but do not form membranes between themselves. They produce one giant cell with many nuclei. This curious structure sprouts roots that penetrate deep into the uterine lining, much as a plant sends roots into the ground.

At four months of development, a skeleton originally formed of rubbery cartilage begins to change into hard bone. Here, the shin bones stand out as dark rods in the legs. These and other long bones grow from the middle toward both ends.

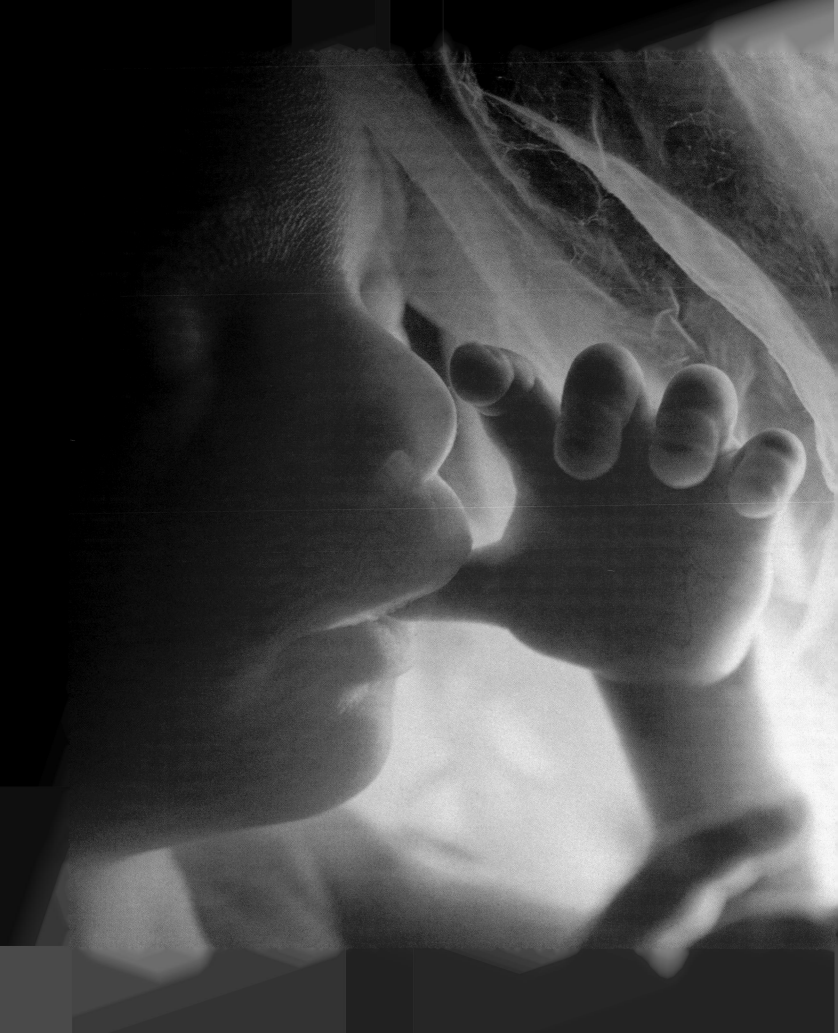

All alone in the amniotic sac yet intimately allied to its mother, a mid-term fetus develops primarily in size. The radical reshapings of its early weeks have produced an essentially finished form that requires only the completion of fine detail work—the wiring of brain cells, for example, or the elaboration of special cell types within the liver—and a steady increase in size. At four-and-a-half months (opposite), the fetus is remarkably whole yet only halfway to birth. Effectively weightless in amniotic fluid (below), another mid-term fetus—hardly larger than a man's hand—drifts serenely as its mother's body supplies the matter and energy that its own genes transform into what will be a fully developed human being.

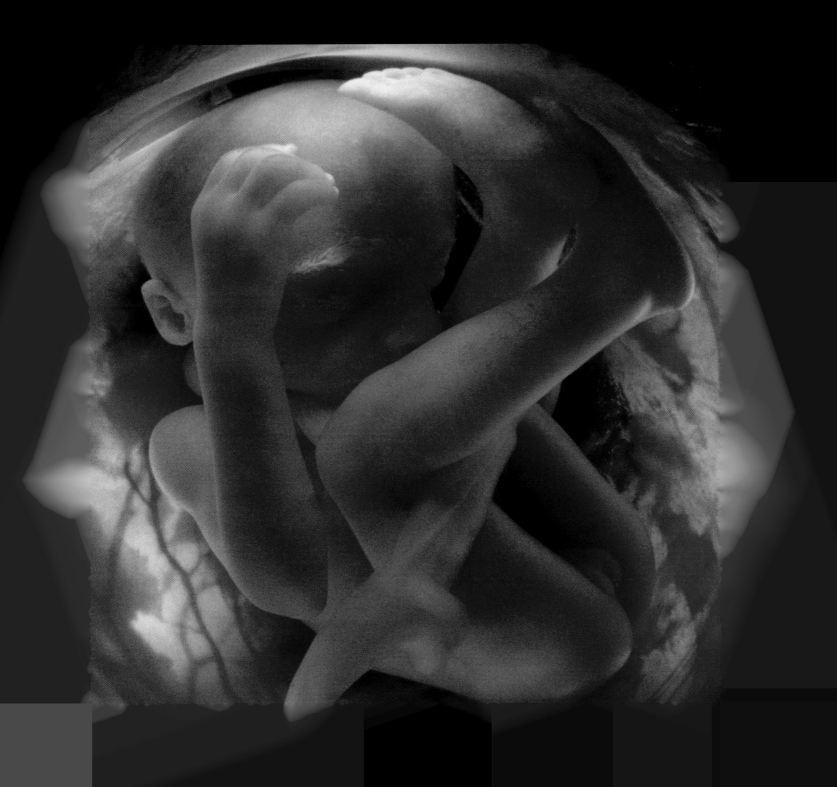

For the next few days, the blastocyst's roots push deeper into the wall of the womb and wrap themselves around the woman's blood vessels. Then the roots make enzymes that destroy the blood vessel membranes, spilling the woman's blood into small spaces inside the increasingly spongy tissue. The roots take oxygen and nutrients from the mother's blood. As development proceeds, blood vessels of the embryo will snake out of its body—through the umbilical cord—into this blood-engorged tissue and back into the embryo proper.

All through this process the outer cells of the blastocyst continue to make chorionic gonadotropin, CG, that enters the mother's bloodstream. Some of it is excreted in the urine. For this reason, the most common pregnancy test involves looking for CG in urine. Plenty of CG remains inside the body, however, and in addition to blocking the menstrual cycle, it sends a message to the ovary that produced the egg. The chemical signal acts on the structure that released the egg, called the follicle. This chemical signal tells the follicle to undergo its own kind of development, changing into a kind of temporary gland called the corpus luteum. In turn the corpus luteum will produce the hormone estrogen, which travels to the uterus and causes it to grow. The growing fetus does not stretch the womb to its expanding size; the uterus, which contains the large, powerful muscle needed to expel the baby at birth must grow on its own to stay large enough. To ensure that it does not contract too soon, the corpus luteum also produces another hormone, progesterone, which inhibits the uterine muscle from acting until the fetus is fully developed.

After the first two or three months of development, the corpus luteum shrinks away and hands off its hormone-making role to the placenta, which makes both estrogen and progesterone for the rest of the pregnancy. This allows the placenta to monitor fetal development and to stop producing progesterone when the fetus is ready to be born and the uterine muscle can be allowed to begin contracting.

From the blastocyst's first puffs of CG that signal the uterus to prepare for pregnancy, to the months-long flow of hormones from the placenta, it is the conceptus/embryo/fetus that calls the shots, commandeering the woman's body for a life-support system. Only after the beginnings of that support are established—by about Day 9—does the modest clump of undifferentiated cells, the inner cell mass, begin to rouse and start the task of constructing a new human being. Yet even at this stage, those 20 or so cells must perform a series of restructurings and differentiations over another five or six days to create the first structural element of the actual embryo—a simple, straight trough of cells that defines the head-to-tail axis of the embryo, the armature of what is to become—over the next 36 to 37 weeks—a baby.

CONSTRUCTION ZONE: MAKING BABY

Although the full course of fetal development takes some nine months in humans, much of that time is spent simply increasing the size of the fetus. The most important and most interesting steps—laying the foundations of the organs in their correct positions—occur during the first eight weeks after fertilization. In fact, they take place in the mere six weeks between the formation of the primitive streak at the end of the second week, when the embryo is the size of a poppy seed, and the beginning of the ninth week, when it has become about as big as a pecan. Developmental biologists use the term "embryo" during this six-week period.

At the start of the ninth week, once the beginnings of all major tissues and organs have been laid down, the embryo becomes a fetus.

Scientists probing the mysteries of embryonic development focus on the earliest steps in the process, for it appears that the mechanisms that control the first steps work much like those that guide subsequent events. The primary effect of each step is to turn on or to turn off suites of genes. The specific sets of genes vary with each step, as do the locations of the cells in which the switching events occur. Thus there are constantly shifting patterns of cells that activate certain genes with the effect that the protein products of those genes help signal other patterns of genes to come into play.

The process is a bit like a symphony orchestra in which different combinations of instruments produce various sets of notes. When one pattern of instruments has sounded its notes, their completion of a task signals the next set of instruments and notes. Molecular biologists speak of genes being induced—roused from dormancy and called upon to send a specific message to the cell, altering its form or function in some way. If one cell produces a certain protein, that induces its neighbors to do something else.

Embryonic development, then, is a constantly shifting pattern of gene inductions and resulting changes in the behavior or form, or both, of the cell. At the same time, the embryo's cells are dividing, often launching the two daughter cells on new careers. Or, sometimes, one daughter pursues one fate and the other a different course. Sometimes a patch of cells coordinates its activities, assembling into a structure that serves for a few hours or a few days and then dies, pruned away as it were by the phenomenon of programmed cell death as it shapes the growing embryo.

For example, in the fifth week of development, the embryo has stubby arms with hands shaped like paddles. The parts that will be fingers are completely joined by flesh. Then, in the following week, four waves of cell death begin at the edges of the paddles and plow furrows toward the wrists. Their mission completed, the cells in the furrows die back, removing tissue that served rather like scaffolding during construction of the fingers.

The fate of every new cell is governed by its genes working in concert with the cell's environment. At various points as development proceeds, some genes become permanently switched off and others are called into action, sometimes never to return to dormancy. At the same times, other genes are activated, causing the cell to respond in various ways. Along with the usual processes of making proteins and dividing at regular intervals, cells sometimes crawl off to new homes in faraway parts of the embryo.

When cells depart one region of the embryo or arrive in another, they alter the local environment for resident cells of that particular area, thus changing the signals that induce genes in those cells. As a result, the resident cells may begin multiplying or changing their behavior or form, or they may even pick themselves up and move somewhere else, sometimes in groups that look like sheets or trains of moving cells.

The early weeks of embryonic development witness an astonishing cellular ballet, cells gliding and turning, moving back or forth, crawling over other cells or diving below them. The choreographer of this dance is, as is nearly always the case with living organisms, a delicate interplay between the entire genome, shaped by evolution over billions of years, and the briefest of chemical signals wafting through the cell's local environment. The genes make a wide range of actions possible, and the chemical environment chooses among the possibilities for each cell.

At about Day 15—roughly the time a pregnant woman misses her first menstrual period—the interplay of genes and environment has led the 20 or so cells of the inner cell mass to multiply and reassemble themselves into structures that look like two hollow balls pressed against one another. The two cellular balls are suspended within the third, larger ball of the blastocyst by a stalk, the primitive umbilical cord. At this stage the "true embryo" is nothing more than the two layers of cells formed where the two inner balls touch, a patch of cells called the embryonic disc. From these two sheets of cells the baby will develop.

The new child's existence begins with the formation of a groove that forms in a straight line in the surface of the embryonic disc. Called the primitive streak, this groove forms as cells from both sides crawl toward the centerline and heap themselves up in ridges on either side. The ridges form first at one end of the centerline, which will be the embryo's tail end—the base of the spinal cord. The ridges lengthen toward what will be the head end of the embryo. As the groove lengthens, sheets of cells from atop the ridge slide down into the groove and then double under themselves, forcing their way between the two original layers and creating a third layer. Embryologists consider these three layers—called ectoderm, mesoderm, and endoderm—to be the founding tissues of the embryo's tiny body. These layers form the bases of all the major organ systems.

The first significant part of the body to develop is the nervous system. About two and a half weeks after fertilization, when the embryo is barely 1/16 inch long, the embryo lays down the foundations of the spinal cord and the brain. The spinal cord is constructed in the groove that formed with the primitive streak. The ridges on either side of the groove build up as neighboring cells crawl atop the long mound. Like waves breaking on a shore, the tops of the ridges then bend toward one another until they touch, forming a hollow tube. This so-called neural tube then sinks below the surface as other cells close the gap where the tube has been. The tube is the founding structure of the spine and spinal cord. Closure starts at the middle of the groove and moves toward both ends. It is at this early stage of development that two of the worst birth defects arise. If the tube fails to close at the tail end, it can cause spina bifida, a birth defect in which some of the spinal cord bulges through an opening in the baby's back. If the failure occurs at the head end, the defect can be far worse. The brain may fail to develop, leading to the birth of a baby with anencephaly, the absence of a brain. That usually fatal destiny is determined between Day 23 and 26.

Formation of the neural tube gives rise to a number of other tissues. One of the most extraordinary classes of these cells develops along the crests on each side of the neural groove. Just before the neural tube closes, cells from its crest crawl away, some to become parts of the nervous system, others to help make teeth, still others to make cartilage and the bones of the face. Some neural crest cells travel to the outermost cell layer of the embryo, to what will become the skin. There they reside and make melanin, the pigment that gives skin and hair their color.

THE DEVELOPING EMBRYO

Embryonic development in this early period is so rapid that each day brings major changes, so much so that embryologists have no trouble telling, for example, an 18-day-old embryo from a 20-day-old. Every time cells move to a new home or welcome immigrants to their neighborhood, the pattern of genes being expressed shifts. And with each shift comes a new pattern of cell form and function. At one point, for example, sheets of cells in the middle layer curl themselves into a network of tubes that become the circulatory system—the first system to function. These arteries and veins do not grow like a tree sprouting branches. Instead, growth originates simultaneously in scattered sites throughout the embryo and placenta. Each tube lengthens and sprouts buds that branch. Then these blood islands, as they are called, meet and fuse to form one large network.

Around Day 22 cells wrapped around one of the larger tubes start expressing their genes for actin and myosin, the two main proteins of muscle cells. These proteins assemble themselves into parallel filaments, and the myosins begin pulling on the actins. The cells

contract and then relax, taking up a rhythmic pattern of action. The heart has begun to form. At first it has only one chamber, like the heart of a worm, but just three weeks after the sperm entered the egg, and barely a week after a woman has missed her menstrual period for the first time, it is beating inside an embryo the size of a sesame seed. Inside the vessels of the new circulatory system, other cells transform themselves into blood cells. As development proceeds, special cells in bone marrow will take over the job of making blood cells.

In the early days of embryological research more than a century ago, scientists made one of the most surprising and haunting observations in biology. Through its first few weeks of development, the human embryo is almost indistinguishable from that of a cow, a monkey, a lizard, or even a bird. At four weeks after fertilization, a human embryo has a tail and even gill pouches, structures that are the precursors of gills in fish. In humans the tail disappears, the victim of a wave of programmed cell suicide, and the gill pouches change not into proper gills but into bones, muscles, and other parts of the face, ears, and neck.

The reason different species look so much alike in early embryonic development is that they all share the same evolutionary past and have inherited similar genes. The genes for gill arches, for example, arise in all vertebrates because the early chemical signals that activate genes are similar in all the species. But as development proceeds, the later signals are different. Humans no longer have the genes that fish use to continue the development of gill arches into full-fledged gills. Those genes have mutated into something different in humans, either rendering them ineffectual or changing their effect on the cells of the gill arches so that they develop into other structures of the head and neck.

In addition to the signals that tell a cell which genes to switch on or off, cells must have other kinds of information to know what to do and, if they are migrating cells, where to go. In some cases researchers have found attractant signals exuded by cells in certain parts of the embryo. Once released, these signals diffuse outward from the source. They work something like the sperm-attracting chemical produced by the unfertilized egg.

Hands, arms, and legs all begin to move at about eight weeks of development, but it is not until around the nineteenth week (opposite) that these motions become powerful enough for the mother to feel—the moment of so-called quickening. FOLLOWING PAGES: Fine hairs, usually shed before birth, cover the face of this nineteen-week-old fetus.

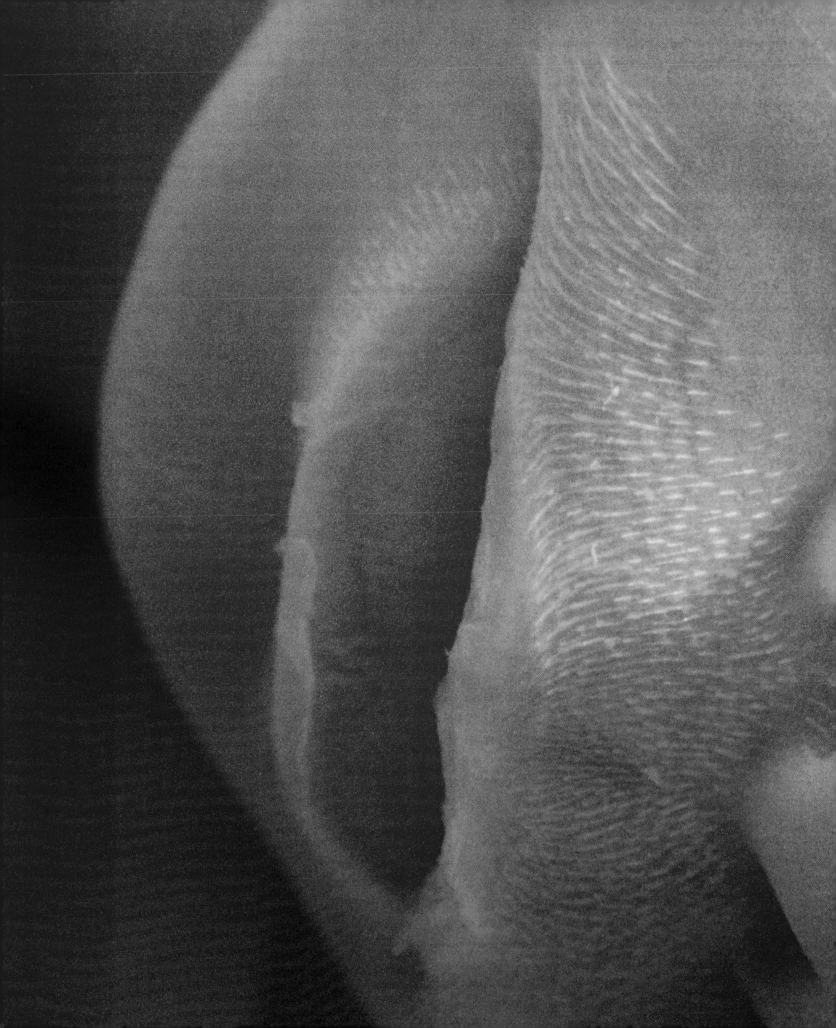

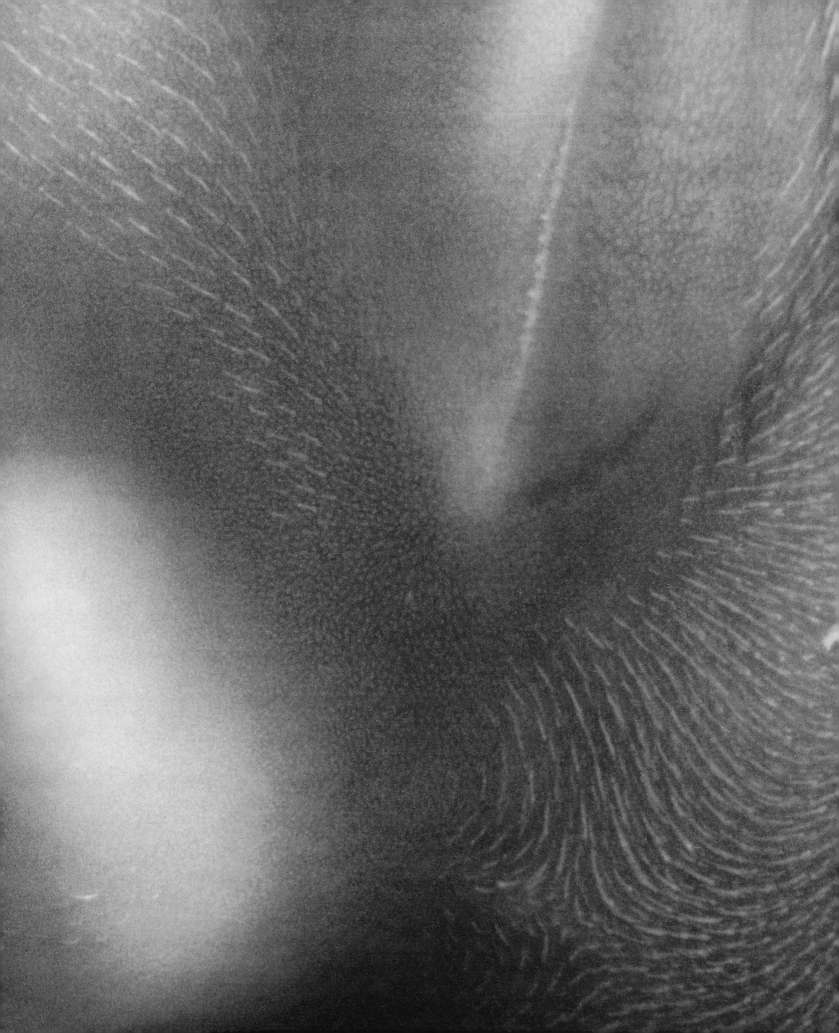

The migrating nerve cell, for example, behaves like a bloodhound sniffing its way to a person; it crawls "upstream" seeking higher concentrations of the attractant molecules. Eventually it reaches the source.

There is evidence that embryos have yet another kind of chemical signaling system, one that works even for cells that do not migrate. This system effectively tells them where they are situated within the embryo. The evidence comes from fruit fly embryos, but because it is well established that all living things share fundamental biological processes, researchers suspect the same system may operate in human embryos. This signaling system is a system of chemical map coordinates; certain cells at one end of the embryonic disc give off specific molecules that diffuse toward the opposite end. In the fruit fly there are signals that spread from head to tail and others that move from tail to head. There are molecules, too, that diffuse from top to bottom. Thus any one region of the embryo receives a specific combination of concentrations of chemicals that specifies its location. Those chemical signals may then act directly on genes or collaborate in some other process. The exact mechanism is not understood. In the human embryo the situation is even less clear. There is no evidence yet that humans use this chemical coordinate system. How a human embryo cell knows where it is remains a mystery.

Yet another coordination system must operate in embryos. In many cases a new functional ability requires not just one gene to come into play but groups of genes, all of whose protein products are needed at the same time. Thus the embryo must have a mechanism to coordinate suites of genes. Much of the coordination is done by what might be called master genes. When stimulated by the right combination of signals, a master gene commands the manufacture of several copies of a protein that can bind to several different, but functionally related genes, and trigger them simultaneously. In insects, where much more is known, one of these related sets of genes dictates the development of a whole leg. Biologists have found instances of mutant fruit flies in which cells that were supposed to develop into an antenna got the wrong signal and instead became a leg. The hapless insects have perfectly formed legs growing out of their heads. The mutation was in a master gene, a type that exists in nearly all species, including humans.

Once the nervous system has begun to form during the third week after fertilization, development accelerates. The circulatory system, as we have seen, begins to form soon after. The heart starts to beat around Day 22, but, curiously, it is not situated in the chest. The heart forms in front of the head. Two dome-shaped bulges swell out from the brain to start construction of the eyes. By the end of the fourth week, the embryo has sprouted four buds that will lengthen into arms and legs. During this same period, the embryo begins to fold, head and tail curling toward each other in a C-shape until they almost touch. This maneuver tucks the heart into the chest.

As the fifth week begins on Day 29, a slit widens across the head, the primitive mouth, and above that slit form pits that will become nostrils. Though the rudiments of a face are forming, a portrait at this stage would appear grotesque. The eyes are on opposite sides of the head, and the nostrils are wide apart. By the end of the fifth week the limb buds are longer and tipped with flat

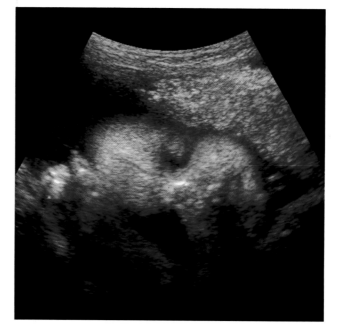

Baby's first portrait, this ultrasound scan of a full-term fetus shows the face, lying sideways. Commonplace today, ultrasound helps doctors detect abnormalities and problem births, often long before labor begins.

paddles. And, perhaps most important for our species, this is the week the brain begins enlarging—though not yet functioning—starting a process that will continue until the child is several years old.

The beginnings of the skeleton are also laid down in the fifth week. Like the circulatory system, the skeletal system arises in islands throughout the embryo. At numerous sites, special cells begin to secret the proteins that become the tough material cartilage, better known to some as gristle. In other islands, similar cells exude a protein material called pre-bone. This is a fibrous matrix that takes up calcium phosphate, a chemical present in the fluid between cells. Pre-bone gradually crystallizes into bone. From now through the rest of development, the tiny spots of bone will grow, and most of the cartilage will be replaced by bone. The enlargement of bone imprisons cells that produce pre-bone, stranding them in tiny compartments inside solid bone. There the cells survive by maintaining long processes that reach through channels in the bone to other bone cells and to sources of nutrients and oxygen.

As the embryo grows, its bones must keep pace. Throughout development, bones are continually being remodeled by two types of cells. One chews its way through bone, breaking down the calcium and returning it to solution. As existing bone is dissolved, the imprisoned bone-making cells are released to secrete new pre-bone. Through a delicate interaction between the bone-destroying and bone-making cells, the tiny bones of the early fetus are enlarged and reshaped to suit the growing fetus and, of course, the growing child. This same mechanism, incidentally, heals broken bones in later life. When the bone breaks, it frees bone-making cells and spurs them to action to knit the pieces together.

....

FINAL STAGE: THE FETUS
....

The embryo phase closes at the end of the eighth week, Day 56, when the foundations of all the major organs and structures have been laid down. Traditionally, this is the time that medicine drops the term "embryo" and begins to call the unborn a "fetus." All major organs are present, and the fetus has changed well away from its early resemblance to an embryonic fish or lizard. Now no

bigger than the last joint of a man's thumb, it looks clearly human. The fetus has a torso with arms and legs. Arms have hands with separate fingers, and those fingers have the beginnings of nails. Leg buds have become jointed legs with feet and toes. The face, however, still looks slightly grotesque, the eyes still too widely set, the nostrils not yet near enough to make a nose. The mouth is simply a slit without lips. With all the specialized tissue types established and properly positioned, the next 30 weeks until birth bring growth in size and an increasing complexity of cell types within tissues. Another month or so must pass before the fetus begins to look "normal."

The fetus also develops in another way: It begins to exhibit behavior. At eight weeks of development, a fetus can move, but its arms and legs swing and jerk without purpose. The muscles that power them are working independently, with no instruction from the brain. The brain is present and has nerve cells, but they are not yet linked to one another. They cannot receive or send signals. That will be possible only after pairs of nerve cells make links with one another. These junctions, called synapses, are places where the tip of a long process from one nerve almost touches the surface of another nerve. Synapses are necessary for even the most basic mental functioning. They are the switches and relays that must function before the fetus can move in any coordinated fashion. Similar contact points form where nerves deliver their commands to muscle cells.

The earliest of these hookups occurs around eight weeks—the time of transition from embryo to fetus. Nerves reaching out from the spinal cord make contact with muscles in the arms and legs, allowing them to twitch reflexively in response to outside stimuli. At this early stage, however, the spinal cord is not yet wired to the brain, so there can be no conscious control of movement. The motion is like that of a knee jerk in response to a thumping below the knee.

The fetus's earliest motions are too feeble for the mother to feel. As the developing human grows in size and strength, however, the day will come—typically between the 17th and 20th weeks—when the woman notices a poking or kicking in her belly. This is the so-called quickening, the event that people long ago believed was the sign that the unborn had suddenly been endowed with the mythical vital force that they believed caused motion and, hence, life. To many in those days the quickening was synonymous with the installation of a soul in the fetus.

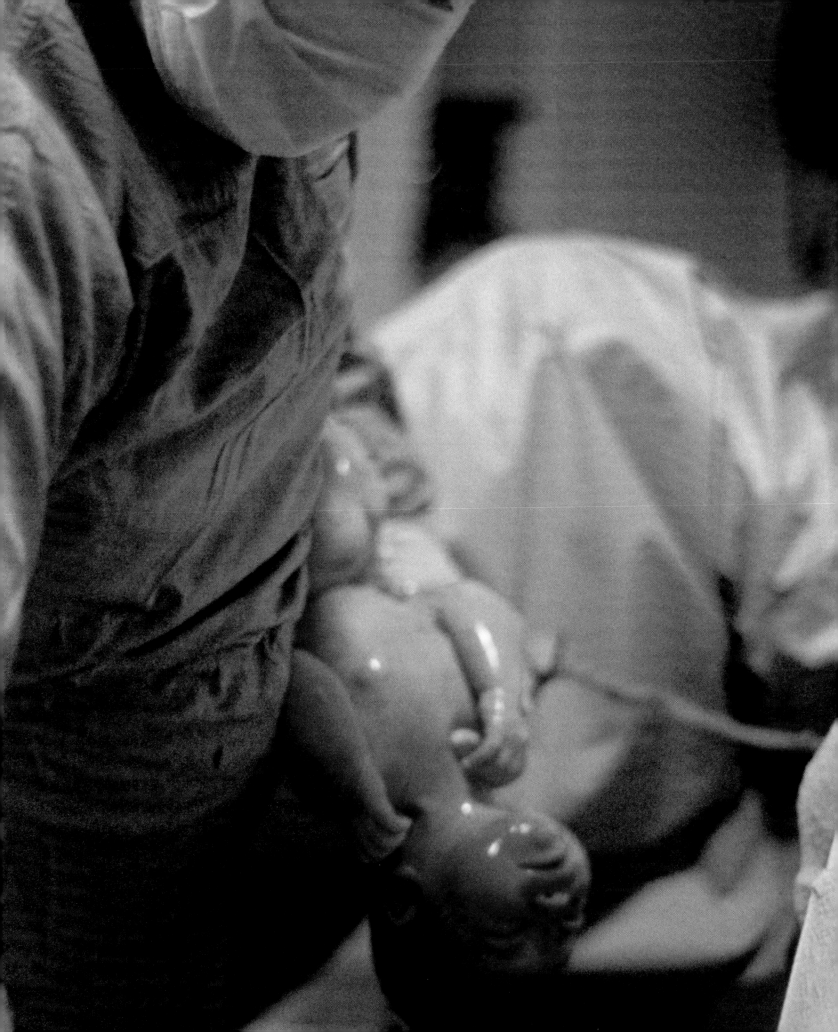

If the fetus can move and respond to outside stimuli, can it feel pain? This question has vexed many people, especially in the era of fervent debate over abortion. The beginnings of answers have come from more than a century of careful dissection of fetuses that died in the course of miscarriage or medically performed abortion. The evidence comes from study of fetal brains, for without a functioning brain it seems inconceivable that a fetus could perceive or feel anything. Thus, any inner experience that requires a brain, such as feeling pain, is unlikely in an embryo younger than 20 weeks, which is the time when the thinking part of the brain, called the cerebral cortex, has formed synapses with the rest of the nervous system.

Pain perception may not even begin to be possible until around 25 weeks. This is when nerve fibers from the brain's central coordinating organ, the thalamus, make connections with cells in the cortex. Still, the fetal brain is quite unlike that of a newborn. Studies of brain waves detected by a machine called an electroencephalograph, or EEG, show that the fetal brain does not begin to resemble that of a newborn until about 30 weeks of gestation.

This means that for its first six months, a fetus lacks the nervous system necessary to have anything that a born human would recognize as consciousness. After that transition, however, there is good physiological evidence that a fetus of seven months does have at least a rudimentary ability to feel.

The unconscious power of the fetus late in gestation is undeniable. Just as the conceptus commandeered the woman's body to turn it into a life-support system, the fetus now instructs her body that it is time for her to give it birth. When the fetus is about 38 weeks old—approximately 8 months and 25 days old—two critical steps occur. One, possibly signaled by the placenta, is the production of a hormone by the woman's uterus and by the corpus luteum, the structure on the ovary where the egg follicle formed nine months earlier. Called relaxin, the hormone acts on the cervix, the bottle-shaped, fibrous opening of the uterus.

For nearly nine months, the cervix, which opens downward, stayed clamped shut to keep the fetus safely inside. Relaxin makes the cervix soften, a process obstetricians call "ripening."

A second event is triggered by another hormone, oxytocin, produced in the mother's brain. Somehow her brain senses that gestation is complete and that it is time for the powerful uterine muscles, which have been developing for months, to take on their brief but heroic job. When the time is right, the placenta gives the final signal by shutting down its production of the hormone progesterone. This is the substance that has been inhibiting the uterine muscle from labor contractions all through the pregnancy. Now released from progesterone's control, the uterus begins contracting in powerful waves that build in strength over a period of hours, finally pushing the baby through the softened, stretched cervix.

The wonders of birth, however, are just beginning. The newborn child's body undergoes several rapid and immediate changes in its internal anatomy and chemistry. The umbilical cord is no longer a lifeline supplying oxygen and nutrients. These must now be taken in by the mouth and nose. Lungs that have waited, filled with fluid for months, must quickly empty themselves and take in air. The intestines, which never held anything but gulped amniotic fluid, must be ready to digest the proteins and fats in mother's milk. The heart and blood vessels, which have pumped blood out of the belly and into the placenta to obtain oxygen and nutrients, must shut off certain valves and open others to reroute blood flow through the lungs and intestines. The umbilical arteries squeeze shut instantly, preventing fetal blood from leaving the infant's body. The umbilical vein, which carried blood back to the fetus, waits a moment for placental blood to return to the little body before itself constricting. The cord as a whole shrivels on contact with air, and a few days later the dried cord falls off, leaving the belly button.

Still other dramatic changes take place at birth. With the baby's first breath, fluid is expelled from the lungs and air is pulled in. Oxygen enters the lungs and stimulates lung cells to secrete a hormone that will act on the circulatory system. Called bradykinin, the protein causes sphincter muscles to contract around various blood vessels and stay clamped until the vessel walls grow together, permanently blocking blood flow. One clamps

Entering a new world, babies' bodies perform a host of unseen changes. Valves, sphincters, and shunts of the circulatory system, for example, reroute blood flow from the umbilical cord—long the pipeline for oxygen and nutrients—to the untried lungs and digestive tract, which now must take over the job of sustaining life from the placenta and the mother.

down on the vessel that carried blood to the placenta and diverts the blood flow to the liver, which now must begin to fine-tune its many biochemical functions.

In other parts of the circulatory system other sphincters tighten, some over a period of weeks, to reroute the pattern of blood flow. The problem of "a blue baby" with "a hole in the heart" can arise at this stage. The hole is a normal part of the fetus's heart, allowing blood to cross from one side of the pump to the other as long as the fetus is to derive its oxygen from the mother's blood in the placenta.

At birth the lungs take over the oxygen-supply role and blood must course through them to take up the oxygen. The switching over in the circulation pattern is caused by the constricting of various sphincters, which reverses the difference in blood pressure between the two sides of the heart.

During the fetal period, a flap of tissue over the hole is easily pushed open to let blood cross over. When the blood pressures reverse, the flap is pushed shut and normally covers the hole. Over time it fuses with the heart wall. But if the flap is too small, part of the hole remains open and must be repaired surgically.

For all the complex events that must happen over the course of nine months to turn a fertilized egg cell into a baby, development is far from complete at birth. The child must, of course, grow and change for many years. The nervous system, one of the first systems to begin development, is one of the last to finish. Even as elementary a structure as the myelin sheathing around nerves is largely lacking.

For another year after birth, the child's nerves will gradually acquire the insulation that allows them to conduct signals properly and show the fine motor control that develops only gradually in infancy. The brain continues growing in size and complexity for several years. The reproductive system, of course, so essential to survival of the species, will not be fully developed for many more years. Not until adolescence will it be ready to renew the cycle of human origins, producing the sperm or the egg needed to relay life to a new generation.

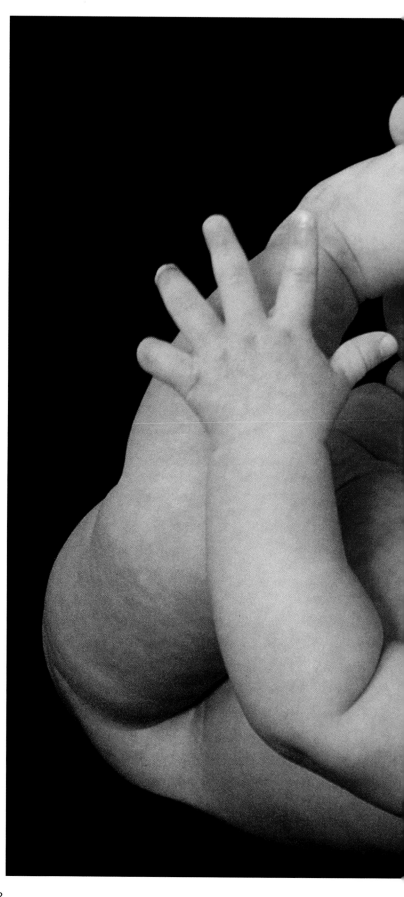

. . . .

Ten toes and ten fingers mark only one milestone in the development of a human being. At five months of age, this baby depends utterly on parents for food, shelter, and love. All are needed to support years of further development, both anatomical and behavioral.

. . . .

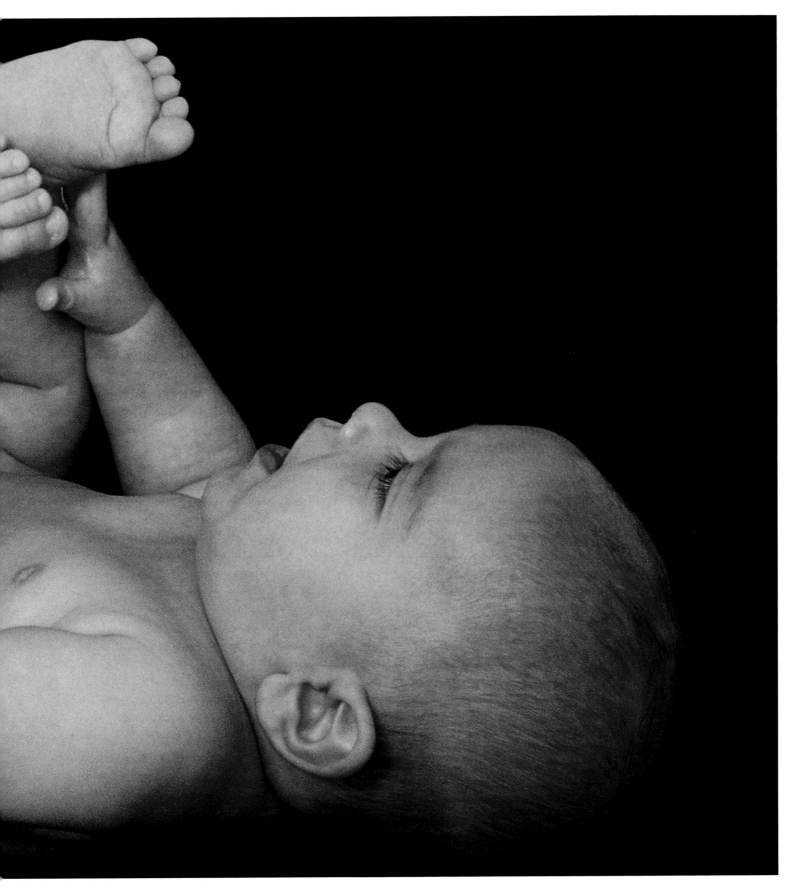

All systems are go in these bulked up—but superbly conditioned bodies of Japanese sumo wrestlers.

Systems

CAROLE J. HOWARD AND

JOEL SHURKIN

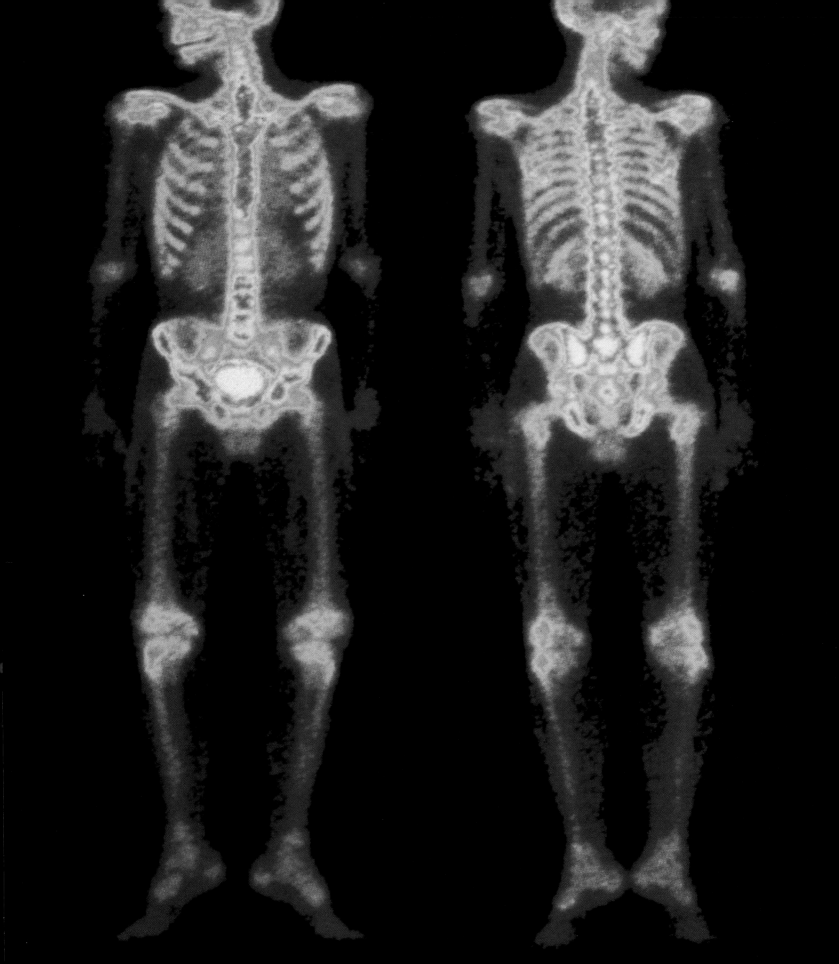

If the human body is a temple, it is surely the most elaborate ever wrought. It is a marvel of architecture, complete with domes, windows, arches, and thousands of miles of intricate passageways.

But this is no placid, subdued temple. Far from it. Every cell of the body, every fiber of its being, blazes with activity. The human body is a bustling place, even when asleep—it is always building, renovating, reproducing, growing. It converts energy from one form into another, it sends and receives messages, it fends off intruders, and it performs the most amazing balancing acts.

This human "temple" is constructed according to levels of organization that increase in size and complexity. Cells form the basic structural and functional units of life; they are the smallest living parts of our bodies. Some entire organisms consist of only a single cell.

Complex organisms—such as humans—require many levels of organization to keep things running properly. Groups of similar cells come together to form tissues; tissues unite to form organs. When two or more organs, along with their associated structures, join forces to perform certain vital functions—such as digestion or reproduction, for example—this group of organs working together is called a system. The complete organism comprises various systems.

From a biological standpoint, the ultimate goals underlying all the body's internal activities are survival and reproduction. With our large, sophisticated brains, we humans can set many goals for ourselves; but our bodies are run largely by systems designed to ensure that we survive and reproduce.

The human body may be considered to have ten systems, each with its own job but all highly interdependent. The main job of the skeletal system is to protect our innards and support us. Muscles help us move and respond to external stimuli. The task of the endocrine system is to maintain order among the body's trillions of cells. Both the digestive and the respiratory systems provide raw materials for our daily lives and for growth, and both carry off wastes. The circulatory system transports nutrient- and oxygen-rich blood throughout the body. The urinary system rids us of liquid waste, while the nervous system interprets and responds to stimuli from outside our bodies as well as those from inside. The job of the reproductive system is to ensure survival of our species. Skin, the integumentary system, holds the whole package together and helps protect us from invading microbes. The tissues that marshal bodily defenses are classified by some scientists as a system—the immune system, described in the next chapter of this book.

If one system fails, the others are affected, either directly or indirectly. Let's take a look, then, at how these ten systems do their jobs. For herein lies the magic that differentiates our living bodies from even the most exquisite temple.

SUPPORT: THE SKELETAL SYSTEM

Our skeletal frames are more than just scaffolding that holds us erect; they serve as the structures upon which we hang all that we are. Our bones are the enduring anchors to which muscles attach; they act as the levers and fulcrums of our daily activities; they are strong and sturdy, yet light in weight. They also cradle and protect our vulnerable vital organs. They house the cellular factories that formulate our blood. And they act as storehouses for phosphorous and calcium.

Bones are far from rigid, lifeless structures. Nerves etch their surfaces; blood vessels interweave them. Bones bustle with metabolic activity. Break one and you will immediately understand how sensitive they can be. By studying a bone from someone who died 100,000 years ago, a physical anthropologist can tell the sex, size, general health, diet, and age of the bearer.

The body of an adult contains about 206 bones. The reason for the "about" is that a few of us have an extra rib or vertebra. The difference is genetic and no one knows the evolutionary reason for the extra bones. A baby's bone count is higher than an adult's because some bones, particularly those of the pelvis and skull, converge and meld together as the child grows.

The skeletons of men and women are generally the same. The major exception is the pelvic girdle, which is

Going or coming, the skeleton serves as nature's armature for the human body. Some 206 bones compose the durable structure that supports our weight, underpins our movements, and shelters our fragile internal organs.

shallower and wider in women than it is in men, due to the needs of childbirth.

Bones start forming in the first two months of fetal life and continue to grow long after birth. Long bones, such as the leg and arm bones, do most of their growing where plates of cartilage separate the bone shafts from their ends. These cartilage cells grow, then die, leaving spaces for blood vessels and bone-forming cells called osteoblasts. As the osteoblasts multiply, they move outward, invading the cartilage and pushing the knobby ends of the bone farther away from each other. Cells left behind begin forming new bone.

The process continues into young adulthood, when the bones stop growing. All else being equal, we are as

renovation. Bone that receives sustained stress is replaced more quickly than bone that does not. Bones are far more complicated than they look from the outside. A cavity that runs along the center of bones holds yellow marrow, fat cells, and blood vessels. The ends of these bones enclose red bone marrow, which produces blood cells. The central cavity connects to the rest of the bone through an intricate labyrinth of small channels. These run through spongy material surrounding the cavity and out to a fibrous membrane that covers the bone, called the periosteum. The periosteum carries the blood supply and the nerves that make the bones sensitive to pain.

Bones come in various shapes and sizes, which reflect their varied functions. Long bones, such as the

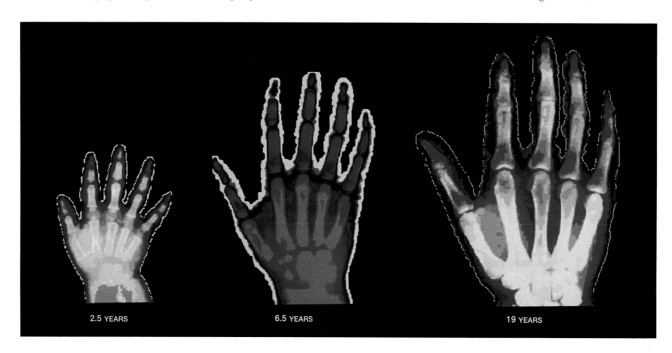

2.5 YEARS   6.5 YEARS   19 YEARS

tall as we are meant to be. When we reach our assigned height, the pituitary gland, working on instructions from our genes, tells the bones to stop growing. The ends of each bone solidify, and bone growth ceases. That's a clue in determining the age of skeletons ancient or modern, and that's why anthropologists inspect the condition of the bone ends. The end of growth does not mean bones become inert. New bone is constantly forming, while other bone is being torn down and reabsorbed—a form of

femur, act as levers; short bones, such as those of the wrists, are rather cube-shaped and provide support and flexibility; flat bones are protective shields. The femur is the body's longest bone; the ossicles of the middle ear, which vibrate to sound, are among the smallest.

The skull, whose crucial function is to support and protect the brain, consists of no less than 29 bones, 8 of which constitute the cranial vault, or cranium. The cranial bones are connected at irregular seams, called

....

From toy-toting toddler to ball-palming teen: Hands grow as cartilage cells near the ends of long bones degenerate and are eventually replaced by bone. Cavernous spongy bone (opposite) forms in areas of low stress; compact bone forms where stresses are greater.

....

sutures, which look like wavy lines on the skull's surface. The lower jawbone, or mandible, hangs from attachments to the cranium, forming the only truly movable joint in the entire skull.

Four bones of the skull contain the sinuses, small air-filled chambers lined with mucous membranes. These sinuses aren't just sources of headaches—though they certainly can be that when an allergy or infection inflames the mucous membranes and makes them swell. Sinuses connect with the nasal cavity to drain fluids. They also make the skull lighter without lessening its strength, and they act as echo chambers that add resonance to the human voice.

The skull rests on the spine, literally the backbone of the whole operation. The spine consists of 26 irregular, drum-shaped bones called vertebrae. These form movable joints that allow us to bend and twist, curl and stretch.

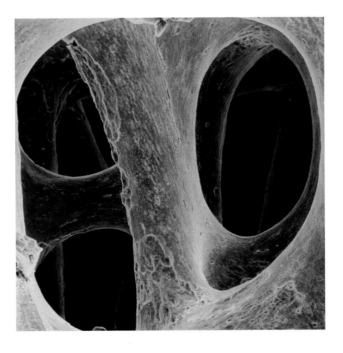

The vertebrae are separated by fibrous, fluid-filled disks of cartilage that regularly absorb hundreds of pounds of pressure per square inch. They are, in fact, the spine's shock absorbers. Whenever you lift something, your back ultimately has to bear the weight.

In addition to providing support, the spine also protects the spinal cord, the body's vital communication link with the brain. Connected to the spine, the rib cage protects the heart and other internal organs, such as the liver and the spleen.

Nature tends to repeat models that work. Most mammal limbs have essentially the same design: one long bone running from the main torso to a complicated joint, either the elbow or the knee. From that joint, two bones run to another complicated joint, the wrist or ankle, and from there to a hand or foot. The bones of the hands and feet are themselves remarkably similar, except that the foot bones are much larger and the big toe is unequipped to swivel as the thumb can.

An old spiritual sings that the foot bone is connected to the leg bone and the leg bone is connected to the thigh bone. In truth, every bone of the body is connected to another, with one exception: The hyoid bone in the throat, which supports the tongue muscle, stands alone.

Bones connect at joints, mechanical works of art that allow us to bend in all manner of ways—simultaneously. The body's most mobile joints—including those joining the various arm and leg bones—are cushioned and lubricated by a slick, clear fluid. Joints are designed in various forms, depending on their function. Hinge joints, such as the elbow and finger, are the simplest. The shoulder and hip joints, ball-and-socket structures, may be the most versatile. They move up and down, forward and backward, as well as around. Some joints pivot, such as the one at the base of the skull that allows us to shake, nod, tilt, or turn our heads. Movement is more limited in gliding joints, in which two nearly flat surfaces slide over one another. Some joints in the foot move this way.

Perhaps the most interesting and important joint of all, however, is the saddle joint at the base of the thumb. Both of the bones meeting here have concave and convex areas, allowing them to rock back and forth. The result is an opposable thumb, a thumb that can grip strongly in a direction opposite from the rest of the fingers. On the strength of this little joint—combined with our agile brains—we have built tools, towns, civilizations.

The body's skeletal system isn't perfect; things can and do go wrong. Bones break, and joints, despite their ingenious design, don't always stand up to the pressures we place on them. Ask any athlete. Bones can fracture in various ways. If the two broken ends are kept in contact, they usually heal, which is why doctors set the bone's edges so that they touch.

Joints suffer myriad ills, ranging from torn ligaments to torn cartilage to dislocation. In addition to the outside

forces of wear and tear, attack can come from within. One auto-immune disease in which the body's own immune cells mount a treasonous campaign on the joints is rheumatoid arthritis, which can result in swelling, pain, and deformity. Certainly, we can have other problems with our skeletal system. On the whole, however, it serves us well—an engineering marvel that supports the miracle that is a human being.

MUSCLES MAKE MOVEMENT POSSIBLE

Muscles provide the force behind every movement. They allow us to run, carry, speak, play music, create art. There are three types: skeletal, smooth, and cardiac muscles. Smooth muscles are also called involuntary muscles because they run without conscious thought. They line the walls of our blood vessels and internal organs, helping move blood through the arteries and propel food through the gut.

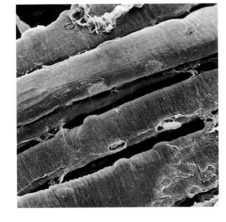

Unlike smooth muscles, both cardiac and skeletal muscles have a striped or striated appearance when viewed under a microscope. Cardiac muscle, found only in the walls of the heart, pushes blood out of the heart and through the body. Its action, like that of smooth muscles—is involuntary. You can't will your heart to stop, although some people who practice yoga or biofeedback can apparently slow it down to some degree. Most of us have strong hearts. They flutter and begin beating around the 22nd day after conception; they never stop, one hopes, until we die, perhaps some two and a half billion contractions in an average lifetime.

Skeletal muscle attaches to bones. Its movement is voluntary—we choose to tap a foot, wave an arm, crook a finger, or smile at a friend. By and large, when we speak of the muscular system, we are referring to skeletal, or voluntary muscles.

The human body moves by the lever principle. A lever is the simplest and one of the most efficient of all machines. Basically, it involves a rigid bar, or lever arm, that pivots on a fixed point, called the fulcrum. When a force is applied to one part of the bar, that force moves a weight elsewhere on the lever. When we move our bodies, bones serve as levers and joints act as fulcrums; skeletal muscles supply the force. Mostly we use what engineers call a third-class lever in which the force is applied to the lever between the weight and the fulcrum. The biceps muscle of our upper arm, for example, exerts pressure beyond the elbow, or fulcrum, on the forearm, which lifts the hand—the weight.

This mechanical system can be amazingly powerful. Naim Suleymanoglu, a Turk only four feet eleven inches tall and weighing all of 141 pounds, lifted 738 pounds in the 1996 Olympics using exactly that biceps leverage—with help from the other levers and muscles in his body.

It takes more than 650 skeletal or voluntary muscles—layered in overlapping, complex patterns just below the skin and fat, or deeper yet—to run our bodies. These muscles account for about half our total body weight; the percentage generally is higher for men than women. Some voluntary muscles, like those in our backs, are built for strength; others, such as those controlling our hands, are gracile cords capable of the subtle motions required to play a Mozart concerto.

Fibrous cords and sheets of connective tissues—tendons and aponeuroses, respectively—tie muscles to bones. Tendons extend from the hands up the arm to the muscles that help control them. Because some of the controlling muscles are separate from where the work actually is done, we can have very powerful but compact hands. They can play that concerto with lightning finger

In a clutch, muscles often hold life in the balance (opposite). The body's 650 muscles work in groups, accomplishing motion by either contracting or relaxing. Each contains bundles of muscle fibers (above), which react to signals from the central nervous system that trigger energy-producing chemical reactions.

strokes with just the right pressure, expressing not just the notes Mozart wrote but the soul of the pianist as well.

The basic unit of the muscle is the muscle fiber, a single cylindrical cell that often runs the entire length of the muscle. Muscle cells are among the longest cells in the body—some in the thigh stretch more than a foot in length. Muscle cells are bound together like wires in a suspension bridge cable.

Each cell consists of between 1,000 and 2,000 even thinner cellular strands called myofibrils, which actually do the contracting. Viewed through a microscope a muscle cell looks striated with alternating bands of dark and light because myofibrils are composed of thick and thin myofilaments of proteins partially overlapping one another.

When your muscles relax, the thick and thin myofilaments overlap only a little. The arrangement looks much as your hand looks when you interlace your fingers up to the first knuckle. When a message from nerves tells the muscles to contract, the thick filaments slide further down among the thin ones and look like hands closely entwined. The myofibrils shorten, and so does the muscle. The shorter the myofibrils get, the more the muscle contracts.

Muscles run on electricity in the form of ions—atoms that have lost or gained an electron, giving them an electric charge. The rush of these charged particles into the muscle cell is commanded by motor neurons, nerve cells coming from the spinal cord. Each motor neuron commands an average of 150 muscle fibers, but some neurons may jolt several thousand fibers.

All a muscle can do when stimulated is contract; it can't stretch or elongate. Muscles can only pull; they can't push. Movement requires an elaborate symphony of muscles to perform even the simplest tasks. Active muscles must work in harmony with those not contract-ing. Imagine, then, what it takes to play that concerto! Want to lift an arm, for example? Just having one muscle contract won't do the job. The brain sends a quick series of orders: The front and back sectors of the shoulder's deltoid muscle balance each other, while the middle section exerts the lifting motion. You can make your arm go up slowly or quickly by moderating the balance between different parts of different muscles.

Muscle fibers cannot contract partially. How much a muscle contracts depends, rather, on how many fibers get stimulated. The lowest level of nerve stimulus required to contract a fiber is called the threshold. Any stimulus above that threshold causes the fiber to contract all the way; with any lesser amount, nothing happens.

We can get subtle motion from a muscle because the brain orders just the right number of fibers within that muscle to contract. Facial expressions are deceptively complex. It takes more muscles—sixteen—to frown than it does to smile—eight.

Muscles differ in how fast they can contract. Some are fast-twitch, others are slow-twitch. Slow-twitch fibers are dark; fast-twitch are light. The difference shows in their color: In a chicken or turkey the type of muscle is evident in the color of the meat. Red fibers, the slow-twitch ones, are relatively small muscles that are full of blood vessels. They need a lot of oxygen and use fatty acids as fuel, the reason dark meat usually is greasier than white meat. These fibers use up energy as they go and are designed for endurance.

Chickens walk much more than they fly. So they need the staying power of slow-twitch muscles in their legs—the dark meat of their drumsticks. Fast-twitch muscles in their breasts accommodate their frequent but brief wing flappings. White fibers—the fast-twitch muscles—are designed for quick bursts of power; they are subject to oxygen deficits and lactic acid build up. They are fueled by glucose, the sugar that serves as the body's basic energy source. Wild fowl, in contrast, have red meat in their breasts because they use their chest muscles for some serious flying, not just flapping around the henhouse.

Muscle fibers for most of us are about half red and half white; the exact proportion is genetically determined. Long-distance runners and swimmers tend to have more red than white; sprinters more white than red.

Any muscle cannot hold a contraction for very long. The fuel that powers it burns up in a matter of seconds, and the body can't replace it fast enough. The body then turns to other sources of energy. One such energy source is a substance in muscles called creatine phosphate, which can boost fuel production briefly. It sustains the muscle for about five to eight seconds. Beyond that period a process called glycolysis allows muscles to borrow from stores of glucose in their cells. Glucose is a form of sugar capable of producing large amounts of energy, which is why eating a chocolate bar can spur you into activity when you're feeling lethargic.

Energy produced by muscle contraction provides most of the heat generated in our bodies. That's why we get so warm when we exercise. We shiver when cold because the brain orders the muscles to contract rapidly to burn fuel, create heat, and drive up the body's temperature. Several hours after we die, the lack of any fuel causes the muscles to go into solid contraction. This makes the body go rigid, in the condition known as rigor mortis (the reason a dead body is sometimes called a "stiff"). Rigor mortis lasts for 15 to 25 hours; then the muscles relax—forever.

Muscles are more forgiving than bones; if injured they usually repair themselves. Overexertion can cause injuries to muscles and tendons, and they may even tear in some cases. You can increase a muscle's endurance and strength with the proper conditioning. The involved bones will become sturdier as well—more minerals will be deposited.

that has delivered its nourishment and contains carbon dioxide and other cellular wastes.

Every cell in the body needs oxygen, nutrients, enzymes, proteins, and hormones to live. Blood delivers these supplies. The waste products of life—carbon dioxide and the toxins normally built up by metabolism—have to be carted away. Blood performs that service as well. It also carries the soldiers of the immune system—white blood cells—which patrol the bloodstream and pounce on invading microbes. Blood provides clotting abilities to prevent its own loss and helps regulate body temperature and the acid-base balance in the body. Only after decades of research and the lure of huge profits, are we coming close to producing an artificial substitute.

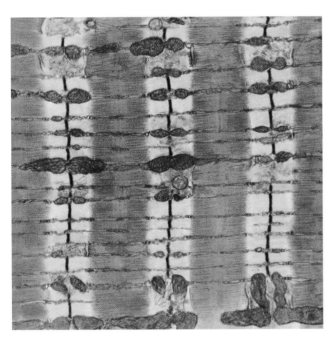 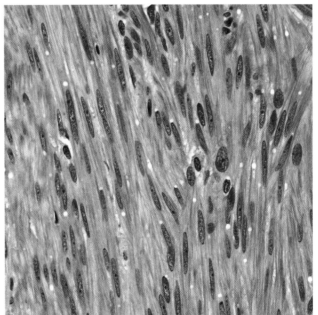

TRANSPORT AND DELIVERY: THE CIRCULATORY SYSTEM

Blood, as the poet Alice Meynell wrote, is the color of life. The circulatory system transports that vivid fluid.

Driven by the heart, blood flows through the body; arteries carry oxygen-laden blood; veins transport blood

Blood makes up about 8 percent of our body weight. Plasma, a straw-colored fluid that is about 90 percent water, is the liquid part of blood. A variety of dissolved substances—mostly proteins, plus various nutrients, gases, hormones, and electrolytes—make up the rest.

The solids in blood include red blood cells, white blood cells, and platelets. These three are largely responsible for performing the blood's main jobs—transportation, protection, and regulation. Platelets,

Striated skeletal muscle (left) contains bands of the proteins myosin (yellow) and actin (pink) that slide across each other as the muscle contracts. The same proteins occur in smooth muscle cells (right), but their arrangement does not produce striations. Smooth muscle forms the walls of many internal organs and is organized to work for long periods without tiring.

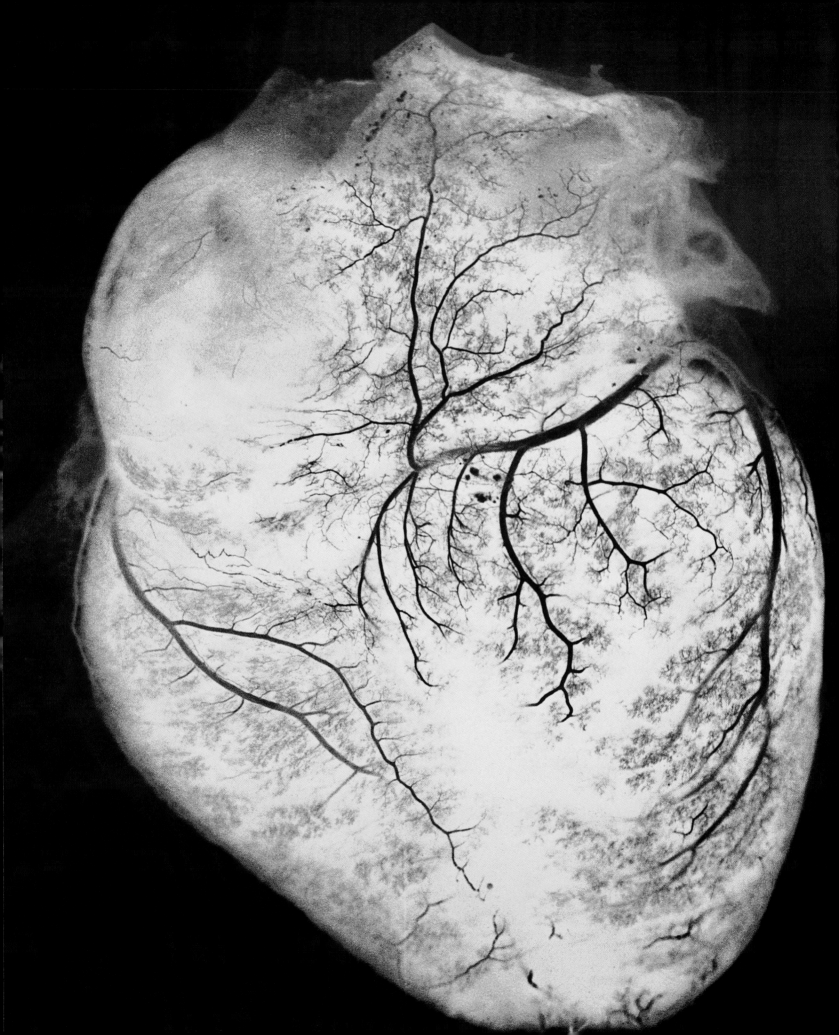

the smallest formed elements in blood, are essentially cell fragments containing chemicals that initiate the highly complex process of clotting. Clotting serves to plug up breaks in blood vessel walls following an injury, thus helping to prevent blood loss.

White blood cells, far less numerous than their red counterparts, represent less than one percent of total blood volume but are the warriors that defend our body against invasion, rushing to trouble spots to challenge invaders and heal wounded tissues. Red blood cells contain hemoglobin, an iron-bearing compound that combines easily with oxygen. Hemoglobin combines with oxygen from the lungs, and the red cells then transport it throughout the body.

When red cells reach an area low in oxygen, the bond to the oxygen breaks, and the gas is absorbed into those tissues. The combination of hemoglobin and oxygen gives blood its bright red color. Red blood cells are concave on both sides, like jelly donuts with the jelly sucked out. This shape increases the surface area of each cell, enabling it to disperse more oxygen to tissues.

Vital as they are, red blood cells are short-lived, surviving only about 80 to 120 days. And, because they have no nuclei, they can't repair themselves as other cells do.

Three million of your red blood cells died as you read this sentence. Fortunately, the red bone marrow in your body produced an equal number of replacements in that interval. And, should your kidneys ever detect that the red blood cells are providing an insufficient amount of oxygen to the rest of your body, they will order the marrow to make more of them.

The heart transports and directs the blood. Commander of the entire circulatory system, the heart is a muscle that pumps throughout a lifetime, driving blood through our bodies. Though the ancient Greeks and Romans were ignorant of the heart's true function, they assigned great importance to it; they thought the soul abided there. That notion is still reflected in our language. When you are in love, you give your heart. Your heart may break—or ache; you might wear it on your sleeve.

Not until the early 17th century did the English physician William Harvey discover that the heart is a pump. And what a wondrous, amazingly durable pump it is. Cardiac muscle fibers, unlike many other cells in the body, do not die and get replaced. The heart muscles you were born with never rest, once they begin work in the third week after conception.

About the size of a grapefruit, the heart is actually not one pump but two; each half is a coordinated pump. The top chamber of each—called an atrium—receives incoming blood and passes it through valves to a larger, lower, V-shaped chamber we call a ventricle. Valves between the atria and ventricles operate like spring-loaded trapdoors, allowing blood to surge through—in one direction only. In some people, the valves do not close properly, allowing some backflow to the prior chamber; this malfunction can be heard through a stethoscope as a murmur.

The two sides of the heart pump different types of blood. Blood coming from the lungs, freshly supplied with oxygen, flows into the left atrium. Oxygen-depleted blood collected from other parts of the body returns into the right atrium.

Surrounded by the coronary arteries and veins that serve its own cells, the heart (opposite) is an indefatigable muscle, contracting 100,000 times a day as it sends blood to the body's farthest reaches. Specialized cardiac muscle cells (above) combine features of both smooth and skeletal muscle types; each cell stands capable of beating on its own.

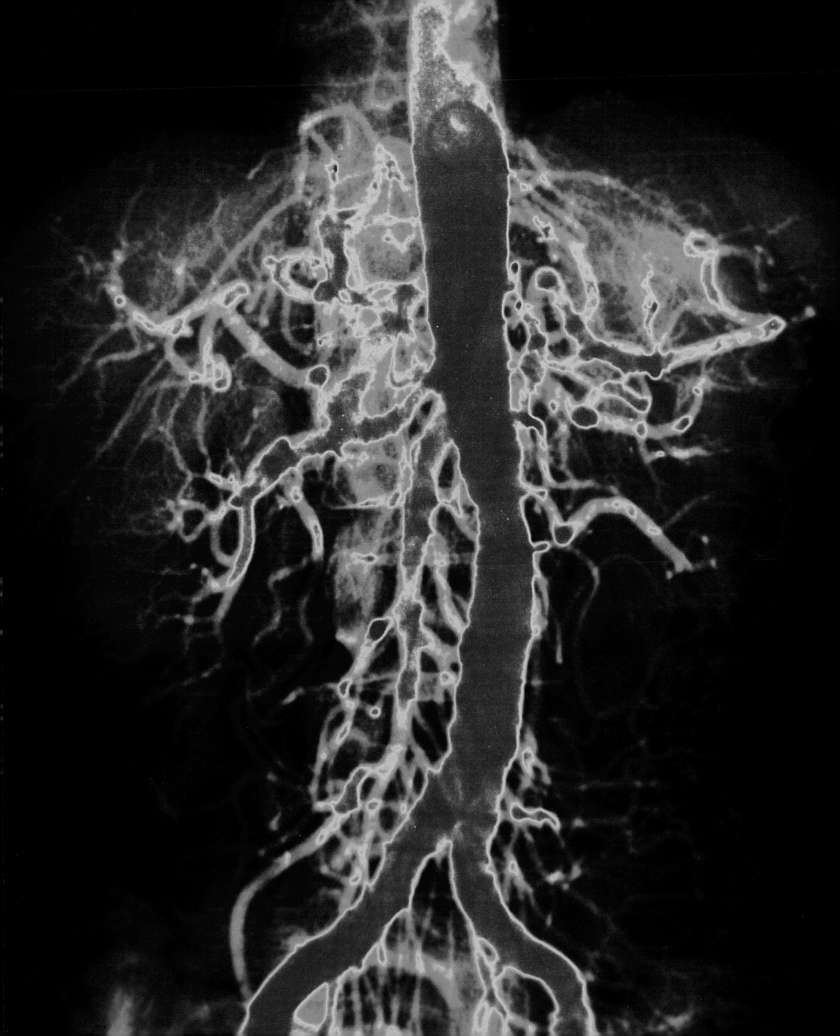

AORTAGRAM OF AORTA (OPPOSITE)

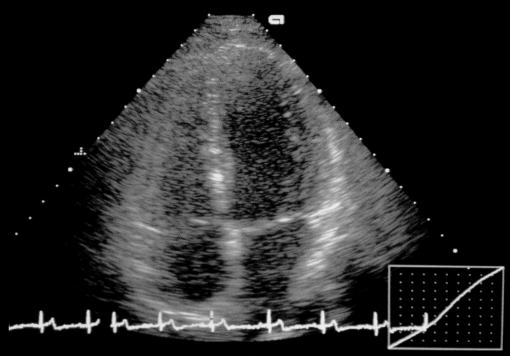

ULTRASOUND OF HEART SHOWING (UPPER) ATRIA AND (LOWER) VENTRICLES

Light years seem to separate the humble stethoscope from modern imaging techniques that reveal the heart's innermost workings. An aortagram (opposite) highlights blood flow in the aorta, the heart's outlet for freshly oxygenated blood. Echoing sound waves create an ultrasound image of the heart, its four chambers outlined in blue (above). Electrocardiogram (ECG) tracings of the heart's electrical activity ride below. PET scans of the left ventricle, in longitudinal section, use radioactivity to map blood flow (below, left). Cooler colors in the middle pair of images indicate a restriction and explain the presence of chest pain, perhaps announcing an impending heart attack. Both ultrasound and ECG combine to evaluate the state of the heart's blood flow over time (below, right). FOLLOWING PAGES: Snapshot of the river of life, disk-shaped red blood cells and blue-tinted white ones course through a tiny section of vein.

BLOOD FLOW

CHEST PAIN

NO CHEST PAIN

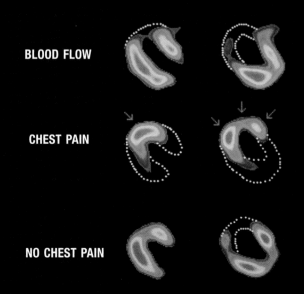

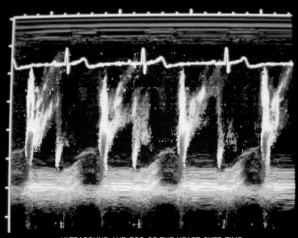

ULTRASOUND AND ECG OF THE HEART OVER TIME

PET SCANS OF BLOOD FLOW IN THE HEART

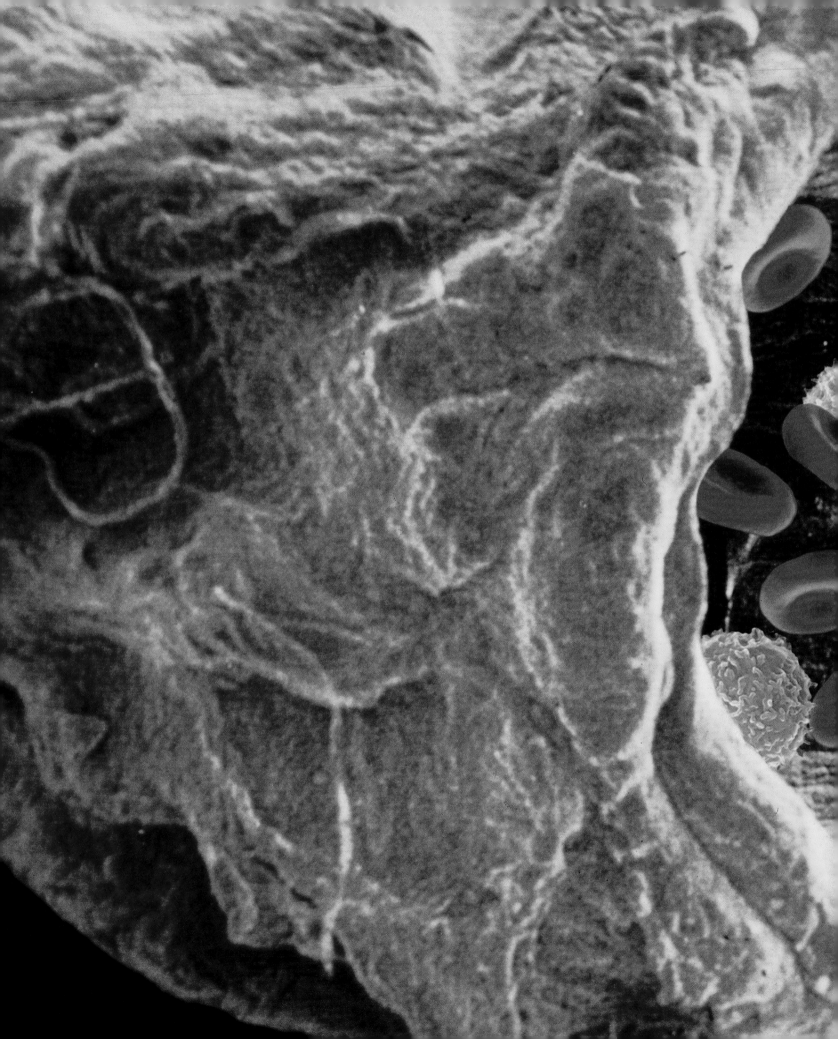

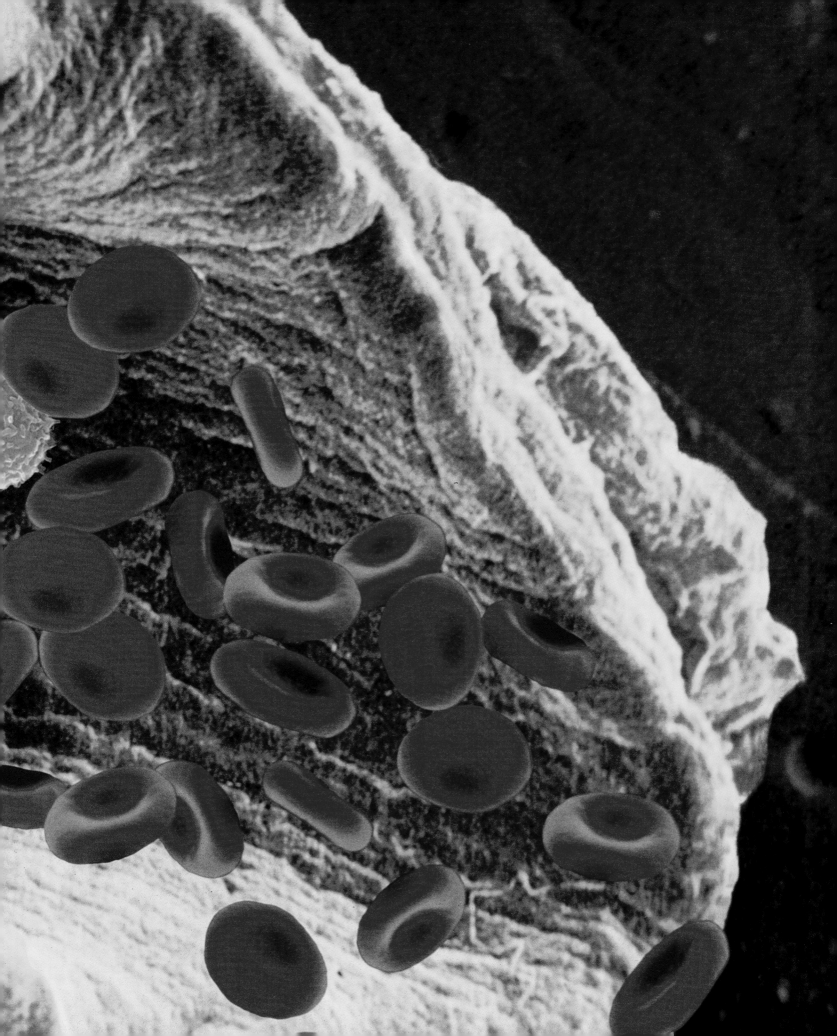

The atria have relatively thin walls and expand as they fill. The two atria then contract simultaneously, pumping their respective loads through the valves and into the ventricles. Filled with blood, both ventricles then contract. The left ventricle sends its bright red, oxygen-rich load out to the body. The right ventricle pushes its cargo of deoxygenated blood to the lungs, where it is replenished with oxygen.

Blood on its way from the heart to the lungs for an oxygen recharge is carried in veins; oxygen-rich blood traveling away from the heart to tissues is transported in arteries. To cope with the high pressures generated by the ventricles forcing blood to the body, arteries have thick, tough, elastic walls.

The main artery, the aorta, is about an inch in diameter when it branches from the heart. Veins, which carry deoxygenated blood from the tissues back to the

blood to exchange nutrients and gases with surrounding cells and tissues. Carbon dioxide and waste materials diffuse from the tissues into capillaries; oxygen and nutrients move from them into the tissues. The blood, darkened by its loss of oxygen and bearing a load of wastes, then moves up the capillaries into venules, the tiniest of veins. These merge to form larger veins, which return the blood to the heart. Nearly 70 percent of the blood in our bodies is in the veins at any one time.

By the time blood reaches veins, the pressure is so low it needs help to return to the heart. Peripheral veins contain one-way valves, cuplike structures something like those in the heart, to ensure that blood travels forward. If the valves weaken and fail, blood pools below them, forming the ballooned, twisted structures known as varicose veins. Venous valves are more passive than those in the heart and need outside help. Skeletal muscle

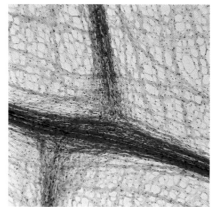 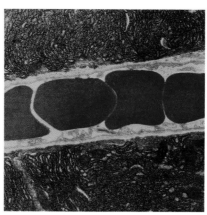 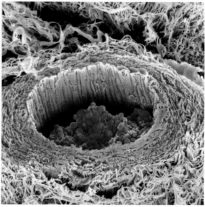

BRANCHING CAPILLARY   RED BLOOD CELLS IN CAPILLARY   CROSS SECTION OF ARTERIAL VESSEL

heart, operate under much lower pressure than arteries do, and so have thinner walls.

As they leave the heart, arteries branch out, growing progressively thinner and narrower, eventually becoming arterioles. Arterioles, as they reach the tissues, shrink further until they lead to the body's smallest vessels, capillaries. Composed of a thin layer of smooth muscle and an outer layer of connective tissue, these microscopic tubes are so narrow that red blood cells must squeeze through them single file. This is where the action is. Capillary walls are only a single cell layer thick, allowing

provides the extra push. Physical activity is crucial to circulation. Doctors recommend moving around some when you are on a long airplane flight, for example, or sitting at a computer. Even the simple act of inhaling can help move the blood in veins on its way; the abdomen swells as you inhale, increasing pressure on nearby veins.

Like any flowing material, blood moves from an area of high pressure to one of lower pressure. The standard instrument for taking blood pressure measures that fluid pressure against a column of mercury, measured in millimeters (abbreviated mm Hg). In the human body,

....

Smallest and thinnest of blood vessels, capillaries bridge the network of veins and arteries, making intimate contact with body cells as they deliver oxygen and nutrients and cart off wastes. Although only 1/3,600 of an inch wide, red blood cells must push through capillaries in single file. Deformed red blood cells (opposite) signal the potential for sickle-cell anemia, a severely debilitating disease that affects blacks much more commonly than whites.

....

maximum pressure occurs as the aorta walls expand to accommodate a surge of blood from the heart. This peak pressure is called the systolic pressure, and in a healthy adult it should average about 120 millimeters of mercury. When the wave of blood leaves the aorta, the aortic walls return to their original shape and the pressure drops to its low point, what doctors call the diastolic pressure. In a normal adult the diastolic pressure should run between 70 and 80 mm Hg. Blood pressure above 139/89—that is, with a systolic of 139 mm Hg and a diastolic of 89 mm Hg—is considered high.

Blood pressure is determined by the amount of blood the heart pumps in a given time, by how much resistance the blood meets as it courses through the system, and by the amount of blood in the system. A change in heart rate or in stroke volume—the amount of blood pumped by each contraction—changes the pressure. The stiffening of the blood vessels that normally occurs with age tends to raise pressure. Damaging deposits may also form within the vessels as we grow older.

The main long-term feedback controllers of blood pressure are the kidneys. When blood pressure or volume increases, the kidneys allow more water to pass out of the blood and form diluted urine. When pressure decreases, the kidneys conserve water, and the urine becomes concentrated. Diuretics, a common treatment for high blood pressure, override the kidneys' control and force them to release water.

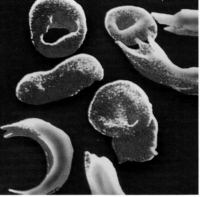

DAMAGED RED BLOOD CELLS

The circulatory system is a marvel of engineering, but things do go wrong with it. The most dangerous and best known of circulatory disorders, atherosclerosis, involves the narrowing of arteries by fatty deposits on the arterial walls. When these plaque deposits occur on the coronary arteries—which serve the heart muscle itself—they can prevent sufficient blood from reaching the hardworking heart. One result can be angina, a dull, pressing chest pain usually brought on by exertion, which is a signal that the heart is receiving insufficient oxygen. Angina may be followed by a full-blown heart attack. This occurs when a narrowed coronary artery becomes fully blocked— usually by a clot—and the part of the heart muscle it supplies then dies. Sometimes such an attack stops the heart from beating; death usually follows unless the heart can be restarted. Risk factors for atherosclerosis include smoking, lack of exercise, high blood pressure, a diet high in fats, and genetic predisposition.

AIR CIRCULATION: THE RESPIRATORY SYSTEM

Without oxygen, we would die within minutes. Every cell in the body needs it to do its metabolic business. The respiratory system—the lungs and connecting tubes— brings oxygen into the body. Our lungs alternately inflate and deflate, usually between 12 and 20 times a minute. When we exercise, we breathe more rapidly because our muscles are using more oxygen, which must be replaced.

The respiratory system starts processing air as soon as we inhale it. The nasal passages immediately filter, warm, and humidify it. The entire nasal cavity is lined with mucous membranes full of blood vessels; heat from the blood warms the incoming air as it passes. Evaporation from the membranes acts as a humidifier, moistening the air. The surface layers of the mucous membranes contain cilia, tiny hairlike structures that literally wave in the wind, creating a flow of mucus down the throat, or pharynx. Small particles such as bacteria and dust become trapped in the sticky mucus and can then be swallowed.

The pharynx, too, is lined with mucous membranes that further filter the air. Below lies the larynx, or voice box, home of the vocal cords. Vocal cords vibrate when air rushes between them, generating sound waves that we can then shape into sounds. The walls of the larynx are mostly cartilage. The Adam's apple—that slight protuberance at the front of the throat—is actually formed by the largest cartilage of the larynx.

The larynx marks an important crossroads in the pathways for digestion and respiration. The esophagus— a muscular tube in the throat—carries food from the pharynx to the stomach. In front of the esophagus, the trachea, or windpipe, carries air to and from the lungs.

When we swallow, the epiglottis, a small cartilage suspended by muscles and ligaments over the opening to the larynx, presses down like a lid, closing off the trachea and preventing any food or liquid from entering it. On the rare occasion that the epiglottis fails as we eat, we cough. If food gets stuck here, we can choke to death.

The trachea measures about an inch in diameter and about four inches long. It is lined with mucous membranes and waving cilia that continue to filter the air and push harmful particles back up toward the pharynx, where they can be swallowed. The end of the trachea splits into two branches, the right and left primary bronchi, each leading to one of the two lungs.

Breathing is automatic. While we can regulate the rate and depth of breathing to some extent, it mostly is run from the brain's involuntary centers. During inhalation, the diaphragm, a sheet of muscles that forms the floor of the chest cavity, contracts, pulling downward to enlarge that cavity. This creates a lower air pressure in our lungs than exists outside the body, so air flows into the lungs. Exhalation reverses the process. The diaphragm muscles relax, the chest cavity shrinks, and pressure within the lungs builds. This causes a pressure differential in the other direction, and air rushes out.

Lungs, which occupy most of the space behind our rib cage, are soft and spongy air-filled sacs. They are made up mostly of the air tubes, cells, blood vessels, and capillaries—all bound together by tissue. The bronchi branch repeatedly, subdividing into innumerable small passageways called bronchial tubes that end in microscopic pouches called alveoli. The arrangement is an intricate network that looks like an inverted tree, with the trachea as the trunk. The actual exchange of gases—the business of breathing—takes place within the alveoli. Like tiny balloons, they are elastic, thin-walled, and filled with air. Since all the body's metabolic processes produce carbon dioxide, a gas that would suffocate us if we had only it to breathe, the respiratory system must remove carbon dioxide the cells produce as well as transport oxygen to the cells.

....

Blur of a runner hints at the intense activity of his respiratory system—a system designed to bring life-sustaining oxygen into the body and remove unneeded carbon dioxide. The exchange takes place in the lungs, twin organs protected by the rib cage. At top performance an athlete may process 300 quarts of air each minute; a couch potato, only eight.

....

Fresh air we breathe into the alveoli is relatively rich in oxygen and poor in carbon dioxide, compared to the blood that is moving through the capillaries surrounding the alveoli. This blood is oxygen-poor and carbon dioxide-rich because it has already made its circuit of the body, depositing oxygen and picking up carbon dioxide waste from the cells it has reached. Due to such differential pressures of these gases, oxygen-rich air diffuses out of the alveoli and into the capillaries, where 95 percent of the incoming oxygen is bound by hemoglobin in the red blood cells. Similarly, carbon dioxide diffuses out of the blood and into the alveoli, for exhalation.

In all capillaries of the body except for the lungs, the situation is exactly reversed; the oxygen concentration is greater in the capillaries, forcing diffusion of that gas into the surrounding cells. Simultaneously, carbon dioxide, which is produced by the cells, leaves them and enters the bloodstream.

We need some 300 million alveoli to provide the enormous surface area required to get enough oxygen into the blood to keep our millions of body cells supplied. Each lung holds only a gallon or two of air, yet its internal surface area rivals the size of a tennis court.

The respiratory center in the brain largely regulates breathing; we have little conscious say in the matter.

Under normal circumstances, two groups of neurons control the breathing process. One causes the chest muscles and diaphragm to contract when we inhale; the other causes them to relax when we exhale.

Breathing is controlled by a sophisticated feedback system through areas in the body sensitive to carbon dioxide, oxygen, and hydrogen. One respiratory center is the medulla, in the brain stem. It is exquisitely receptive to high levels of carbon dioxide in the arteries of the brain. Should levels reach a certain point, the medulla orders faster and deeper breaths. You can change this process somewhat—emotions can have an influence on how you breathe, for example. But breathing is too important to be left to free will.

You cannot hold your breath beyond a certain point, no matter how hard you try. Even if you fall unconscious, when the carbon-dioxide level in your blood gets high enough, your medulla will order more breathing, and you will inhale. This happens to drowning people: They try not to breathe so as not to inhale water, but sooner or later carbon dioxide builds to the maximum permitted level, and they inhale.

Some mammals are better at holding their breath than we are. Weddell seals, for instance, can stay underwater for an hour and dive as deep as 2,000 feet. Although their lung-to-body size is

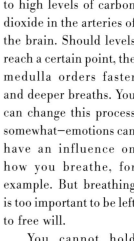

Inside the lungs (top), numerous treelike branchings of blood vessels and air passages greatly multiply the surface area. Millions of tiny alveoli (above) allow oxygen from a bronchus (blue) to diffuse into a capillary (red) while carbon dioxide takes the opposite route. Fingerlike cilia lining the windpipe (opposite) waft harmful particles away from the lungs.

similar to that of humans, the seals have proportionally more blood in their bodies than humans as well as more red blood cells.

The combination of adaptations a seal calls into play when diving deeply is called the diving reflex. Humans have one, too. Several years ago a Michigan college student in a car accident crashed through the ice on Lake Michigan and was submerged for more than 30 minutes. When he was retrieved from the frigid water, he had no pulse and no heart beat. His skin was cold and blue, and his eyes were fixed in a deathly stare. To all outward appearances he had drowned. His rescuers, however, did not give up and began CPR, breathing and pumping his heart for him, and warming his body.

The "drowned"student revived and finished his next semester at college with a 3.2 grade average. He was lucky. Normally just four minutes without oxygen means brain damage. But because the water was extremely cold, his metabolism slowed considerably. More importantly, oxygen was shifted to those organs that needed it the most, particularly his brain.

The respiratory system— specifically the tissues of the nose and throat—is in direct contact with the outside world. Despite all the filtering that goes on along the way, the system is vulnerable to disease, insult, and injury.

Most of us get respiratory infections, especially in the upper reaches: the sinuses in the skull, the tonsils, the larynx, and the pharynx. The common cold, caused by 200-odd viruses, is the best known. Pneumonia attacks the bronchioles and alveoli, usually due to a bacterial or viral infection. Lifestyle and occupations can contribute to lung diseases, too. Dust and other particles may attack the lungs, resulting in diseases such as silicosis. The world's most common occupational disease, it is caused by silica dust in quarries and coal mines.

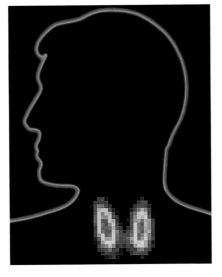

FOOD TO ENERGY: THE DIGESTIVE SYSTEM

Like the respiratory system, the digestive system works to provide our cells with the nourishment they need. In this case, the supplies come from what we eat and drink. The digestive system performs its own little miracle, turning a loaf of bread and jug of wine into the nutrients and water we need to stay alive.

Some people take minimal interest in food. "Life is too short to stuff a mushroom," cracked one British journalist. Others live to eat. Even the formidable Virginia Woolf advised, "One cannot think well, love well, sleep well, if one has not dined well."

Far beyond the chemical reactions their bodies make of food, it is important to most people. Their mood, their self-worth, even their national pride, may depend on what's on the dinner plate. They judge others by what they eat as well. The French and Italians particularly glory in their cuisines. On the other hand, hell has been described as a place where all the chefs are British. Some vegetarians think eating meat is barbaric; carnivores often think being a vegetarian is effete, and everyone disagrees on how long to steam asparagus.

Our bodies are far less discerning. They want food, need it, and will as eagerly down a greasy hamburger as broiled halibut in chardonnay. Foods—all foods—contain chemicals that your body uses to live and thrive. If you do not get all the nutrients that you need—the fatty acids, the amino acid units of proteins, the carbohydrates and minerals and vitamins—you can get sick and even die.

No matter how long and how lovingly prepared, once a meal passes your lips the mechanical and chemical

Someone to look up to: Mikhail Gorbachev greets Lithuanian basketball star Arvidas Sabonis. Differences in height often result from varying levels of somatotropin, one of more than 200 chemical messengers we call hormones. Created by the endocrine system, somatotropin stimulates growth. The double-lobed thyroid gland (above) manufactures hormones that help to regulate the body's rate of metabolism, the sum of the processes of cells.

processing we call digestion begins. Your teeth begin the mechanical part of digesting, grinding food into small pieces. Three pairs of salivary glands secrete a mixture of mucous, water, and a digestive enzyme into the mouth that immediately starts breaking down starches in the food. Swallowing begins as a voluntary process: You decide to move the food out for the next taste, and your tongue rises as it pushes the food to the back of the mouth. Reflex action takes over as the food, now shaped into a moist, soft lump called a bolus, passes into the pharynx and embarks on a stormy trip through the rest of the digestive system. Muscles in the pharynx contract, pushing the bolus to the esophagus, a muscular, ten-inch-long tube leading to the stomach. One wavelike constriction of the esophagus moves the bolus onward.

At the entrance to the stomach, a structure called the cardiac sphincter acts as a one-way valve to prevent food from returning to the throat once vigorous stomach contractions start. The stomach does some mechanical processing—churning and mixing—and some chemical processing. Shaped much like a fat letter *J*, it produces about six pints of gastric juice a day, mostly corrosive hydrochloric acid that kills microbes and begins to break down proteins. The most important digestive component of gastric juice, an enzyme called pepsin, works best in acid and begins taking apart all the protein in the food. Mucosal lining on the stomach walls keeps the pepsin and acid from eating away the stomach itself—unless you have a peptic ulcer, in which case the defense mechanism has failed—a situation, doctors are now realizing, often brought about by a bacterial infection.

The stomach absorbs few nutrients, some water, sugars, and salts. But the agitation and churning now have turned your most recent meal into a semifluid paste called chyme. Peristalsis—successive waves of involuntary contractions—pushes the chyme to the base of the stomach. A muscle called the pyloric sphincter serves as the gatekeeper to the duodenum, the upper part of the small intestine. The pyloric opens and squirts small amounts of chyme through at a time, then closes to prevent too much stomach acid from getting into the intestine. About four hours after eating, the food has passed into the small intestine, where the real business of digestion is done.

Often the body knows what it needs. For example, pregnant women eat different diets than they did before they conceived. Children undergoing growth spurts alter their food intake to supply the fuel they need. When you are sick, you eat differently. Complicated signals shuttle among the brain's hypothalamus, the gut, and fat cells to regulate our preferences.

Food provides not only the energy all cells need to power their activity but also the raw materials for growth and repair—the stuff of new cells, neurotransmitters, hormones, and more. Carbohydrates, fats, proteins, nucleic acids, trace minerals—all are fodder for our digestive system. Most of the nutrients the body needs are extracted in the small intestine. Although it's only an inch in diameter, the small intestine takes up most of the lower abdominal cavity. It measures about 20 feet in length, all coiled in place like sausage and held to the back of the abdominal wall by a thin sheet of connective tissue. Besides absorbing nutrients, the small intestine also extracts two-thirds of the water from food.

The small intestine gets help from other organs. For instance, when food enters that organ, the brain signals the pancreas to secrete pancreatic juice, a mix of enzymes needed to break down proteins, fats, and carbohydrates. Pancreatic juice also contains sodium bicarbonate, which neutralizes any remaining stomach acid.

Also essential is the liver, the body's primary chemical-processing plant and toxic waste facility. The liver also produces bile, a solution the small intestine needs to digest and absorb fats. Bile works the same way detergents do: It breaks fats and oils into smaller pieces. Once processed, the fats are shipped to cells that need them. In addition, the liver detoxifies the blood of many potential poisons, from metabolic by-products of our own bodies to dangerous substances in our food.

Once the small intestine completes its processing—about three to ten hours after eating—the material that is left moves into the large intestine, or colon. Coiled atop and around the small intestine, the colon is about five feet long and three inches in diameter. Food stays here for from ten hours to up to several days, depending on circumstances and a person's regularity.

Not much of any use is left in the food you ate by the time it reaches the large intestine. The small intestine already has extracted most of the usable carbohydrates, fats, and proteins, as well as most of the vitamins and minerals necessary for healthy life. The colon reabsorbs water from all the spent digestive juices and returns it to circulation. The colon also harbors millions of bacteria. These "bugs"— harmless and, in fact, beneficial as long as they stay in the intestine—produce several vitamins as they feed on our wastes.

With every bit of useful material now absorbed, the concentrated remains pass into the rectum and the anal canal. There two circular sheets of sphincter muscles act as valves, controlling elimination of intestinal wastes.

LIQUID WASTE REMOVAL: THE URINARY SYSTEM

Removing liquid waste is the task of the urinary tract, which regulates the content of our body's fluids. The key organs of this system are the kidneys. These two reddish brown, bean-shaped organs sit in the back of the abdominal cavity. They are the body's sanitary engineers. The kidneys essentially filter and condition the blood, eliminating waste materials, balancing the salts and liquidity of bodily fluids, and keeping the acid-base level just right.

Fortunately, we have two kidneys. Lose one and life will go on. Its mate enlarges to assume the function of both organs. Lose function in both kidneys, and you need a dialysis machine or a transplant to survive. Dialysis machines filter the blood through a membrane that takes out larger molecules, such as proteins.

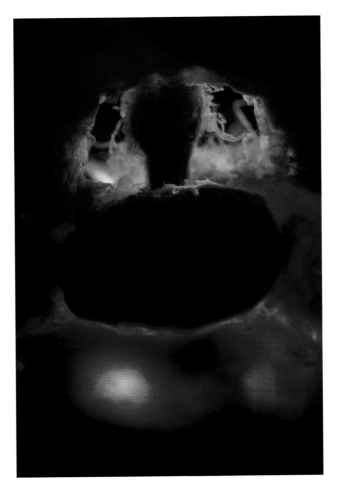

The human body is mostly water. Our remote ancestors came from the sea, and we still carry that heritage with us. About 60 percent of our body weight is water, salty like the sea. You need to drink two quarts of water a day just to make up for what you breathe out or excrete. Water is stored mainly in blood, muscle tissue, and the skin. Moving through the body, water distributes nutrients and chemicals, helps balance body temperature, dilutes toxins, and is the medium in which the complex chemical reactions of our metabolism occur.

The urinary system regulates the amount of water in our bodies, processing what we take in and getting rid of what we don't need. Through coiled masses of capillaries in the kidneys called glomeruli, substances such as water, glucose, and certain minerals are returned to the blood.

Nitrogenous wastes—from proteins—as well as excess water and mineral salts exit the kidney as urine, which passes through a duct called the ureter into the large and muscular reservoir that is the bladder. Periodically, urine is disposed of through the urethra.

Men and women have slightly different urinary systems, a function of differences in their pelvic areas. Men's urethras are about eight inches long. The urinary system in women accommodates their childbearing capacity. To make room for the uterus, women's bladders sit lower than men's, and women's urethras are only about two

Dangling on a stalk from the brain's hypothalamus (purple), the pituitary gland (black)—once called "the master gland"— produces ten hormones, including growth hormone. It also directs the production of hormones in other glands.
FOLLOWING PAGES: Not a second lost for two men in danger thanks to the "fight-or-flight" hormone, epinephrine, sent coursing through their bloodstreams courtesy of the adrenal glands, which sit atop the kidneys.

inches long. Because the shorter passage to the outside world is more likely to admit bacteria, women tend to have more urinary infections than men.

At any given time, about one-fourth of the blood being pumped by the heart is being processed in the kidneys. They process about 45 gallons of blood every day. The kidneys contain microscopic structures called nephrons, about a million of them in each kidney. Nephrons consist of complex, convoluted balls of blood vessels, each encased by a capsule. Glomeruli surround the capsule. Blood flowing into the nephrons backs up against this capsule, creating pressure. Blood cells and even large molecules are too big to pass through the filtering membrane of the capsule, so these are retained in the bloodstream. Pressure forces fluids and dissolved substances through.

Typically, one percent of a person's blood volume, with various salts and smaller molecules, is filtered in the nephrons and passed out of the kidney's draining tubes as urine. Urine consists of mostly water and potentially harmful by-products of protein metabolism, such as urea and uric acid. The accumulation of either substance in the body could be life-threatening. Tissues filled with waste products cannot absorb oxygen or food, and they die by poisoning, starvation, or suffocation.

The composition and volume of urine depend on what is happening throughout the body. Eat a salty meal, for instance, and you produce more concentrated urine, with high salt content. Your body continues to produce concentrated urine until you answer your thirst response to the excess salt, and drink. Exercise, and your urine volume is lowered—it becomes more concentrated because your active body needs more fluid. The metering of fluid is regulated by a series of hormones from the adrenal and pituitary glands, and by an enzyme called renin produced in the liver.

Urine passes to the bladder through the tubular ureters, which connect each kidney to the bladder. The bladder is a storage vat; it is extremely elastic and shrinks when empty. Normally, it holds about a pint of urine, but it can store more if necessary.

When the bladder is full, nerves send a message to the smooth muscles in the bladder to relax. Muscle contractions then push the urine into the bladder's drain, the urethra, from which it is excreted when we relax the sphincter muscles.

Proof of the efficiency of the urinary system—and the body as a whole—urine is a sovereign diagnostic tool, its collection painless and convenient. Chemical tests can tell doctors much about our health. Blood in the urine can mean the existence of a cancer or a hemorrhage; glucose can indicate diabetes.

WRAPPING UP THE HUMAN PACKAGE: THE SKIN

The human body is an amazing package encased in an equally spectacular wrapping—skin. The largest of our organs, skin helps regulate body temperature, excess moisture, and salts. It is full of sensors that measure temperature and that, maybe best of all, empower us with a sense of touch. It is also our main shield against infection by bacteria and viruses. Normally, we wear our skin well. And we are always changing it.

Our skins have an outer layer, the epidermis, that is constantly shedding cells. The epidermis consists of epithelial cells, platelike and flat, that serve as our body armor. Underneath the epidermal skin layer are the basal cells of the dermis, which constantly divide, replacing our outer wrapping. The innermost dermis consists of tissue penetrated by blood and lymph vessels, nerves, hair follicles, and sweat and sebaceous glands. At its base, dermis connects to the underlying tissues.

Our outer wrapping keeps inner tissues from drying out. And it allows for the excretion of wastes in sweat. Our skin provides crucial temperature control as well. Underneath it, a layer of fat serves as insulation. Also, when temperatures drop, muscles in the skin contract, causing goose bumps, which trap some air for added warmth. At the same time, blood flow to the skin is reduced so that the blood's warmth can be used as an internal radiator, limiting heat loss to the outside.

First stop on the digestion tour, the mouth begins both the physical and chemical breakdown of food. Within 24 hours or so, more than a dozen different actions throughout the gastrointestinal tract will reduce these flakes to their essentials, wringing out every last bit of nutrition to provide fuel and other necessities for all the body's cells.

When body temperature rises, blood vessels in the skin dilate, letting more blood stay near the surface so heat can radiate out. Evaporation, too, helps dissipate heat from the skin.

Skin may be telling us even more about our origins. Exposed to sunlight, the skin manufactures vitamin D. According to one theory, the earliest humans came from the tropical area of Africa and were dark skinned. As some moved north, where sunlight was weaker, the screening effect of dark pigment against ultraviolet rays became less essential for survival. In time lighter skin, which produces more vitamin D than dark skin, prevailed in the limited sunshine of northern latitudes. Tropical areas still favor dark, screening tones; the farther north one goes, the lighter the skin.

Skin contains nerve endings—small, simple, and exquisitely sensitive. Hair embedded in skin is coiled with similar, unelaborate nerve endings. These terminals are all directly wired to the central nervous system, and their activation gets immediate attention. The deeper or harder the touch, the more rapidly nerve impulses are produced. Fingertips, palms, nipples, lips, and any places where hair is abundant contain the highest density of nerve endings and are especially sensitive. Using this acute sensitivity, the blind can read with their fingertips, discriminating the fine points of Braille, and violinists and cellists can put just the right pressure on the strings.

Going from the inner to the outermost layers of the epidermis, one finds increasing amounts of a tough protein called keratin. Forms of skin themselves, hair and nails also are composed of keratin. Hair is basically dead skin cells packed in a column; nails are tough scales of keratinized dead skin cells cemented together.

Skin is so sensitive that it sometimes acts as a barometer of mood. Embarrassed, you might blush as blood flows to the surface. Or you might flush with anger. No one knows why skin does these things, but it is known that skin provides an indication of general health. A rash, or irritation of the skin, comes either from allergies or infections. Some rashes, such as chicken pox, are spread from one person to another; most are not.

Odd collections of skin cells can form warts, moles, or tumors, benign or otherwise. Skin cancer can be the least harmful form of cancer a human can have, or one of the most aggressive. Exposure to ultraviolet light is the leading cause. The most common form of skin cancer is basal cell carcinoma, affecting the basal layer. It rarely spreads and is easy to remove. Squamous cell carcinoma and melanoma are more dangerous. Malignant melanoma occurs in the pigment-producing cells of the skin. These tumors are usually very dark, irregular in shape, and have indistinct borders. They must be removed at once because they spread quickly.

COMMUNICATION CENTRAL: THE NERVOUS SYSTEM

Each of the body's systems must do more than just function well on its own. The systems have to work together, too. The job of running the show, of helping maintain order among the body's trillions of cells, falls to the nervous system. Its main task is communication. It receives and stores information about the outer and inner worlds, initiates activity, and carries impulses throughout the body. The nervous system works in tandem with the ductless glands of the endocrine system.

Our nervous system never sleeps, even when we do. Our bodies are under continuous stewardship, with data constantly gathered, shuttled up to the brain for analysis, and sent down through the spinal cord to incite action when and where it is necessary. The action of the nervous system is life. Death is simply what happens when that system finally shuts down.

Because our world can be dangerous, the nervous system must work fast. It does, thanks to some remarkable engineering. Analogies between the brain's functioning and the fastest parallel-processing supercomputer, say, or the biggest telephone switchboard imaginable work only to a certain point. For the human body is far more complicated than anything we are ever likely to build, and the nervous system that drives it defies our engineering capabilities.

Primitive animals use a simple tube of nerve cells to help shove food into an oral cavity, or they contract a ring of muscles to snap away from noxious encounters with enemies. Human beings evolved spinal cords that process and coordinate our responses.

In the spinal cord, knots of nerve cell bodies send commands along gossamer threads cabled together, some coated with glistening myelin, the fatty sheath that insulates nerve cells and speeds electrical signals on their way. The brain, with its own organized groups of grayish cell bodies and gleaming white fiber bundles, asserts control over the cord, commanding action by the muscles, glands, and the other systems of the body.

Our nervous system is brilliantly organized. A basic hierarchy exists in which the brain and spinal cord, which make up the central nervous system, serve as the apex. Nerves branch from the central nervous system, interlacing tissue and bones of the body. These nerves are considered the peripheral nervous system. The basic unit of the system is the nerve cell, or neuron, a cell capable of taking in information from all over the body and granting an appropriate response.

Within the complex web of the peripheral system are two subsystems—the autonomic and the somatic, or voluntary, system. The latter is concerned both with the nerves that register incoming stimulation and the body's voluntary muscle responses. The somatic system performs conscious actions. Decide to scratch your ear, and that ear gets scratched. Slam the door to make a point, and it closes with a bang.

As its name implies, the autonomic system takes care of automatic, involuntary processes, such as the beating of the heart and glandular secretions. The autonomic system operates on autopilot, running quietly in the background, revving up instantly to respond to challenge. It helps regulate heart action and the activity of other organs, causing the heart to pound with fear. Our glands charge our muscles for action, our lungs fill and exchange oxygen rapidly to fuel flight—or prepare to fight. The pupils of our eyes dilate to increase the acuteness of the image of what we see.

Basically an extended and convoluted tube, the 30-foot-long digestive tract packs itself efficiently into the abdominal cavity (opposite), largely walled off from the rest of the body. The mere taste of a peach (above), or even the aroma of food, often is enough to stimulate production of saliva and other juices that start the digestive process.

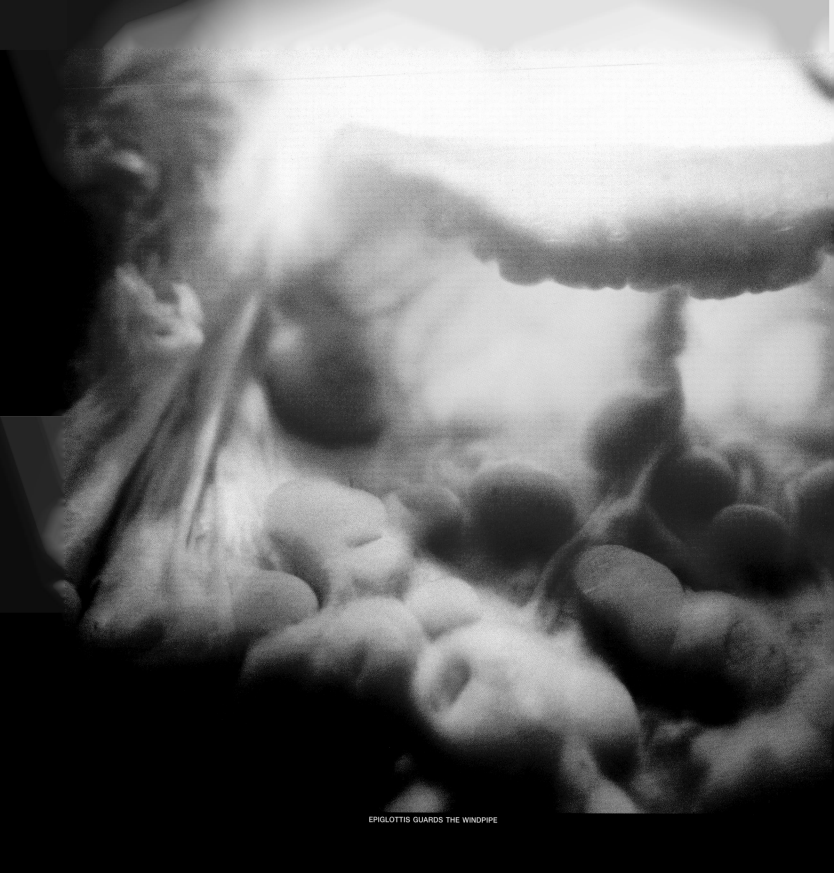

EPIGLOTTIS GUARDS THE WINDPIPE

Trapdoor to the trachea, a flap called the epiglottis (above) remains open during breathing but clamps down during swallowing to prevent food from entering the windpipe. The tongue propels food toward the throat; projecting papillae on its surface (opposite, top) help turn food into a bolus, or ball, for easy swallowing. The esophagus (opposite, bottom) consists of an expandable inner layer surrounded by smooth muscle that pushes food along with strong, sequential contractions we call peristalsis.

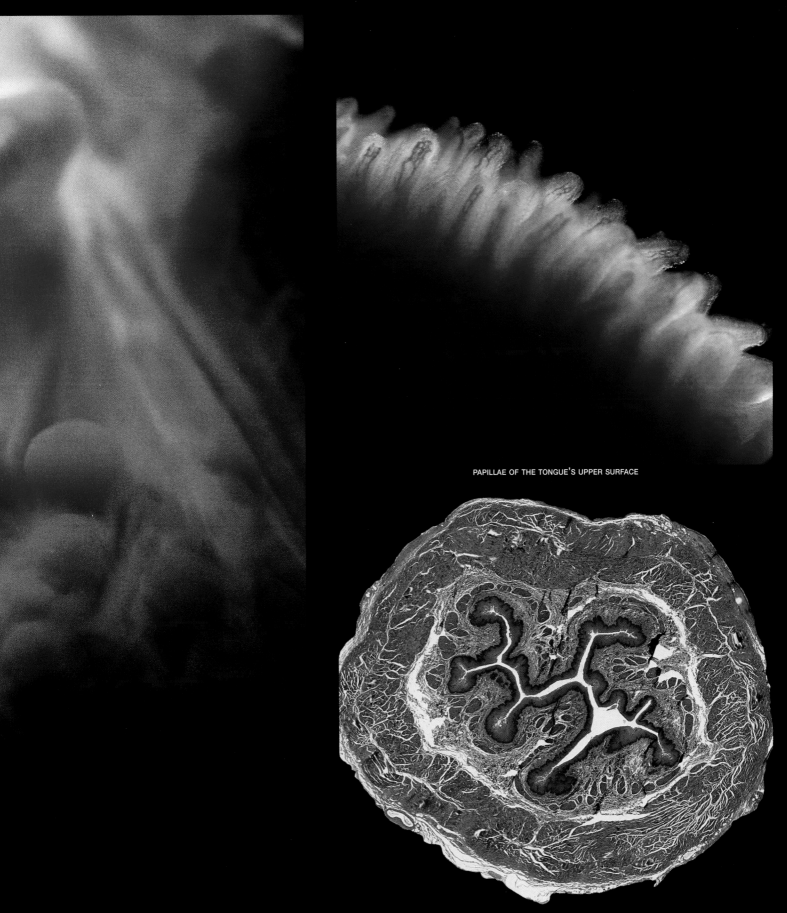

PAPILLAE OF THE TONGUE'S UPPER SURFACE

CROSS SECTION OF THE ESOPHAGUS

The autonomic system is further divided into the sympathetic and parasympathetic systems, which control internal organ reflexes. The sympathetic system responds to stress or emergencies. The parasympathetic system, in an opposite manner, helps conserve and restore body resources—generally working to calm it down.

The nervous system sorts its tasks into sending or receiving, with different pathways maintained for each function. The sensory system transmits information from both the outside world and from the body itself—impulses from the skin and from sense organs—to the brain, data that tells it what is happening within us and outside us. Impulses travel from a sensory nerve to the central nervous system. The sensory system transmits the images of our children, registered through our eyes; the smell of roses through our noses; the touch of the wind, the sear of a burn, the reflections in our mind.

The motor system, largely voluntary, sends impulses from the spinal cord and brain to muscles and glands of the body. When you decide to walk or eat or curl a pinky, impulses travel from motor nerves in the brain through the spinal cord to the appropriate muscles.

A neuron, the essential unit of the nervous system, is a specialized cell designed to receive messages from and transmit them to sensory organs, glands, muscles, or other neurons. Its chemical and electrical characteristics give it the most complicated combination of functions ever stuffed inside a cell membrane. Two types of neurons— sensory and motor—serve the two nerve systems.

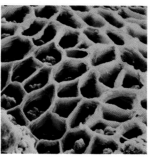

A nerve cell may be round, star-shaped, or oval, with a single nucleus to supply working instructions. Many short arms and one much longer arm extend out from the body. The shorter arms, dendrites, are studded with a multitude of knobs that receive incoming signals.

An axon, the single long extension leaving most neurons, transmits outgoing signals to the dendrites of other, nearby nerve cells. The longest axons in the human body, which run from the legs to the spinal cord, can reach more than three feet in length; in some whales, axons can measure thirty feet, from brain to tail.

Neurons are grouped into nerve fibers, which, like their component cells, can be sensory or motor. Each nerve is made up of a bundle of nerve fibers, and it can contain both motor and sensory nerve fibers. For example, sensory nerve fibers carrying messages from the taste buds to the central nervous system are packed into a nerve containing motor nerve fibers that send impulses to the scalp and face. Within the nerve, sensory and motor nerve cells—the neurons—communicate with each other.

A nerve, whether it runs along a bone between bunches of muscle, or traverses the delicate lining of a body cavity, consists mostly of a cable of axons, each strand responsible for carrying either a sensory or a motor message. The message is electrical as it is sent along the fiber, but neurons never quite touch their target dendrites. Tiny gaps exist between neighboring neurons, and they must be jumped, the signals transmuted from electrical to chemical at junctions called synapses.

Chemicals flow across this cleft only when given the right electrical stimulation. Special knobs at the end of an axon contain vesicles, tiny packets filled with chemicals known as neurotransmitters. When an electrical signal arrives at the axon terminal, the vesicles instantly fuse to the cell membrane and spew their neurotransmitters into the cleft.

These neurotransmitter molecules, one kind for each kind of neuron, cross the synapse instantly. At the receiving neuron they encounter receptors, specific recognition sites embedded in the receiving nerve cell's membrane. Each kind of neuron complements a particular kind of molecule; a given neurotransmitter molecule docks only with a molecule of complementary shape. This proper fit then triggers the receiving cell to recognize that an appropriate signal has arrived.

Meeting a challenge at every meal, the highly muscular stomach sits atop yards of folded intestines (opposite). On the inside, acid, enzymes, and other digestive juices break down foods and kill microbial intruders. A latticed abnormality in the stomach lining indicates an ulcer (above), likely the nefarious work of the bacteria *Helicobacter pylori* (top).

So far, scientists have found at least 50 different kinds of neurotransmitters, with new ones identified regularly. Many of the medicines we take, from pain pills to Prozac, are modeled on neurotransmitter messages. Each signal must be swept rapidly away almost as soon as it is sent, in order to ready the neurons for their next exchange. And the machinery designed to carry out these crucial cleanups has been the target of everything from anti-itch creams to chemical warfare.

Chemical transmission across a synapse provides only part of the picture of the working nervous system. The poet Walt Whitman could "sing the body electric," and he was correct. Neurons act like tiny electric batteries. Like all cells in the body, neurons have a cell membrane separating the innards of the cell from the outside environment. Each neuron has ion pumps, which move positively charged sodium and potassium ions through channels in the membrane. While a nerve idles, there are ten times more sodium ions outside the membrane than inside, and about ten times more potassium ions inside than outside. Inside the membrane, the presence of other, negatively charged ions results in a net negative charge; the outside environment has a positive charge.

The charge separation, what electrical engineers call a potential difference, is the same stored electrical energy found in a battery. The difference is minute but measurable—70 millivolts or 1/1500th of a D battery.

When a signal arrives from an adjacent nerve cell, the potential difference reverses in a flash. The membrane snaps open its channels to permit the sodium outside to enter and the potassium trapped inside to escape, changing the charge nearly instantly from -70 to +30 millivolts. The charge immediately outside the membrane is now negative. The switch in charge lasts far less than a blink, a thousandth of a second; then the ions reestablish their previous separation, and the potential energy is restored until the next time.

Nerve impulses must move quickly—about 390 feet a second in the faster nerve fibers—if we are to survive our world's dangers. A sudden confrontation with a speeding car leaves little opportunity for prolonged thought; it is the quick who survive. Curiously enough, what makes nerves so fast is not their electrical conductance but rather the insulative qualities of the axon's covering sheath. Simply sending signals down the nerve cables wouldn't work—this method isn't fast enough. So, the fastest neurons are wrapped in a myelin sheath so ingenious that it awes electrical engineers. It is sectioned

by gaps called the nodes of Ranvier (named for a French scientist). The nerve impulses skip from node to node, rather than traveling straight down the nerve. The result is that these impulses go 50 times faster.

Depending on the sum of the various signals and neurotransmitters received, the end result is a neuron that is either excited—that is, ready to signal the next neuron down the line—or inhibited from doing so. Like muscle cells, neurons work on an all-or-nothing basis: They are either on or off; there is no middle ground.

Nerves that lead from the brain through the spinal cord are contained and protected in the bony spinal column. The spinal cord itself, from the brain stem to the lower back, is only about as wide as a finger, and about 17 inches long. Thirty-one pairs of spinal nerves branch out from between the vertebrae.

In addition to transmitting voluntary impulses, the spinal cord acts as the body's 911 emergency center by implementing reflexes. Burn a finger with a candle, and a reflex kicks in. Instead of waiting for one message to travel up the cord to the brain for a decision, then for another to go back to the muscle for a response, the spinal cord routes the signal directly to a local motor nerve, and your hand immediately jerks away from the flame. No thought was necessary, no time was lost.

A more benign example of reflex action is the common knee-jerk reflex, tested when a doctor taps the tendon below the knee. The tap stretches the front of the thigh muscle. That sends a signal to a neuron whose stretch receptor lies embedded in the muscle, which then relays its message to the spinal cord. Instead of asking the brain what to do, the cord quickly sends back a message to a muscle to contract. The simplest creatures still get through life solely by reflex. Our brains provide much finer control.

Beyond the spinal cord, the peripheral nervous system spreads a multitude of nerves. Many of these nerves connect to muscles anchored to bones, and are used for directing voluntary actions. Other peripheral nerves beyond the spinal cord are responsible for involuntary functions. They monitor and regulate internal organs, heart rate, blood pressure, digestion, and such through a complementary set of subsystems— the sympathetic and the parasympathetic systems. These systems use distinct sets of nerves that employ their own brands of neurotransmitter molecules.

The parasympathetic system reigns when the body is at rest, maintaining or storing energy. It relies primarily on the neurotransmitter called acetylcholine. When you eat a heavy meal, the somnolent feeling that results is the parasympathetic system at work, conserving energy so the body can efficiently digest its food. Consider siestas a victory for your parasympathetic system. Your pupils constrict, your heart rate and the strength of cardiac contractions decrease, your liver stores glucose. You are in a low-maintenance, restful mode.

The sympathetic set of nerves, on the other hand, fires up with stress, preparing the body instantly to fight or flee, and it is unconcerned where the energy to do so comes from or how long it will be required. The job of the sympathetic system is to ensure survival. Its commands are carried out primarily by the neurotransmitter norepinephrine, closely related to the stress hormone from the adrenal gland, epinephrine (also called adrenaline). When you are enraged or in imminent danger, your sympathetic system takes over. Your pupils dilate, your heart rate and blood pressure increase, your hair stands on end (a relic, perhaps, of hairier ancestors who wanted to appear bigger instantly). At such times, you are tapping the most primitive parts of humanity, an instinct that harks straight back to a primordial time.

Under normal circumstances, these two systems—the sympathetic and parasympathetic—are in balance without any conscious effort, achieving what is called homeostasis. This state is largely maintained by reflexive action originating in the core structures at the brain's base, the hypothalamus and the medulla oblongata. In these basic brain centers, which evolved early in vertebrate history atop the spinal cord, reflexes still dictate internal responses for optimal functioning. If something happens to change conditions—internally or externally—these structures automatically adjust the metabolic levels to maintain stability.

Humans of course do more than rely just on instinct. Our brains can direct a response to an emergency on a conscious level—should I get out of here or not? The convolutions of the brain that, over evolutionary time, enabled more and more neurons to be packed into a given cranial space now provide us with the power of decision-making strategies: How do I get out of here?

But humans are not just a basic vertebrate nervous system capped by the human brain. Our entire nervous system evolved together. And in human beings, the whole assemblage meets complex emotional needs that now extend into realms far beyond mere survival. We write poetry, build cities, compose symphonies, heal the sick, contemplate the universe. Due to the complex inner lives our brains enable us to create, people find that damage to the nervous system ranks among the worst disasters that can happen. Peripheral nerves sometimes can regenerate if they are crushed or partially cut, as long as the myelin sheath remains basically intact. In an injured nerve, a neural fiber beyond the break no longer gets the proteins and enzymes it needs, so it shrivels and dies, leaving behind a hollow sheath. Other neurons in the nerve send out sprouts in an attempt to reconnect, and sometimes some of those sprouts will find their way through the hollow sheath, will connect to the target, and will restore the functional connection.

Such regenerations do not occur with nerves that begin in the brain or spinal cord, which is why strokes or severed cords can cause death or paralysis. Injuries to the middle or lower part of the spinal column can produce paralysis in the legs and possibly parts of the trunk, affecting such things as breathing and bladder and sphincter control. Injuries higher up on the spinal cord can paralyze all the trunk and limbs. Even today, when a spinal or brain nerve is severed we know of no way to put it back together again.

Just as injuries can destroy the nervous system's ability to function properly, so can various diseases. In multiple sclerosis, the body's own immune system destroys the myelin sheath, causing short circuits in the nerve fiber. Recent research shows that the nerves themselves also may be attacked. Multiple sclerosis is the most common neurological disorder, striking one in every thousand people. Drugs that protect the nerves might help those who suffer from what is now an incurable ailment. A far rarer disease, amyotrophic lateral sclerosis (ALS), better known as Lou Gehrig's disease for the famous baseball player who succumbed to it, destroys motor neurons in the spinal cord and brain. It, too, is incurable and largely untreatable, although biotechnology companies are hard at work seeking new therapies.

....

CHEMICAL MESSENGERS: THE ENDOCRINE SYSTEM

....

Nerves alone do not suffice for all the fine tuning of our bodies' processes; many of the chemical controls that

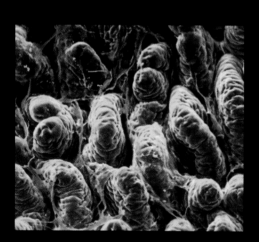

Richly laced with blood vessels, the small intestine (right) accomplishes digestion's most important work. Enzymes continue processes begun in the stomach, further breaking down carbohydrates, proteins, and fats. Dense villi, projections in the folded intestinal lining (above), greatly increase the organ's surface area, increasing its ability to capture and absorb glucose, amino acids, fatty acids, vitamins, and other essential nutrients. Capillaries reach into each villus and transfer its catch directly to the bloodstream.

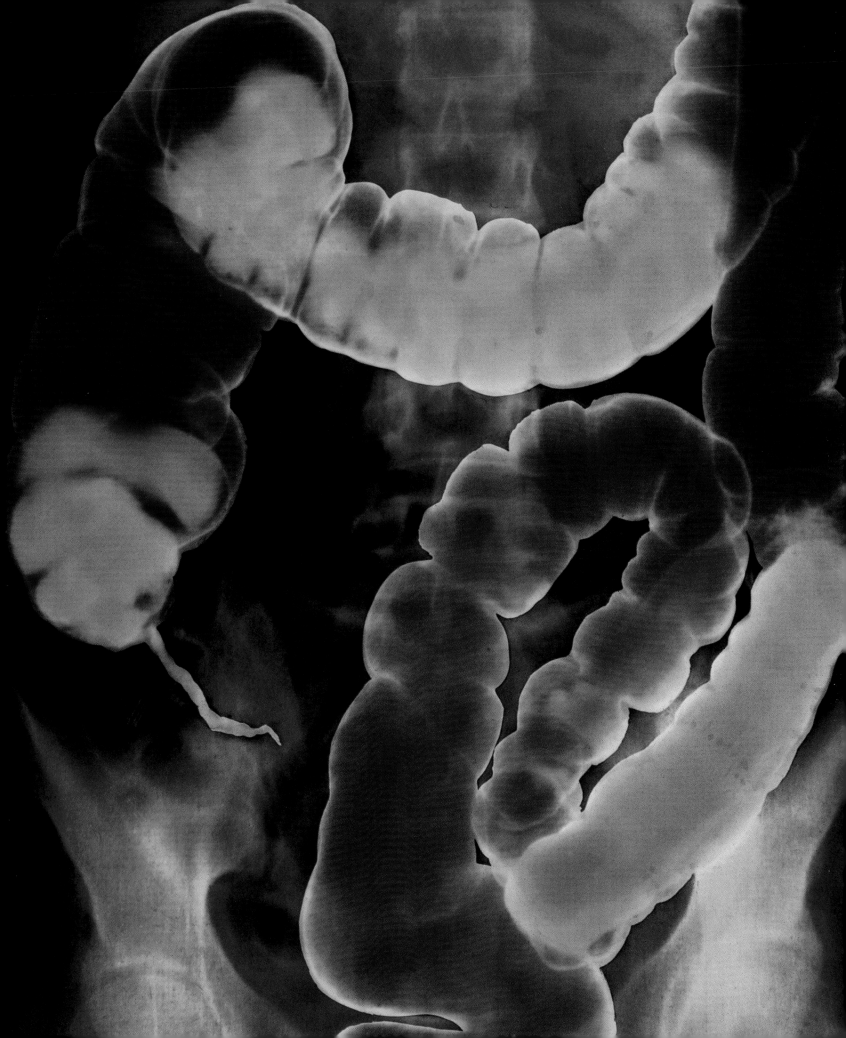

keep us healthy and fit are instigated by extraordinarily powerful chemical messengers called hormones, which enter the bloodstream and exert general control over many vital bodily activities. Their mere presence in the fluids bathing certain organs can trigger remarkable action. Other glands, called exocrine or ducted glands, are not members of the broadly based endocrine system. Rather, they work locally, secreting substances into their cell-lined ducts, which empty either into spaces between organs or directly onto organ surfaces. Examples of exocrine glands include the sweat glands, saliva glands, mucous glands, oil glands, and mammary glands.

Endocrine glands are ductless; they simply release their products, which are eventually swept up by the

compared to neurotransmitters allows regulation of continuing processes in our bodies and concerted influence over large areas. Hormones circulate through the body in the bloodstream until they find the organs they are to influence. As a result, the glands that secrete hormones do not have to be near the organs they control. The endocrine system thus consists of glands tucked into various nooks about the body. Some are close to their spheres of influence; others far away. One of the most crucial glands of the system, the pituitary, is the size of a pea; it dangles from a tiny stalk at the base of the brain. The efficiency of the system almost defies belief.

In 1849, the German scientist A. A. Berthold removed both testes from young roosters, then implanted

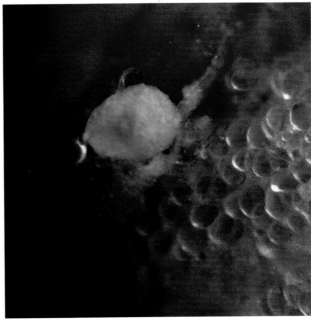

bloodstream and taken to where their message is read. Examples of endocrine glands include the pituitary, adrenal, thyroid, parathyroid, pineal, thymus, pancreas, and the sex glands—the testes and ovaries. Although this ductless messenger system is much slower than, say, the nervous system, it is exquisitely accurate—and effective over a much longer term. Its actions last minutes to days, not milliseconds. The longer action of hormones

one testis inside each of the birds' bodies. The animals grew up to be normal male roosters. The glands could not be discouraged. Because the hormones of our endocrine glands are potent, only minute amounts are required. To effectively influence heartbeat and blood circulation, for example, the adrenaline in your body need only be as concentrated as a teaspoonful of dye dissolved in a large lake.

....

End of the line, the large intestine (opposite) extracts the last bit of value from food that now consists mostly of bacteria, indigestible material, water, and mucus. The wormlike appendix, a vestigial organ, no longer participates in digestion. Absorbed nutrients travel the bloodstream to the liver, where specialized cells (above, left) carry out some 500 vital functions. Pancreatic cells (above, right) make insulin and glucagon, hormones that regulate glucose level in the blood.

....

Substances this powerful must be under tight control, so the body has feedback systems—both positive and negative—to make sure there is just the right amount of hormone when and where it is needed. There is also a safety feature: Enzymes in the blood and the liver quickly dismantle any excessive amounts of hormones.

Almost every cell in your body is the target of at least one hormone. Each hormone works only on cells it recognizes, and then only at certain times. All cells bear specific recognition sites for hormones. These sites, called receptors, are protein molecules embedded within the cell envelope. Some hormones, unlike neurotransmitters, actually enter the cell to trigger metabolic changes. A cell accepts certain hormones based on an exact molecular fit with the receptors, and will reject all others.

Once inside a cell, a hormone begins changing the cell's metabolic processes. Some messages alter the rate of protein assembly, perhaps as a way of preparing the cell to divide. Other hormonal messages, scientists have recently discovered, simply tell a cell to drop dead.

Most of the body's control over hormones is through negative feedback. When the amount of a hormone gets too high, the excess in the blood tells the secreting gland to slow production. For instance, the parathyroid hormone regulates how much calcium is incorporated into bones. If the level of calcium in blood gets too high, the parathyroid gland simply shuts down production for a while.

A few endocrine glands rely on positive feedback. For instance, the hormone oxytocin, released by the pituitary, stimulates uterine contractions during childbirth. The more oxytocin in the blood, the more the pituitary releases, increasing uterine contractions and speeding the birth process.

A few other glands are regulated by the nervous system, which sends out signals directing when and how those glands should function. If you are facing grave danger, for example, the nervous system instructs the adrenal gland to start production of epinephrine and norepinephrine, which increase available blood supply and make your heart beat faster.

While the pituitary was long considered the "master gland" that directs all other endocrine glands, we now

....

Down and out: Potty training goes communal in this Russian nursery school. As the large intestine finishes its work, waste descends into the rectum. Strong sphincter muscles at the anus keep elimination a mostly voluntary action.

....

know that it gets its directions from the real dictator, the hypothalamus. Located just above the pituitary at the base of the brain, the hypothalamus regulates the quantities of each hormone that the pituitary produces. The pituitary, in turn, directs the actions of other glands.

Some glands, such as the adrenals, produce more than one hormone. The two adrenal glands, each perched on top of a kidney, contain different regions. The innermost heart, called the medulla, produces epinephrine and norepinephrine, both of which stimulate the heart and cause the coronary arteries to dilate in response to emergency situations.

The layer of the adrenal gland just above the medulla produces sex hormones that influence sperm production in men, ovulation and menstruation in women, and growth of body hair in both sexes. Yet another layer of adrenal cells produces the steroid cortisol, which reduces inflammation and controls how the body uses fats, proteins, carbohydrates, and minerals. The outermost layer of the gland produces aldosterone, which works on maintaining blood volume and blood pressure, and helps regulate the amount of sodium excreted in urine.

The acorn-size pituitary, a two-lobed gland at the base of the brain, is also multitalented. In addition to generating hormones that stimulate the activity of other endocrine glands, it produces a growth hormone called somatotropin, which provokes the cells in the body to multiply. Somatotropin also increases the rate at which we metabolize carbohydrates. The same lobe of this gland makes the hormone that encourages production of melanin, the chemical that gives skin color, and produces the hormone that maintains milk in nursing mothers. The other lobe of the pituitary gland manufactures oxytocin for birth contractions and an antidiuretic hormone that influences fluids in the body, plus a few others for moderating the chemicals in our blood.

The thyroid, which lies in the neck, consists of two lobes that wrap around either side of the trachea. A major thyroid hormone—thyroxin—controls metabolism and influences heart rate. Too little can lead to mental retardation and dwarfism. Sometimes, low thyroxin level can be due to iodine deficiency, and can be treated by adding iodine, required to make the hormone, to the diet.

This knowledge led to the widespread use of iodized salt as a mineral supplement, which has dramatically reduced the incidence of hypothyroidism and practically eliminated goiter, the enlargement of the thyroid gland that occurs when the gland is attempting to increase hormone production. Four tiny parathyroid glands, embedded in the thyroid, regulate calcium and phosphates, and also balance the effects of calcitonin, another hormone produced by the thyroid.

The pancreas, next to the adrenal glands, produces two hormones that have opposite effects. One, glucagon stimulates the change of glycogen, a stored form of glucose, into glucose, the simple sugar that is our main source of energy. The other pancreatic hormone, insulin, assembles glucose back into glycogen when sugar concentrations get too high in the bloodstream. The two hormones have a negative feedback system to keep in balance. The amount of glucose in the blood determines the rate of insulin production. After you eat a big meal, your glucose level is very high, so the pancreas releases insulin to stimulate the liver into taking some glucose out of the bloodstream, hoarding it as glycogen. After the body has coped with the meal, insulin production shuts down. When blood glucose levels drop to a certain point, glucagon is released from the liver to free up more glucose. Diabetes is the disease that results when the pancreas cannot produce enough insulin, destroying the balance between the two hormones.

Seated in the chest by the bronchial tubes, the thymus does most of its work in children, producing the hormone thymosin necessary for nourishing parts of the immune system. The thymus grows until puberty, then gradually shuts down. Apparently, its role is to give children extra protection against infections.

The pineal gland, deep within the brain, produces melatonin, the substance that regulates our daily rhythm of sleep and wakefulness. Melatonin production builds at night and falls in the daytime, one reason travelers may take melatonin supplements when they try to adjust to travel between time zones. Exposure to light during the dark cycle causes melatonin production to halt, part of the reason for jet lag.

The testes of males and the ovaries of females have a function not only in reproduction but also in the

Liquid waste-makers, the kidneys (green) filter some 425 gallons of blood daily, creating about three pints of urine, a mix of water and metabolic wastes. Urine travels to the expandable bladder via the ureters (red tubes) for temporary storage.

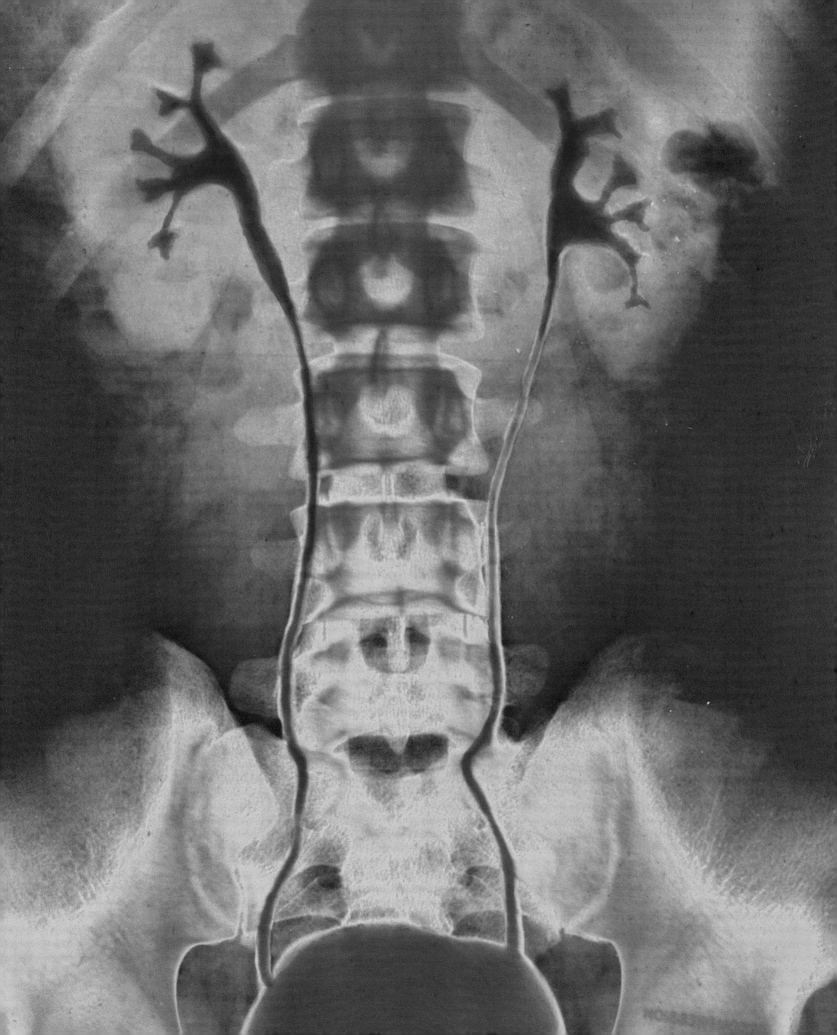

endocrine system. Both organs produce hormones that are integral to the normal development of men and women. Cells in the ovaries secrete estrogen; those in the testes produce androgen, from which testosterone derives.

Circadian rhythms—that is, those that occur in 24-hour cycles—govern much of our hormone production. These seem to be in response to day-night cycles. Adrenal hormones rise and fall daily, following the waxing of dawn and the waning of dusk. But the rhythms are not tied tightly to light cycles. If you were locked in a pitch-black cave, the rhythms would not end. Instead, they would disengage somewhat, perhaps expanding to a longer cycle, one that might drift. Daily alteration of day and night thus appears to modify rather than determine the cycles.

There are other circadian effects. We are more likely to be born between 3:00 and 4:00 a.m. than any other time of day. We are more likely to die then, as well. Just as the production of various hormones flows to a circadian rhythm, so does our body temperature, pain tolerance, respiration, and heart rate.

Our bodies also contain hormonelike substances called prostaglandins, which are produced all over the body and found in almost all tissues. More than a dozen have been discovered, and they are even more powerful than hormones. Prostaglandins are involved in the prevention of blood clots and help stimulate uterine contractions in instances when labor is too long. Together with the hormones produced by the endocrine sex glands, prostaglandins play crucial roles in reproduction.

ANOTHER GENERATION: THE REPRODUCTIVE SYSTEM

All the systems discussed so far work together to ensure our survival as individuals. The reproductive system allows the survival of the human species. We reproduce with remarkable fidelity across generations,

A boy and his amphibian share contact of their largest organ—their skin. Our sense of touch resides in the skin's outer layer, the epidermis, which also regulates body temperature, hoards precious internal fluids, and serves as first defense against hostile bacteria and other invaders.

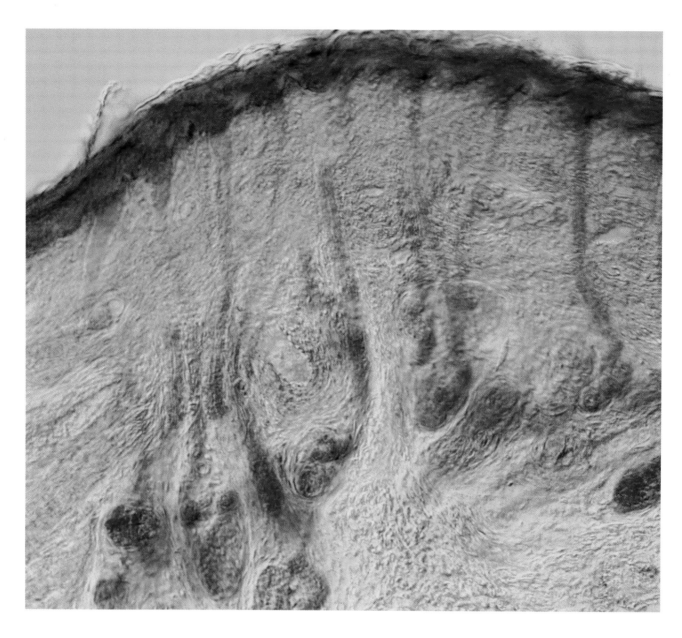

making new beings that are identifiably like us. As each of us entered the world, one of the first things anyone wanted to know about us was our gender—girl or boy? Not all that long ago, one never knew until the doctor made the portentous pronouncement in the delivery room. Now, of course, the question can be answered even before the mother begins to show her pregnancy, by sonograms or blood tests. Human reproduction may not be quite the mystery it once was, but it remains a miracle. The human reproductive system is a chemical, mechanical, even a hydraulic marvel.

Boys and girls start out much the same in the womb. In the fetus, a set of primordial tissues in the pelvis sort themselves either into testes—which descend into the scrotum before the time of birth—or into ovaries, which stay put. Either way, the sex organs stay quiet until puberty, when signals from the maturing hypothalamus order the production of sex hormones. Testosterone directs the boy's body to become a man; estrogen and progesterone turn girls into women.

The physical differences are obvious—and evolution wants it that way. Secondary hair growth, deepening voice, and broadening shoulders in a boy, developing breasts and widening hips in a girl tell that the body is ready to reproduce itself. Sperm develop in the testes. A girl's first ovulation releases an egg cell, or ovum, that has been waiting for the right signal since it formed in the fetus.

To prepare for all of these events, the reproductive system carries out an elegantly orchestrated sequence of changes. These changes are mostly cyclical in a woman, keeping her body primed with the cellular and hormonal supplies she needs to create a new being.

A woman's ovaries release an egg usually every month, from puberty to about age 50. The uterine lining must be prepared monthly in anticipation of pregnancy, becoming enriched with blood and nurturing tissues during the hormonal state induced by ovulation. This lining is shed in menstruation should an embryo not arrive, and then is rebuilt according to renewed endocrine instructions. The average cycle is every 28 to 30 days.

As a girl matures, pubic hair grows. Hair also appears in the armpits at puberty. The opening to the vagina is usually covered by a thin sheet of tissue called the hymen, which breaks at first sexual intercourse—if it hasn't been broken already in strenuous activities such as horseback riding.

The rising load of hormones at puberty tends to rouse sexual interest in both sexes, often emphatically.

Progesterone readies the breasts to produce milk—which is why it is not uncommon for women to have tender breasts during the latter stage of their menstrual cycle.

Physically, the breast is composed of fat intermixed with a rounded collection of milk-producing glands, all of which rise on the developing girl's chest with the hormonal changes of puberty.

The glands of the breast begin producing true milk a couple of days after a woman gives birth and continue for as long as a nursing baby demands it. The reflex that delivers milk is begun by the stimulation of sucking and activates hormonal signals from the brain and pituitary gland to eject milk through the ducts and out the nipple. But the signals can come under more complicated reflexive control as well. Many nursing women find that even the sound of a baby crying, any baby, in any place—a bank, a supermarket—can stimulate lactation in them.

The uterus, about the size of an adult's fist, is a muscular organ that expands many times its original size to accommodate a growing fetus. Its lining, the endometrium, is able to respond to the reproductive hormones whose levels wax and wane over the course of about a month. The paired ovaries produce the eggs and secrete the female hormones estrogen and progesterone.

Extending from the uterus on either side are two fallopian tubes, which widen to enfold each ovary and spread like a baseball mitt to catch an egg. The uterus opens into the vaginal canal through a ring of thick tissue called the cervix. It must dilate to many times its resting, pinhole state to allow a baby to pass through.

If a fertilized egg does not arrive by about Day 24 in a woman's cycle, progesterone levels fall. Over the next few days, small arteries in the uterine lining constrict, cutting off the blood supply. The damaged capillaries ooze blood that pools beneath the endometrial surface. Dying, the outer layer separates from the uterine wall, sloughing out through the cervical opening of the uterus to be shed as menstrual flow. About 20 to 200 ml of blood are lost with each menstruation—roughly between a tablespoon and about three-quarters of a cup.

The production of too much prostaglandin, a messenger that acts on blood vessels and other smooth muscles in the body, can cause problems during menstruation. Prostaglandin made in the degenerating

Sweat glands in the dermis, or inner layer of skin (opposite), churn out droplets of sweat (above). Sweat cools the body through evaporation and provides food for "good" microbes that inhabit the skin's surface and help thwart the "bad" kind.

uterine lining can result in headaches, cramps, even nausea and vomiting in women who suffer during their periods. Aspirin and ibuprofen decrease prostaglandin production, which is why they work so well to relieve menstrual cramps. Premenstrual syndrome, or PMS, can also complicate a woman's cycle. This condition may result from an imbalance in the hormones preparing the uterus to receive the fertilized egg. These hormones act on behavioral control areas in the brain, including regulators of mood. Therapy for PMS has included progesterone and drugs that alter serotonin, a mood-influencing brain messenger. Endometriosis is associated with menstrual cycles as well. It occurs when a tiny bit of endometrial tissue escapes from the fallopian tubes and attaches to surrounding organs. This hormonally receptive speck can then begin to grow and bleed with each passing menstrual period.

A girl baby is born with a quarter of a million potential eggs, though she will send only about five hundred of them in search of a partner. The male reproductive system lacks the cyclical nature of the female. It relies instead on continuous and massive production of sperm, producing millions each day once sexual maturity is achieved.

Before the time a boy is born, his testes normally descend into a pouch of skin called the scrotum. Its placement outside the body keeps the testes slightly cooler than normal body temperature, crucial to the survival and fertility of sperm.

In addition to making sperm, of course, the testes secrete testosterone, the hormone responsible for the widening shoulders, bulking muscles, deepening voice, and growth of facial and body hair in men. Testosterone is now recognized to play certain roles in women as well. Minute amounts may help, for example, to alleviate a loss in sex drive women sometimes experience at menopause, when their reproductive system shuts down.

Sperm cells develop heads and tails as they form. They swim like tadpoles into a 20-foot-long, threadlike coiled duct, whipping their tails and maturing as they move slowly through the encircling prostate gland and into the urethra. The prostate gland, a walnut-size tissue that surrounds the urethra, supplies the seminal fluid that is released with sperm. It is thick, milky, and alkaline,

compared to the acidic environment of the vagina. Acidic conditions tend to slow sperm, so the fluid has to counter that to allow the sperm to meet the egg.

Nerves signal arteries in the penis to dilate, allowing blood to rush in and cause the penis to grow in size and harden. With ejaculation, sperm and seminal fluid are released into the urethra.

Many men over fifty have enlarged prostates. This constricts the urethra and makes it hard to empty the bladder during urination. Drugs or surgery can help. Cancer of the prostate ranks second only to lung cancer as a killer of men. Until recently, the only way to screen for it was by manual manipulation. Now, men can be checked for levels of prostate-specific antigen. A component of prostate cells present normally in low amounts in the blood, it becomes elevated with cancer of the gland.

THE SYSTEMS AS AN ENTITY

A human being is the product of all its systems, each working in harmony with the others. These systems do not exist in isolation; each is an integral part of an almost unbelievably intricate entity.

Scientists and lay people have for centuries likened the human body to an incredible machine—and for good reasons. Our disparate parts work together efficiently to produce force, motion, and energy. The human body's systems work in concert much as the various mechanisms in any machine do. The heart is the engine, driving the body onward; the skeleton is the framework, supporting and containing the different mechanisms.

When we are in optimal health, the timing is perfect, and the machine runs smoothly. When an aspect of any body system fails, we seek to be cured—essentially to restore the workings. No single system can stand alone; none is greater than the sum of its parts. Working together, systems inform and maintain the human body, which is, after all, the ultimate machine—and the final masterpiece.

Pushing through the skin's upper layers, hair takes root in the dermis layer. Beneath the dermis, subcutaneous fat pads the body's inner tissues. Blobs of oily white sebum attach to the hair follicles, lubricating both hair and skin.

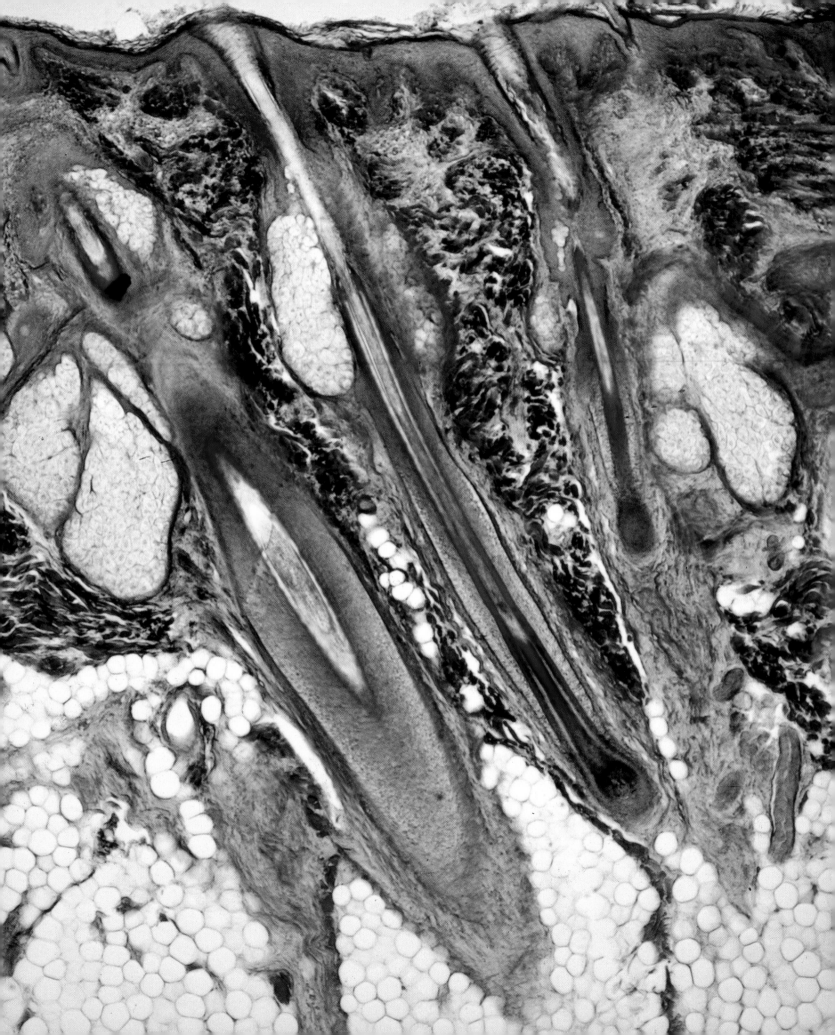

Bodily Defenses

LISA M. KRIEGER

A moment of pain, a lifetime of protection. Eight out of ten U.S. youngsters are immunized against measles, a highly contagious and common childhood disease.

In 1968, while working at a Boston hospital, Steven A. Rosenberg admitted a 63-year-old man complaining of abdominal pain, then conducted surgery to remove a bad gallbladder.

It was a routine case, except for one extraordinary feature. Twelve years earlier the man had come to the same hospital with abdominal trouble of a different kind: stomach cancer. The tumor in his stomach had been excised to ease the pain, but since the cancer had already spread to his liver, it could not be completely removed. Sadly, the man was sewn up and sent home without further treatment—presumably to die within several months.

But over the following months, he began to gain strength and feel better. He continued to improve and soon stopped returning for medical evaluation. Nothing more was heard from the patient—until, more than a decade later, Rosenberg opened him up to remove his gallbladder and discovered that all evidence of cancer was gone. The cancer had simply disappeared. Spontaneous disappearance of cancer, although one of the rarest events in medicine, is also one of the most awe-inspiring, proof of the power of our immune system.

Day in and day out, the immune system quietly goes about its job protecting us from intruders, routinely shielding us from death; in the case of such a dramatic cure we stop and take notice. It is evidence of the immune system's enormous versatility: The cells that attacked this man's cancer are the same cells that thwart the common cold or the deadly Ebola virus. They are also the same cells that, when confused, attack the very body they should protect—provoking sneezing in spring, asthma around certain animals, or the pain of rheumatoid arthritis with every step. Should Dr. Rosenberg's patient ever need a transplant, the same cells that saved his life from cancer might destroy the new life-giving organ.

Friend or enemy? Evolution has created a delicate balance between protection and self-destruction. By defining and defending the self, the immune system makes life possible; when it defends too vigorously, inappropriately, or not at all, it causes illness and death.

In the Book of Exodus, the Bible says that God cast a plague on Egyptians, delivering "sores that broke into pustules on man and beast" because they held Hebrews in bondage. Epidemics were divine vengeance; cures, divine intervention. Everything lay in the hands of God as to whether a person recovered or not.

In this century, we have slowly learned that the activity of the immune system depends on a symphony of highly specialized types of cells in the blood and the tissues, each cell type performing a unique function. Together, these cells and organs form a network to guard the body against attacks by invaders: tiny, infection-causing organisms such as bacteria and viruses, as well as parasites and fungi.

EXPANDING OUR UNDERSTANDING

As the conductor of this symphony, the immune system must keep all the cells working in harmony, coordinating the defense. If it becomes confused, there is dissonance—in the form of immune disorders such as allergies and arthritis. If it becomes disabled, outside enemies move in and eat us alive.

Once science recognized the existence of the immune system at the end of the 19th century, immunology became a new field of study. The word "immunity," based on the Latin term *immunis* for exemption from taxes or the military, was created by early 20th-century scientists to denote freedom from infection. Boosting the body's capability of responding to threat ushered in the age of immunization, when many widespread diseases were brought under control through vaccination. Epidemics were averted; millions of lives were saved.

The mechanisms and processes of immunity are so sophisticated that scientists only recently have begun to unravel the complexities of this system. Physicians could—and did—administer vaccines that promoted immunity to specific infectious diseases. The discovery of penicillin in 1943, for example, saved millions of people from a variety of infections. Beginning in 1954, polio vaccines rescued tens of thousands from the debilitating effects of polio. Two hundred years after

Gaining protection from microbes it has never met, a breast-feeding infant receives not only nutrition from its mother's milk but also disease-fighting antibodies. At birth, the immune system has not yet developed an immunologic memory.

Edward Jenner discovered a vaccine against smallpox, inoculations have wiped that disease from the Earth. Immunologists have accomplished a great deal with little knowledge of the workings of the immune system.

For many years, however, immunology was mainly a public-health tool: practical to be sure, but not one of the most exciting of medical curricula. The roster of newly discovered immune cells grew larger and larger as researchers studied the immune system, but it was not clear how this unwieldy multitude of cells was able to orchestrate its activities into a single selective defense against disease.

Many factors—among them instrumentation, the development of the field of molecular biology, the discovery of the structure of DNA, the boom in biotechnology, and the growing number of researchers attracted to it—have transformed immune research by offering detailed chemical, molecular, and cellular explanations of how the immune system works. Just within the past three decades, scientists have been able to build a complex theoretical scaffold onto which they can hang their many new findings.

In the 1980s, for instance, the cloning of critical immune cells, their receptors, and their genes led to an explosion in understanding the molecular basis of immune regulation. Suddenly researchers could study, under rigorous experimental conditions, both normal and abnormal responses.

Other techniques of molecular biology permitted genes not only to be studied but also to be moved around, at will, from one cell to another, offering yet more experimental possibilities. Closer scrutiny of the chemicals known as interferons and interleukins is improving our understanding of how the immune system turns itself off and on. The practical applications of the new findings are immense. The first payoff came with the successful transplantation of human kidneys, hearts, and other organs necessary for life. On the horizon are immunologic tools to treat cancer and perhaps slow the alarming spread of AIDS. Research has been further spurred by the rise of antibiotic-resistant bacteria—and the threat of newly emerging infectious diseases.

The immune system, we now know, consists of millions and millions of cells organized to perform various functions. The system is a dynamic one in which the cells constantly communicate with each other. The field of immunology is not so different from the study of neuroscience: Like the brain, the immune system learns, remembers, and communicates. Just as our eyes and ears respond to light and sound, our immune system responds to invaders.

When invasion threatens, the cells undergo critical changes, producing powerful chemicals that allow the cells to regulate their own behavior and growth, to recruit reinforcements, and to deploy other cells to problem areas, destroying the enemy.

Although the immune system is intricate, its basic strategy is simple: recognize, mobilize, and attack. Future research will further clarify the components of the immune system, better defining how they work. One of the goals of investigators is to turn down an overreactive immune response, thereby preventing auto-immune disease and the rejection of transplanted organs.

Another goal is to achieve just the opposite: to turn up the response to destroy outside invaders such as malaria or the AIDS virus. Cracking the more subtle secrets of immunity has become one of the central challenges of modern medicine.

Despite huge progress, scientists still struggle mightily to make sense of the underlying mechanism. How does the immune system thwart the seemingly infinite number of invaders? What has this meant for our survival as a species? And why, when the body is so busy working miracles, do we ache, moan, and yearn to climb into bed?

STRUCTURE/MECHANISM

Who am I?

This existential question is the central challenge of the immune system: distinguishing self from nonself. Without this recognition mechanism, as the late scientist and author Lewis Thomas observed, all living matter would become one huge ball of protoplasm. Happily, we are all very distinct individuals. Some molecules are mine. Others are yours—yours alone and possessed by no other.

Your immune system has an exquisite ability to discriminate between you and the outside world. It coexists peacefully with "self" cells, because each of your cells carries on its surface a protein, or a "marker molecule," that is distinct to you. Your immune system recognizes that, and leaves it alone.

Immunity is based on this ability to recognize things that do not belong. Thus, when immune system components detect a marker on the surface of a nonself cell, they immediately take action. Any nonself marker, no matter how harmless it may actually be, is enough to trigger an alarm. So in many cases, the immune system protects us quite blindly, whether we need to be protected or not.

Any substance that prompts an immune response is called an antigen. An antigen can be as harmless as a cat hair or as harmful as a deadly poison. The invader can be an innocent speck of dust or pollen. It can also be a piece of foreign tissue—an organ transplant, for example—that could save a life. It may even be a nourishing piece of food, identified as an enemy until it is broken down by the digestive system.

Occasionally, this detection system goes seriously awry, mistaking self for nonself and launching an attack. Auto-immune disease is the result—and it can be deadly. The ability to distinguish self from nonself seems to have arisen early in evolution. Immunity is an ancient phenomenon, with some forms present in all organisms.

Even creatures as far down the evolutionary ladder as amoebas can both resist bacteria and refrain from eating themselves. The lowly yeast has genes that code for proteins that prevent mating between similar cells. Sea urchins and sponges are able to distinguish their tissue from the tissue of other species and even from other individuals of their own species—and reject it.

But about 450 million years ago, the "big bang" in immune system evolution occurred. Vertebrates began to evolve, and their immune systems became more sophisticated. The modern immune system emerged, with many key players already in place.

The shark represents a great divide in the evolution of immunity. Skin from a hagfish, a creature that evolved before sharks, can be grafted onto another hagfish without causing a rejection response. Sharks, however, resist such grafting. Before sharks, there is no trace of antibodies or other pivotal immune proteins. In sharks and fishes that evolved after them—amphibians, reptiles, birds, and mammals—antibodies and other pivotal immune proteins are present.

Primitive immune systems are perfectly serviceable in most cases, doing an adequate job of protecting against generic invaders. But they can't target any particular invader. Nor do they have a memory, so they can't offer proactive protection should an invader someday decide to return.

Higher creatures have more complex needs, and the immune system appears to have incorporated the defenses seen in invertebrates, then added on, growing more and more elaborate. One has only to look at the illnesses suffered by people with AIDS—caused by a devastating virus that destroys one entire arm of the immune system—to see how the human species would get along without our more sophisticated defense apparatus.

It was Peter Medawar, then a zoologist at Oxford, who identified the importance of the issue of recognition. In his autobiography *Memoir of a Thinking Radish*, he describes the 1944 crash of a Royal Air Force bomber just down the road. Medawar offered to help a local

Patron saint of plague victims, St. Roch (opposite) points to a bubo, a lymphatic swelling characteristic of bubonic plague. One-fourth of Europe's population succumbed to plague in the 1300s; a 1665 death census (above) tallies London's weekly victims. FOLLOWING PAGES: Post-plague desolation marks Pieter Brueghel's "The Triumph of Death."

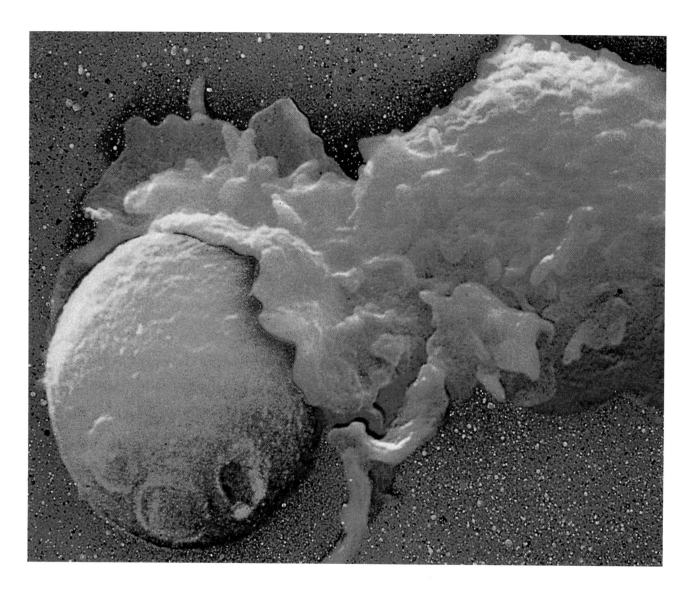

surgeon treat the seriously burned pilot with skin grafts. He observed that the pilot not only rejected the donated tissue but also rejected it more vigorously the second time around. The implication: the immune system perceived donated skin as if it were a virus or harmful bacteria. Moreover, it remembered the donated skin—and disliked it even more on second acquaintance.

Medawar's pivotal observation didn't help the pilot, who nonetheless recovered. But it did lead to research that, 16 years later, brought him the Nobel Prize in medicine. Australian Frank Macfarlane Burnet, who earlier had formulated the first theoretical model of self and nonself, shared the prize.

We now know that the human body has evolved one of the most elaborate mechanisms for distinguishing self from nonself. It can recognize foreign substances whose chemistry is only slightly different from its own.

The force and duration of the immune response depends on the type and amount of antigen and how it entered the body. Immunity also depends on genes; compared with other individuals, you might respond vigorously, weakly, or not at all to a particular antigen.

Our bodies have developed a multi-tiered network of protection against foreign invaders. The first line of defense is the anatomical barrier created by skin and the mucous membranes that line the body's openings.

Swallowing and digesting an infectious yeast cell (green), a macrophage (blue) serves as foot soldier in the body's nonstop war against infection. Other types of white blood cells use varying strategies with different invaders.

The second line, the so-called innate response, is inborn and rapid, prompting an immediate attack on all microbes in the same way. The innate response makes up in speed what it lacks in specificity.

The third line of defense is the immune system's adaptive, or acquired, response. It is much slower to respond, but it is elegant in its sophistication. It also is highly specific: It can recognize, match, and respond to an invader even if it has never faced it before. Moreover, it has a memory, so after once being exposed to a microbe, it is capable of responding much more quickly and forcefully to a subsequent exposure.

The second and third levels of defense work together, in complementary fashion, depending on the type of assault mounted by the invader. A defect in either one creates a huge vulnerability.

These response systems can be seen at work any time a deep cut causes your finger to turn hot and red, when a throat infection makes it painful to swallow, or a cold causes sneezing, coughing, runny nose, and headache.

NATURAL DEFENSES

To penetrate the body, a microbe must first get past the skin and the membranes lining the body's openings. Skin provides a formidable barrier to the vast majority of infectious agents unless it is damaged by injury, bites, or burn. This physical barrier is rich in chemical deterrents. For instance, sweat contains bacterial inhibitors such as lactic acid and oleic acid. Experiments show that if a culture of typhoid germs is smeared on the skin, the number of living microorganisms is reduced greatly in 20 minutes. An identical smear on a glass plate survives a long time.

The numerous mucous membranes of the body—those that line the mouth, throat, eyes, intestines, vagina, and urinary tract—form another anatomic defense. They produce slippery secretions that make it difficult for bacteria to adhere and multiply.

Mucus is sticky like glue, so it traps dirt and germs that get into the nose and throat. When germs get inside and make you sick, the mucous glands work extra hard, manufacturing large amounts of mucus that clog up your head and run out your nose.

Membranes of the eye, respiratory organs, intestines, and sex organs are covered with secretions containing the substance lysozyme, which dissolves and breaks down a bacterium's protein sheath. In the stomach, bacteria are annihilated by acidic gastric juice.

The upper airways have hundreds of tiny hairs called cilia, which trap bacteria before they can enter the lower respiratory tract. But even if this filtering system is penetrated, the cilia sweep mucus laden with trapped bacteria away from the lungs and up toward the throat, where it can be removed by coughing. Mucus in the nose can be sneezed out at speeds of up to a hundred miles an hour—a vacuum cleaner in reverse.

Tears also protect. Like water from a constantly dripping faucet, tears wash the surface of the eye and clear away irritants such as smoke, dust, and sand. Most tears simply evaporate or drain into the tear ducts and are reabsorbed, but excess fluid caused by foreign objects may spill down the face or drain into the nasal cavity.

If a microbe is able to penetrate these barriers, it soon encounters a different kind of resistance: the immune system's second and third lines of defense. One is general, the other is highly specific; their common goal is to destroy the microbe before it multiplies. This two-pronged approach enables the body to recognize and respond to any possible foreign molecule, or antigen.

All immune cells have their origin in the bone marrow, the soft tissue in the hollow center of bones; some move to the thymus, a two-inch gland that lies behind the breastbone, to mature. Scientists first understood the importance of bone marrow after the atomic blasts at Hiroshima and Nagasaki, when they discovered that nuclear radiation had destroyed the marrow, its infant immune cells—and all resistance to infection. People there died within 10 to 15 days.

Where does this cellular line of resistance hide? The cells are transported in a pale, thick fluid—lymph—carried through the immune system's own unique and massive circulatory pathway, the lymphatic system. These cells, called leukocytes from the Greek *leukos* (meaning white light, brilliant, or clear), travel throughout the body in the lymphatic vessels, permeating every organ except the brain.

The lymph system's vast collection of organs, vessels, ducts, and tissues offers sites where these infection-fighting cells can be recruited, mobilized, and deployed at a moment's notice. It is a circulatory system second only to our bloodstream.

Thousands of small lymphatic capillaries feed into larger vessels and pass through one or more lymph nodes. At the base of the neck these vessels empty into a large duct, which returns its fluid contents to the bloodstream, and eventually to tissues throughout the body.

Each lymph node contains specialized compartments where immune cells congregate and where they can encounter antigens. Lymph nodes act as barriers to the spread of infection by destroying or filtering out bacteria before they can pass into the bloodstream. The lumpy, swollen glands in your neck when you get a sore throat are simply the lymph nodes at work, packed with immune cells gobbling up invaders.

Strategically located clumps of lymphoid tissue—residing in the lymph nodes, tonsils, bone marrow, spleen, liver, lungs, and intestines—are packed with white blood cells, ready to be deployed. They patrol everywhere for foreign antigens, then gradually drift back into the lymph system to begin the cycle anew.

The ingenious structure of this system ensures the ready availability and quick assembly of an immune response—anytime and anywhere it is needed.

....

INNATE AND ACQUIRED RESPONSES

....

Say you get a splinter lodged in your skin, breaking through your outermost line of defense. With it comes a tiny world of microbes. When the microbes rush in, they encounter your second line of defense: A natural, or innate, system that is present from birth and is the body's protection against the majority of infectious agents. It is this secondary immune response, characterized by inflammation, that holds infections in check during the first several days. Sometimes it does its job so well that no further response is needed. Its troops are generalists, trained to take on all comers quickly. The cells react similarly to all foreign substances, no matter the size or shape. This innate response, a primitive but effective piece of biological hardware, is uniform in all humans. The ill-fated bacteria are promptly grabbed by large cells called phagocytes—from the Greek, "cells that eat"—whose job it is to entrap and digest microorganisms in an effort to contain an infection in a small space.

Phagocytes, like tiny garbage collectors, also scoop up bits of cellular debris and other cast-off materials. They are the scavengers that waft back and forth, keeping wounds and damaged tissues clean, devouring everything that has no role to play, gobbling up worn-out cells and other debris. They can do this because their cytoplasm contains granules, or packages, filled with potent chemicals that allow them to digest an ingested microbe, destroying it.

Some phagocytes, called neutrophils, circulate in the blood; others, called macrophages, reside in specific tissues. Macrophages and neutrophils work as a team. Macrophages jump in first, then signal neutrophils to join them in battle. When the neutrophils arrive, they digest the invaders. Pus, the unpleasant, yellow fluid in wounds, is the accumulation of neutrophils, digested microbes, and cellular debris from the involved tissue—proof that the immune system is hard at work.

Another important component of the innate response is a system of more than 30 blood proteins known as the complement system, which assists—or complements—the action of macrophages and neutrophils in destroying bacteria. When the first protein in the complement system is alerted to trouble, it triggers a chain reaction. One by one, the subsequent proteins hook together and become lethal—punching holes in the invaders so that fluid flows in and causes them to burst. The complement proteins also dilate blood vessels and cause them to leak, which leads to the pain and tenderness of an injury.

As effective as it is, such innate immunity cannot guard against all infections. Microbes evolve so rapidly that they can devise means to evade the inherited immune defenses of humans and other species that evolve more slowly. When they succeed, the third and final line of immune defense—the acquired or adaptive immune response—comes into play. Thanks to adaptive immunity, the body can identify and react in a sophisticated and form-fitting fashion to any invader, even one it has not previously encountered.

Microbes that have succeeded in crossing the nonspecific barrier are suddenly faced with specialized weapons, lymphocytes called T cells and B cells, tailored just for them. These cells each have their own duties, but are similar in their specificity. Although slower than the initial innate response, adaptive immunity is anything but subordinate.

The cells in the innate and acquired armies have separate functions, but they interact in many ways and can actually regulate one another's activities. Sometimes they communicate by direct physical contact, sometimes by releasing chemical messengers. This signaling system, by coordinating events, is the key to how the immune system works.

Microbe-fighting lymphocytes are crucial to our protection not only because they have an uncanny ability to target specific microbes but also because, incredibly, they retain a memory of each invading microbe and can ward off trouble next time it ventures near. Unlike other immune cells, which live no more than seven to ten days, lymphocytes can live for years, even decades.

Although lymphocytes are white blood cells—as are macrophages and neutrophils—they are smaller. They also are more specialized, not mere garbage collectors. They consist primarily of T cells and B cells, each of which comes with its own different lethal strategy: T cells generally punch and poison, while B cells produce antibodies.

The wonder of lymphocytes is that they have surmounted the huge challenge posed by the microbial kingdom: how to respond to the diverse forms and life cycles of germs. Basically, they've divided up roles and have taken on different but complementary functions and purposes.

The T lymphocytes destroy invaders such as viruses and some types of parasites that have penetrated the host cell, where they are hidden and inaccessible. B lymphocytes and their antibodies go after germs such as bacteria and free-floating viruses that wander outside the cells in the blood or fluids.

The most effective immune response involves both T and B approaches. When a person catches the flu, for example, T cells are needed to destroy cells infected with flu virus, but antibodies are essential to prevent the spread of the virus through the blood. T cells' tasks are divided into yet smaller categories. "Helper" T cells identify the invader, and they stimulate other cells to fight it. "Killer" T cells specialize in destroying cells that are infected or cancerous. "Suppressor" T cells, triggered later, slow the immune response after a microbe has been eliminated.

As a germ enters the body, it encounters one of the literally millions of helper T cells that patrol ceaselessly, night and day. These helper cells aren't aggressive; they merely wander, patrolling the body. Everywhere they go, they inspect cells for special surface markers, then return to the lymph nodes to report their findings. If a cell has the right password— if it is familiar, part of the self and not a threat— the helper cell simply meanders on.

The password is unmistakable. It is a distinctive molecular marking on the cellular surface. Scientists call it MHC (major histocompatibility complex), and consider it the linchpin of the immune system. Found on the surface of nearly every cell in the body, MHC acts like a package label, revealing the cell's contents. Because the structure of MHC is different from person to person, it is a marker of self or nonself.

The MHC password of a germ rings false to the challenging T cell and is not accepted; nonself surface markers on the germ give it away. The helper T cell then signals that the invader is an enemy, and worth battling.

Angel of death works the pump for 19th-century London's poorer residents, dispensing cholera as well as drinking water. Conveyed through human wastes, cholera bacteria continue to devastate communities lacking proper sewage disposal.

FEMALE *ANOPHELES* MOSQUITO TAKES IN HUMAN BLOOD TO NOURISH ITS EGGS

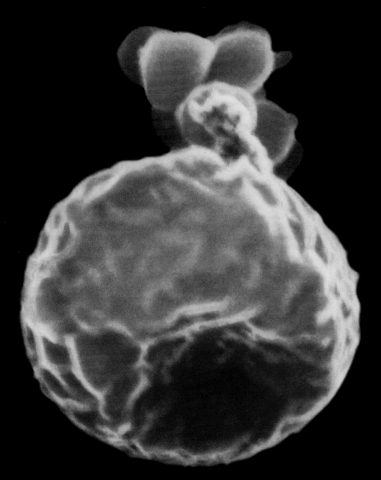

SINGLE RED BLOOD CELL, INFECTED WITH MALARIA

Loading up on a blood meal, this mosquito could also be taking in malarial parasites, if its host is infected. The parasite multiplies in the insect, as seen here in its stomach wall (opposite), eventually dispersing throughout the body. When the insect bites another victim, it transmits the parasite. Once ensconced in the human body, malaria invades the liver and red blood cells, replicating itself at the cells' expense. Prevalent throughout the tropics, malaria affects up to 300 million people worldwide.

OPPOSITE: MALARIA PARASITES CLING TO MOSQUITO'S STOMACH

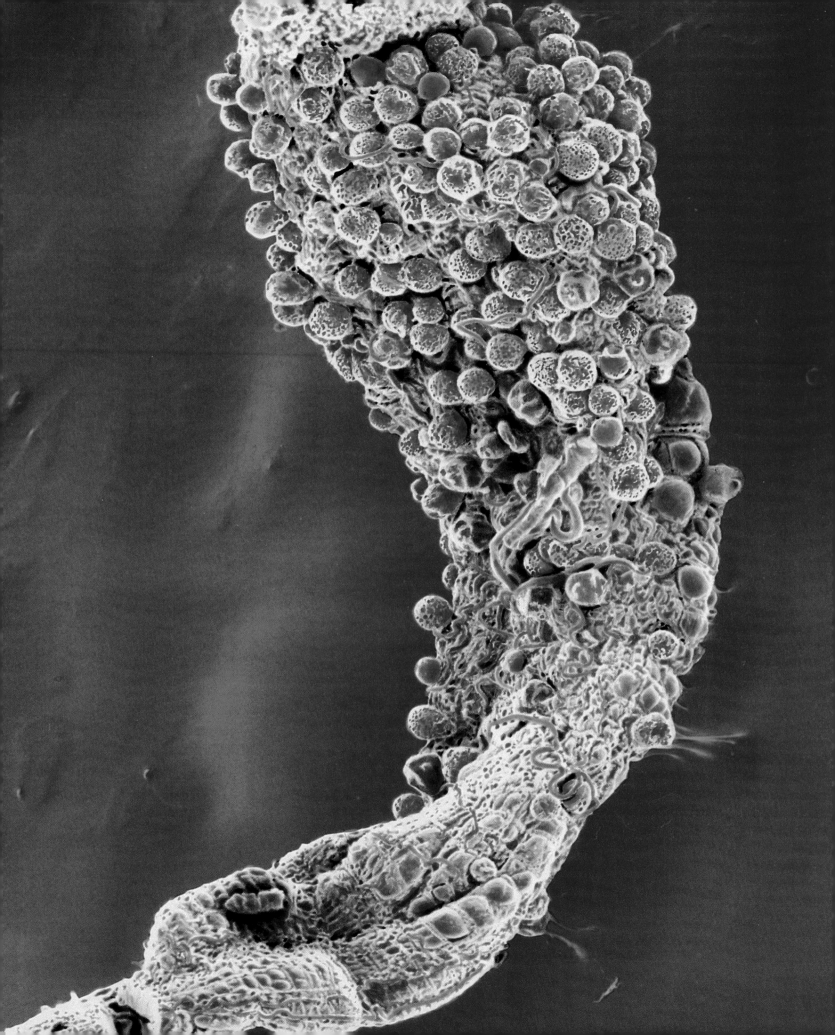

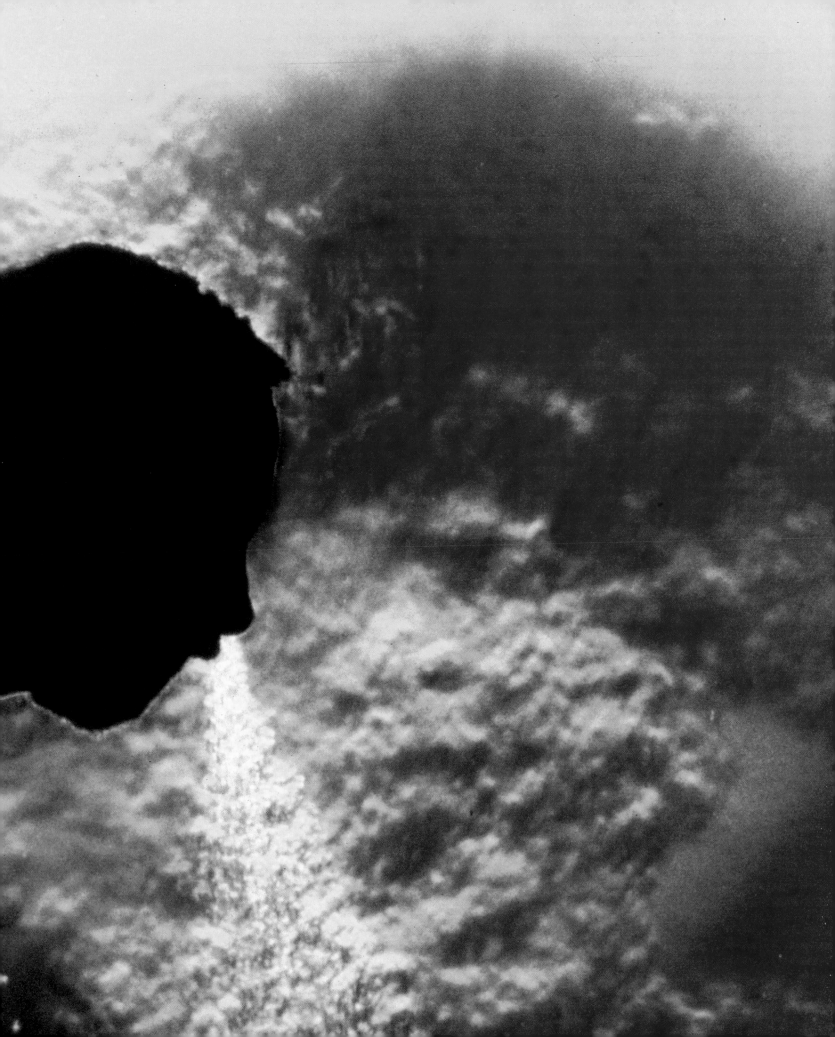

That's when the hatchet men of the immune system, killer T cells, arrive on the scene. They bind to the germ, punch a hole, and deliver a lethal burst of chemicals, destroying it. How do T cells identify specific invaders? Receptor structures on the T cells enable them to identify and counterattack these invaders. The structure of the receptor is determined by an arrangement of genes. Scientists Philippa Marrack and John W. Kappler, both at the National Jewish Center for Immunology and Respiratory Medicine in Denver, found that it takes random combinations of five different gene segments to build a single T cell receptor.

The T cell receptor's specialty is enemies that are clever and evasive about the way they infect, such as viruses that have hidden within healthy cells, or normal cells that have turned cancerous. The AIDS virus, for instance, leaves traces outside a cell that indicate it has been around, but buries itself inside normal cells. Though all seems normal, the invader is hidden inside.

But T cells are only part of the picture. Alerted by news of the invader, B cells and their progeny, called plasma cells, arrive on the scene and manufacture millions of identical antibody molecules— about 2,000 per second—and send them pouring into the bloodstream. Antibodies act as molecular scouts; they don't, as a rule, kill. Their job is to seek out and latch onto intruders before they can infect other cells, staying with them until more destructive immune forces arrive.

B cells have another job to do as well. They activate the complement system, that class of deadly proteins that are present in the bloodstream all the time, awaiting marching orders. Alone, each chemical is impotent. But when all are assembled together in a chemical chain reaction, they become a powerful weapon against germs.

Finally, scavenging phagocytes are brought in to clean up the mess.

Scientists have found the precise physical shape of antibody molecules to be very significant: In each different microbial invasion, antibodies are form-fitted— like a key to a lock—to match the enemy antigen. And whenever antibody and antigen interlock, the antibody targets the antigen for elimination.

Just as one key fits only one lock, B cells are programmed to make one specific type of antibody in response to a specific invader. After infection, B cells rush back to the nearest lymph node carrying an accurate measurement of the invading cell's surface markers. Then, without delay, plasma cells begin manufacturing antibody proteins to fit that specific invading antigen and pour them into the bloodstream as quickly as possible.

The 10 trillion or so B cells in our body can produce more than 100 million distinct antibodies, allowing us to intercept a fantastic range of microbial interlopers. One B cell might produce an antibody that attacks a pneumonia-causing bacterium, while another might counter a cold virus. Still other antibodies fit only molecules of the polio virus or of a particular strain of amoeba. How could such a huge and diverse system be so specific at the same time? Scientists were puzzled for years.

A single sneeze, as shown in this Schlieren photograph of air turbulence (opposite), can broadcast millions of infectious particles. Flu viruses spread easily in airborne droplets; cold sufferers (above) are more likely to be infected by direct contact.

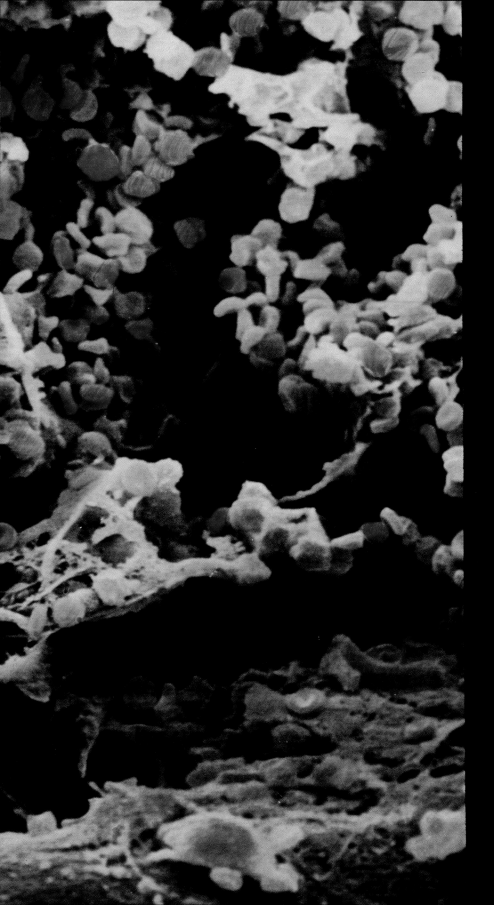

Birthplace of the immune system, bone marrow (left) gives rise to both white and red blood cells. The importance of this soft, fatty tissue occupying bone cavities was discovered only after the World War II bombings of Hiroshima and Nagasaki; studies showed that nuclear radiation destroyed marrow, causing death by infection and internal bleeding.

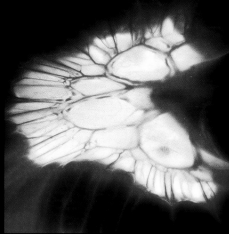

Once mature, immune cells migrate to the lymph nodes (above), where they act as a barrier to infection. The nodes filter out dangerous bacteria and viruses, keeping the bloodstream safe.

Given that millions of antibodies are made, imagine how much space in the genome would be required if every antibody were determined by its own gene. Such a feat is mathematically impossible. There are only 100,000 genes to run the entire human body, most of which are concerned with metabolic and other housekeeping duties, tasks far afield from antibody production.

For years, scientists presumed that an invading antigen acted as a template—so when the antibody molecule confronted it, the molecule would mold itself into a complementary form/shape around the antigen. We now know that the body performs an incredible trick to generate variety, as discovered by Susumu Tonegawa, now at the Massachusetts Institute of Technology. Tonegawa, who won a 1987 Nobel Prize for his work, showed that antibody genes are inherited as gene fragments, which join together to form a complete gene only in individual B cells as they develop. Scientists believe that as the B cell matures in the bone marrow, it randomly picks and chooses fragments of antibody genes like parts from a Tinkertoy set, then joins them to form a unique and functional antibody. The result is that we wind up with a hugely diverse population of B cells, each capable of producing antibodies against one type of antigen.

By waging such an elegant battle, the body keeps the upper hand against the lowly germ. Unfortunately, that war can make us feel pretty crummy. Virtually every infectious disease shares similar symptoms: Fatigue. Fever. Achy joints. Apathy toward sex, food, and newspaper headlines. This is because chemical messengers, called cytokines, relay a warning of infection to immune cells throughout the body. Traveling from macrophages to T cells, they carry this alert. According to Robert M. Sapolsky and his team at Stanford University Medical School, cytokines also influence the brain. For instance, they tamper with the body's thermostat, raising your normal 98.6°F temperature so that you run a fever. They make your head throb, your eyeballs ache, and your eyelids droop with fatigue. They're also known to decrease appetite and sex drive—not a bad idea when your body is busy fighting to survive.

The cardinal characteristic of innate immunity is its ability to distinguish—quickly—between microbes that are good guys or bad, harmless or harmful. In contrast, the acquired immune response is slow to arrive on the scene, but it is far more nuanced and almost infinitely adaptable. Because it actually can rearrange elements of the T and B cell receptor genes, the acquired immune response

can create multitudes of distinct antigen receptors, thus fending off the vast universe of invading microbes.

These two cellular defense systems, innate and acquired, once were thought to be independent of each other. Now it is known that there is a back-and-forth communication between the two. Innate immunity starts the battle quickly and wages it until reinforcements of acquired immunity arrive to help. Then the innate response guides the attack by the B and T lymphocytes of the acquired system, helping it determine which microbes should be responded to—and the nature of that response. Innate immunity, once dismissed as the body's inarticulate and clumsy "first strike" bodyguard, has gained respect. It is not merely a vestige of ancient microbial systems made redundant by the evolution of acquired immunity. Indeed, it dictates the conduct of that more sophisticated system.

REINFECTION AND IMMUNIZATION

Say that the same germ appears a second time. The immune system can remember, even for a lifetime, the identity of a pathogen. As far back as 430 B.C., the Greek historian Thucydides described the plague of Athens, puzzling over the observation that "the same man was never attacked twice."

An 18th-century natural experiment that occurred on the remote Faroe Islands gave us our first real insight into the mechanism of immunologic memory. This experiment began in 1781 with a measles outbreak on these isolated islands in the North Atlantic north of England. For the next 65 years, the Faroes remained measles-free—until another major outbreak in 1846 that affected 75 to 95 percent of the population.

Danish physician Ludwig Panum, studying the two epidemics, observed that "of the many aged people still living on the Faroes who had measles in 1781, not one was attacked a second time." He also noted, astutely, that the people who had escaped measles in the first epidemic acquired it in the second.

Panum's study made two conceptual breakthroughs: First, that the immunity to measles persists for years, and second, that the immune system, once it has met the measles microbe, needs no reminder on reencountering it. Decades after the first epidemic, islanders remained protected. Panum's findings—supported by later observations of epidemics of yellow fever in Virginia and polio among Eskimo villagers in Alaska—showed that the immune system could remember an encounter that occurred years before and that there was an inherent mechanism, unknown at the time, that sustained this long-term memory.

Panum could not explain how the immunity to a reinfection worked. But modern medicine now has an explanation for how the immune system operates—and how it can be manipulated to create vaccines and treatments.

At birth, a person's immune system has not yet encountered the outside world or started to develop its memory files. But over the span of a lifetime, the immune system learns to respond to every new antigen it encounters, either through exposure to microorganisms or as a result of immunization. Moreover, it adapts and remembers. So the next time that a person encounters the same enemy antigen, the immune system is set to demolish it. What happens is that whenever T cells and B cells are activated by an infection, some of them become memory cells and recognize part of the invading organism or tumor cells as an antigen they've met before.

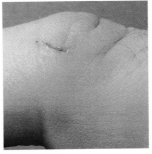

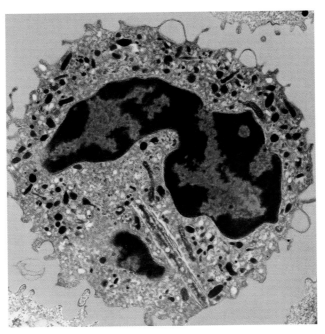

T and B cells, although both players in immunologic memory, differ in the longevity of their responses and in the ways they confer protection. The relative importance of each cell type varies, depending on the nature of the intruder and the type of disease it produces.

B cells, whose plasma progeny produce antibodies, provide the first line of memory defense against a repeat infection. The presence of preexisting antibodies at the site of entry is the most effective mechanism for blocking infection and is the key aspect of protective immunity against many viral and most bacterial illnesses. Since antibodies can last for decades, long-term immunity is already in place against many systemic infections such as measles, polio, mumps, and rubella, as well as smallpox.

The T cell's memory, more complex, happens in several stages. When you first encounter an enemy antigen, there is a 100- to 1,200-fold explosion in the number of T cells. Then, within days, most of them die off, leaving only remnants of themselves. But the remaining 5 percent of those cells offers a stable pool of memory that can persist for many years. You don't need to keep a lot of T cells—because, at a moment's notice, they can again multiply and overpower the infection. This rapid-recall response by these long-lived cells is critical to long-term immunity to many common diseases.

Hard at work even when it seems that nothing's going on, the immune system protects in many ways. Skin provides an excellent barrier to infectious microbes; when it is breached (top left), infection-fighting cells such as neutrophils (above) rush to the defense, speeding recovery (top right). FOLLOWING PAGES: Inside a broken blood vessel, tangles of fibrous protein (green) ensnare red blood cells to form a clot that prevents further bleeding.

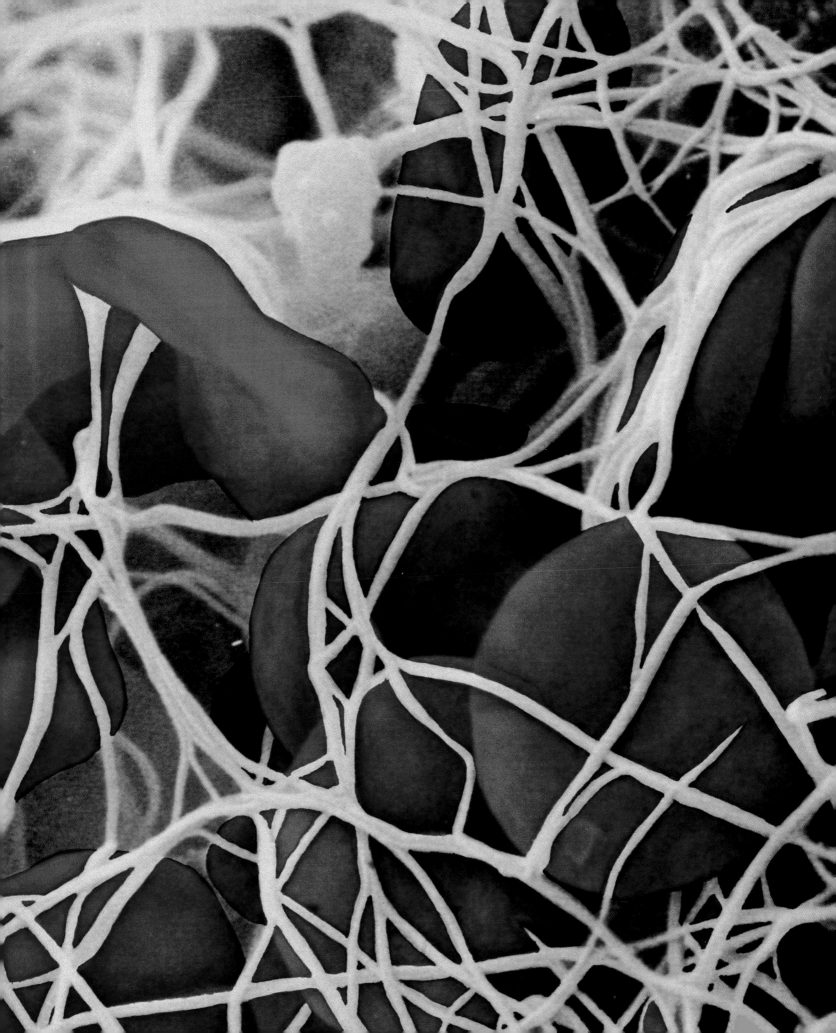

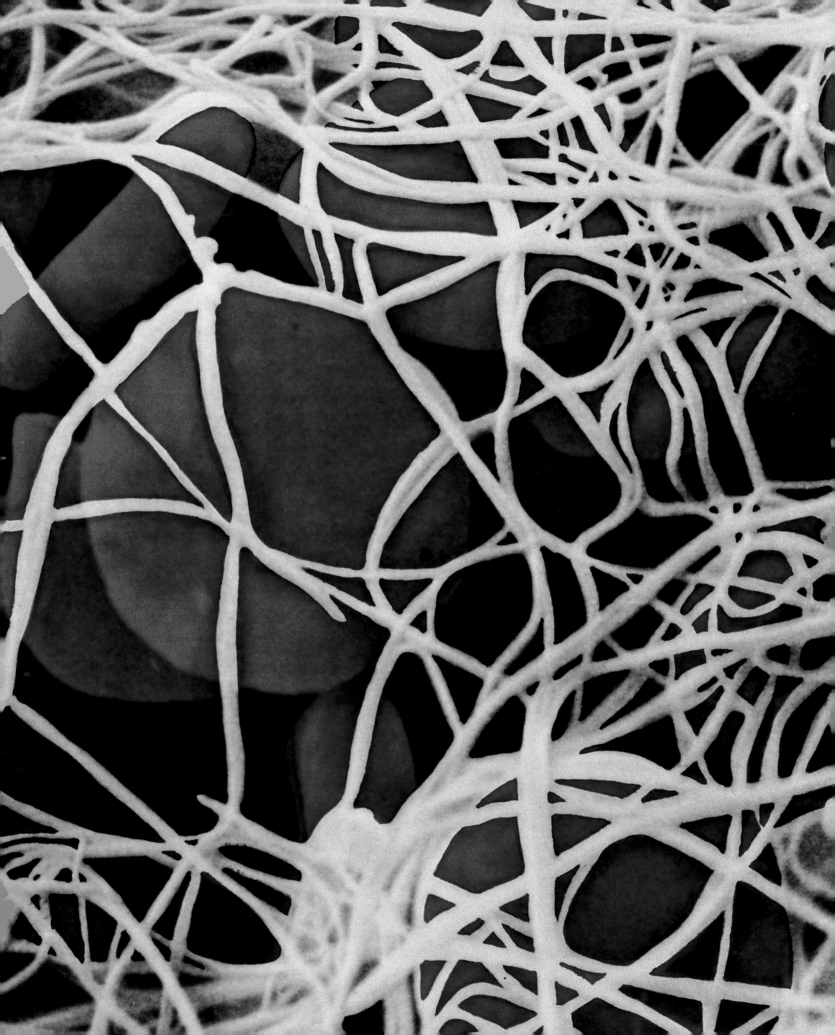

Surrounded on all sides, a sprawling cancer cell becomes the target of killer T cells (below). Normally round, T cells elongate as they press on, altering the enemy's membrane and reducing the cell to a shapeless blob (opposite). The human immune system fights viruses, bacteria, and fungi as well as such cells. Sometimes, however, a cancer cell slips beneath the immune system's "radar," and a tumor results. After years of frustration, researchers are starting to gain deeper understanding of the body's anticancer immune response, prompting hopes that improved treatments for today's most dread disease will soon result.

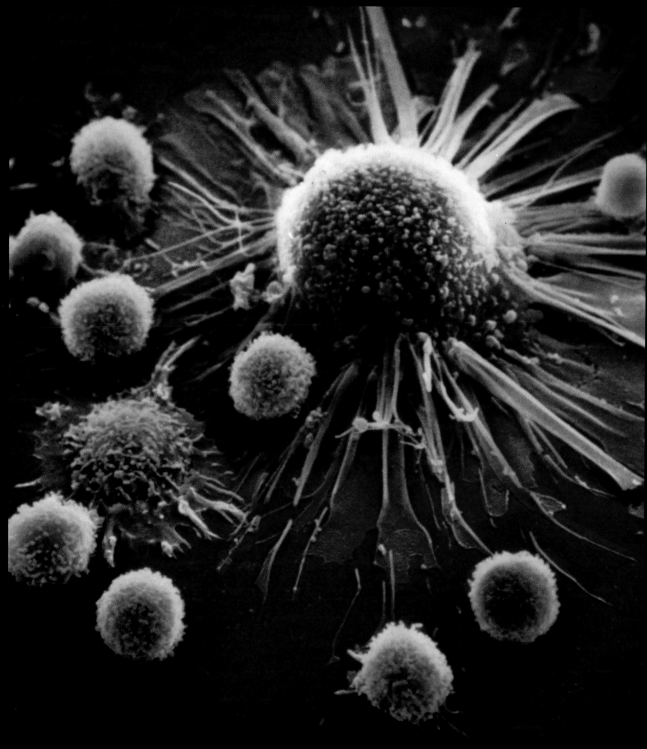

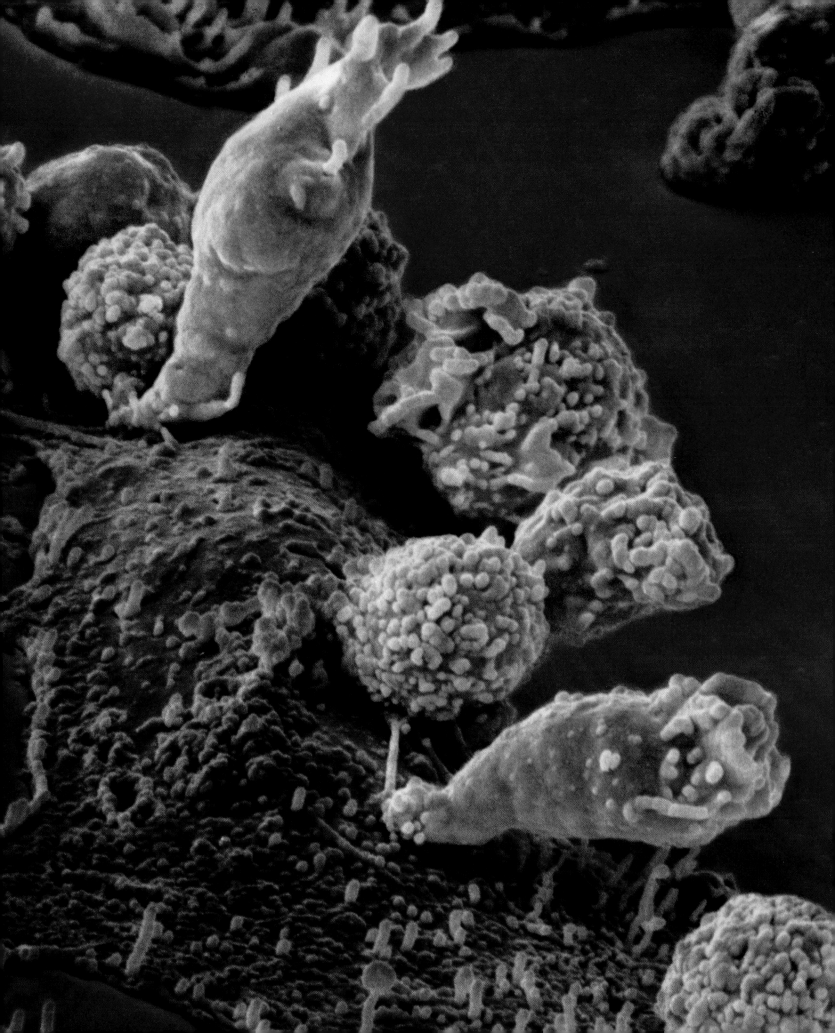

Memory responses are best at preventing diseases that are due to systemic infection—say, by viruses such as measles and polio—because in these diseases, the immune system has a bit of a head start; illness doesn't begin where the virus enters, but only after it starts spreading to other tissues. This allows time for the memory T and B cells to gear up, proliferate, and control the illness.

But there is much less opportunity to intervene against microbes that do their damage right at their point of entry in mucous membranes, such as rotavirus, respiratory syncytial virus, and the rhinoviruses that cause the common cold. In these mucosal infections, memory cells offer too little, too late. The other bad news is that the antibody response at the site of these infections tends to be short-lived, usually a few months or a year. Thus protective immunity to these diseases wanes over time—and we're vulnerable to them throughout our lives.

INFECTIOUS DISEASE AND INDUCED IMMUNITY

With all our defenses, the odds are stacked very much against microbes ever gaining much of a foothold in our bodies. But they don't give up easily—and our changing place on Earth has put us at growing risk of infection. For much of human evolution, the main cause of death was not infectious disease but wounds due to accidents. Primitive hunter-gatherer societies unknowingly limited their exposure to germs, by not having domesticated animals or cities. Microbial transmission was also restricted by geography: Early settlements were so far apart that distance short-circuited disease outbreak. Everyone either died or developed immunity.

Plague expert Charles Gregg suggested how *Homo erectus* first fell victim to bubonic plague: "The first human case may have occurred when the earliest hominids began to vary their diet by running down small game. A sick animal is more easily captured than a healthy one, but the captor risks taking diseases, as well as sustenance, from his prey."

It was during the era of agriculture, with the domestication of animals, that infectious disease became more entrenched. Scientists speculate that wild animals may have harbored viruses for millions of years. When humans began to capture and domesticate animals, the viruses leaped to people. In some cases, viruses jumped back and forth between tamed animal species. Domesticated birds and pigs, for example, are known to exchange mutating and recombining influenza viruses with humans, in never ending variants for which no immunity exists.

As humans evolved and expanded into far corners of the globe, they discovered new animal prey, new microbe-filled environments—and new infectious diseases. Civilization brought benefits, to be sure, such as more efficient farming, better nutrition, and improved hygiene. But as humans moved around, formed family groups, and created tightly clustered villages, the transmission of disease accelerated.

During the population explosion of the Bronze Age 6,000 years ago, farmers and villagers crowded into cities, offering a veritable feast to germs lurking in domesticated animals, wastes, and filth. Plagues and cities developed together, historians say. Evidence from the mummy of the Egyptian Pharaoh Ramses V suggests that he may have died from smallpox more than 3,000 years ago. From the germ's perspective, here were novel environments, rich with opportunity. A germ entering a virgin population—a community with no history of prior exposure, and hence no immunologic memory—caused acute disease in people of all ages, with tragic consequences.

For many populations, centuries of geographic isolation acted as a germ filter. But these populations are isolated no longer. One example concerns an epidemic of measles, spread through the air when its victims cough or sneeze. In the island nation of Fiji in 1875, the son of a chief came down with the disease after a ceremonial trip to Australia. Within four months, more than 20,000 Fijians were dead from the accidentally imported germ.

Like Fijians, Native Americans also led a remarkably disease-free existence, suffering few of the ailments that plagued Europe. New England colonist William Wood described their "lusty and healthful bodies, not knowing the catalogue of those health-wasting diseases which are incident to other countries." The New World had no smallpox, no measles, no plague, no leprosy, no influenza, no malaria, and no yellow fever. Without germs, the immune systems of its people were as vulnerable to invasion as a peacetime army; once germs arrived, those people perished. Disease helped make it possible for mere handfuls of conquistadores to topple the Aztec and Inca empires. After contact with Europeans, the population of North American tribes fell 50 to 90 percent.

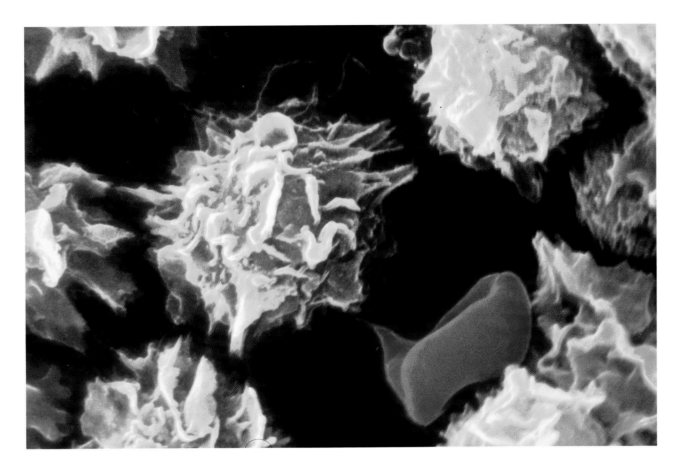

Air travel, population growth, global commerce, the crowding of people into megacities, and the accelerating destruction of ecosystems worldwide have inadvertently opened a Pandora's box of emerging microbial threats.

The human immunodeficiency virus (HIV), the source of AIDS, is a vivid example. Most investigators think the virus originated in chimpanzees or monkeys in Africa; somehow, HIV managed to cross the species barrier from primates to man. Spreading slowly at first, it picked up speed as it reached urban areas. Transmitted through sexual intercourse, the killer moved deep into cities, was transported by rapid travel across nations—and was exported around the globe.

"It was a natural agent in Africa for centuries, probably in chimps. The source of the virus goes way, way back," said Donald Francis, one of the first epidemiologists with the U.S. Centers for Disease Control and Prevention to study HIV. "It was very stable in rural African populations, but got into more urban populations, where there was more sexual activity, and spread, becoming quite widely distributed. It exploded because of an ecological change in humans, not a change in the virus."

Other new threats have emerged: The *Legionella* bacterium, the Lyme disease spirochete, hepatitis C, the Rift Valley fever virus, and the newly recognized, rare strain of Creutzfeldt-Jakob disease that has been linked to bovine spongiform encephalopathy—more familiarly known in England as "mad cow disease."

Epidemiologists have counted more than 28 new or emerging microbes in just the past quarter century. The risks of infection have increased—but fortunately, so have our abilities to combat them. Neither side has rested in the war between microbe and man. Microbes and our ability to combat them have evolved, resulting in, as medical historian William McNeill noted in his book *Plagues and People*, an escalating arms race between "ills and skills."

Spiny projections on these white blood cells characterize hairy-cell leukemia, a cancer affecting immune cells. Leukemia results when cells grow out of control, suppressing normal blood cell formation and infiltrating organs and bone marrow.

VACCINES

Long before microscopes and blood tests, people have been putting the immune system to work in an effort to protect themselves. As early as A.D. 1000 people in parts of Asia, India, and the Arab world were practicing inoculation, scratching matter from smallpox sores into the skin of uninfected people to induce a mild and survivable illness that protected them from a more devastating future infection. In 1700 an English trader described the practice of "opening up the pustules of one who has the Small Pox ripe upon them...and afterwards

mild dose of inoculum could be much stronger than suspected, causing severe illness or death. And people infected with even a small dose of live virus could transmit it to others, causing major epidemics.

English physician Edward Jenner introduced a safer method of vaccination in 1798. He inoculated a farm boy with serum from a cowpox sore on a dairymaid's hand—and proved that he was protected from the related, but far more dangerous, smallpox. No one had yet heard of the germ we now call a virus. And no knew why the procedure protected against smallpox. But Jenner's experiment protected people even when those all around were dying. Within a decade, the practice had spread to much of the world.

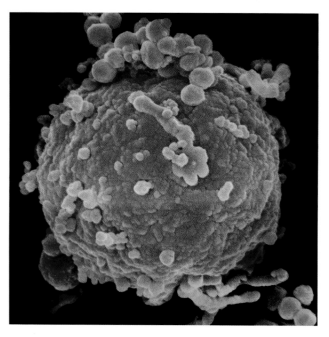

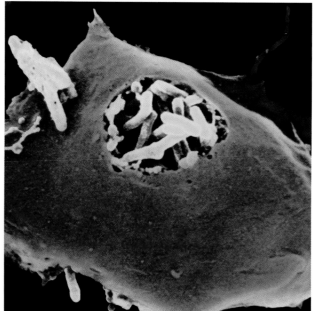

put it up the nostrils of those they would infect." In Turkey the technique gained popularity as a way to ensure unblemished complexions in girls bound for harems.

The practice of inoculation is thought to have reached England with Lady Mary Montagu, wife of the British ambassador to Constantinople. To much consternation, she insisted upon the inoculation of her six-year-old son, and later, during the smallpox epidemic of 1721, her three-year-old daughter. Both sickened slightly, recovered, then were protected from future infections. The practice spread through the Western world, largely via the military. Both Washington and Napoleon had their troops inoculated against smallpox. But the practice was far from perfect. A supposedly

Eighty years later, French chemist Louis Pasteur made a chance discovery that launched the field of modern immunology. In the spring of 1879 he began experiments on chicken cholera; in the fall he found that a culture of the cholera bacillus kept over the summer months did not cause infection when injected into healthy chickens. Clearly, something in the bacteria had altered over time, yet it still offered protection. Pasteur immediately realized the connection between his findings and those of Jenner.

Based on this observation, Pasteur concluded that a live organism can be changed—scientists now call it attenuated—so that it can provide immunity without causing disease. Honoring Jenner and his cowpox experiment,

Pasteur named his discovery "vaccination," after the Latin word for cow—*vacca*.

Pasteur also took a historic gamble—and won. He injected inactivated infectious material from a rabbit that had died of rabies into a boy who had been bitten by a rabid dog. Although Pasteur didn't completely understand the mechanism, he made another discovery—that exposure to a microbe could protect against a full-blown fatal case, even after known contact with the disease had occurred.

Many years would pass before scientists understood why these ounces of prevention were worth pounds of cure. But various experiments helped open the door to the eventual banishment of many of humankind's most serious infectious diseases. During the 20th century, vaccination has rid the world of smallpox and has tamed diseases as virulent as rabies and as widespread as diphtheria, tetanus, whooping cough, polio, and measles.

Using the building blocks devised by Jenner, Pasteur, and others, physicians now rely on two main avenues of defense: passive and active. In what is called passive immunotherapy, antibodies are injected into patients who do not produce them on their own. These antibodies are pulled from the blood of people who have previously been exposed to a disease—and thus have antibodies to spare. This approach offers immediate, but short-lived, protection against specific germs. The antibodies help destroy the microorganism if it is present in the blood or enters it in the next several days. One example is immune globulin, or gamma globulin, antibody-rich proteins that can prevent or treat infections such as

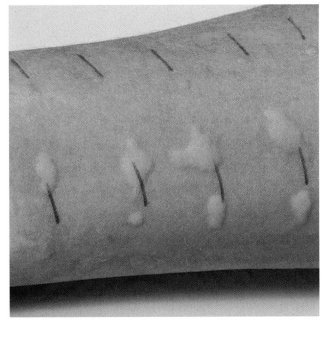

hepatitis. This method, passive immunotherapy, lasts just weeks or months.

The other defensive strategy, active immunotherapy, uses vaccines rather than antibodies, in order to prime the body to make its own antibodies against specific organisms. It confers longer-lasting immunity. It is called "active" because it doesn't merely hand over protection, but arouses a defensive immune response in the body.

Modern vaccines contain microorganisms, or parts of microorganisms, that have been treated so they will provoke an immune response—a mock infection—but not the full-blown disease. The injection of the harmless antigen isn't a real threat and doesn't make you sick. But the immune system doesn't know that and reacts as if the infection were a real one, cranking out immune cells against the modified microorganisms. Years later, even when the antigen from the vaccine has been neutralized, the immune system remains on the alert. It "remembers" the organism, so that if the real invader pays a second visit, it is destroyed before it has a chance to infect.

To cause disease, a germ must beat pretty steep odds. Despite the hundreds of thousands of species of microbes in the world, very few enter humans. Fewer still can be transmitted from one human to others. And modern vaccines, by interrupting transmission, reduce the threat of a successful invasion even further.

To be sure, the battle is not over; effective vaccines are available for only a relatively small proportion of the infectious diseases that afflict humans. But so far, in the evolutionary relationship between man and microbe, man still retains the upper hand.

Fringed with bacteria, a B cell (opposite, left) manufactures specific antibodies designed to destroy the invaders.
A macrophage (opposite, right) engulfs and digests microbes such as these tuberculosis bacteria. Allergic patients develop
red swellings (above) when pollen and other allergens are injected under their skin to test their susceptibility.

ALLERGIES AND ASTHMA

Our arsenals for fighting microbes are so powerful and involve so many different defense mechanisms that we can be in more danger from them than from invaders. "We live in the midst of explosive devices; we are mined," wrote Lewis Thomas. Because of the reflexive way the body reacts to trouble, there are times when the immune system harms more than it helps—sometimes, seriously. In such situations the immune system works so hard to protect the body from a foreign substance that it actually makes the body far sicker than the invader would have; it is our response to the presence of invaders, or perceived invaders, that makes a disease.

Genetic influences and environmental triggers can cause the immune system to overreact, overdoing its job of protecting us against invasions of relatively minor foreign substances. Sometimes that system gets confused, attacking not just foreign invaders, but the body's very own tissues.

An allergy attack is a sign that the immune system is fiercely overreacting to contact with an alien invader. Allergies and asthma affect between 10 and 20 percent of human populations, most of them in North America and Eurasia.

Worldwide, surveys suggest that allergies are becoming more common, with industrial nations showing the steepest rise. They are the most common cause of lost work and school time; treatments for these diseases require billions of dollars each year in the U.S.

Some people, it appears, have immune systems that are abnormally sensitive to otherwise harmless substances that are inhaled, ingested, or even touched. Allergies are simply responses to faulty information. Ordinary, non-harmful particles—pollen or dust, for instance—are misidentified as threats and attacked with antibodies. A similar sort of false alarm can also be triggered by environmental conditions such as cold temperatures or bright light. Each antibody attack is specific; cat hair might provoke one, for instance, and mold another.

Imagine a field of ragweed that sends an invisible cloud of pollen grains wafting into a nearby town. The pollen is inhaled by a child whose body has never been exposed to the substance before. The child's immune system kicks into gear. It cranks out vast numbers of immunoglobulin E (IgE) antibodies, all designed to target ragweed pollen. Yet some inexplicably attach themselves to cells in the nose and throat called mast cells. There they sit until the next pollen bits float by. At that point they release powerful chemicals called histamines, which trigger the familiar symptoms of allergy. The little pollen particles are destroyed—but at the expense of many tissues and a large amount of wheezing, sneezing, sniffling, and red, itchy eyes.

Sophisticated as they are, immune systems can make errors. Those of people with allergies somehow learned something that isn't true—that an innocuous foreign substance is dangerous. Normally, the body learns to defend itself through experience—by encountering, battling, and remembering one enemy after another. Allergic reactions occur when this memory fouls up, and an innocent foreigner is misidentified as a threat.

Different allergens can cause the body to react in different ways. Sneezing, wheezing, nasal congestion, or other symptoms may surface. Allergic reactions to foods, insect bites, or medicines may produce a variety of problems. One is a skin reaction called urticaria, better known as hives. The skin suddenly turns red, swells, and itches—then, just as abruptly, the symptoms disappear. Another skin allergy is eczema, in which patches of skin become thickened, leathery, reddened, and dry.

For most people, allergies are just a nuisance. For others, however, they can become serious health problems; on occasion they may even threaten life.

Exposure to an antigen that previously caused sensitization, such as the venom of a bee sting, may lead to anaphylaxis, or even death. Anaphylaxis is a sudden, massive over-reaction of the body's immune system to an allergen that makes its way into the blood-stream. Within several minutes of a bee sting, the lips and tongue of a sensitized individual grow puffy; itchy bumps appear over the body, blood pressure drops, and breathing becomes difficult. Without immediate treatment to quiet the attack, a simple bee sting can be deadly.

In one of the best known serious allergic reactions, asthma, the body's airways narrow in response to stimuli that don't affect the airways in normal lungs. The tiny air tubes in the lungs squeeze shut, trapping air inside and causing the person to wheeze as he tries to push the air out again. First, the chest feels tight. A few moments later, he is fighting for breath, coughing, and wheezing. Characteristic of the disease is a "twitchy" airway: In asthmatics, even low levels of histamines and other

Invisible to the unaided eye but offensive to the immune system, dust mites (opposite) are a leading cause of chronic allergy. In their droppings, cockroaches (above), give off proteins that can trigger the coughs and wheezes of asthma. FOLLOWING PAGES: Diversely shaped pollen grains, magnified 3,000 times, spark varied reactions in different people.

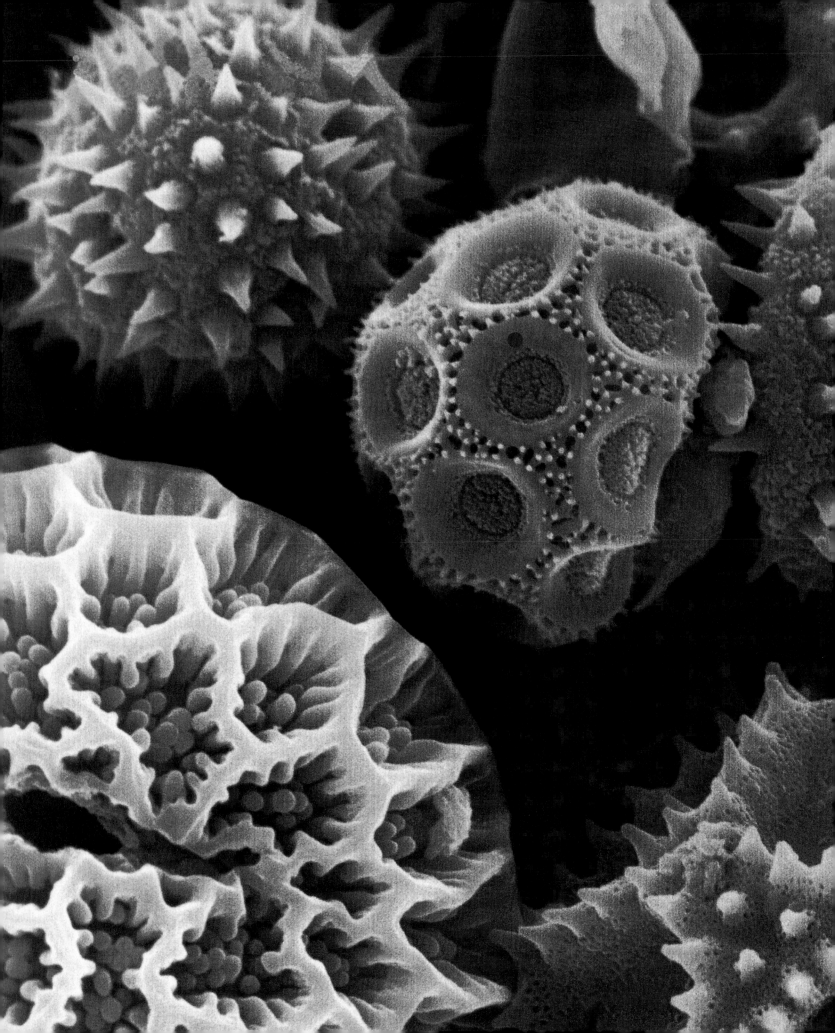

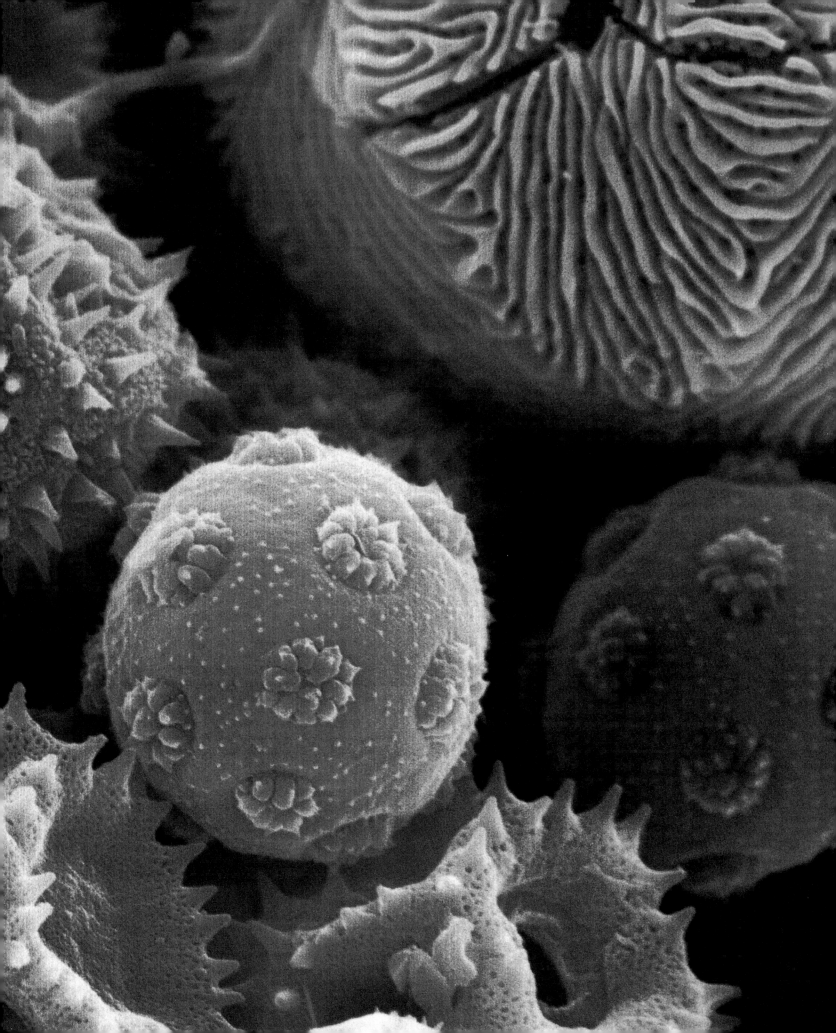

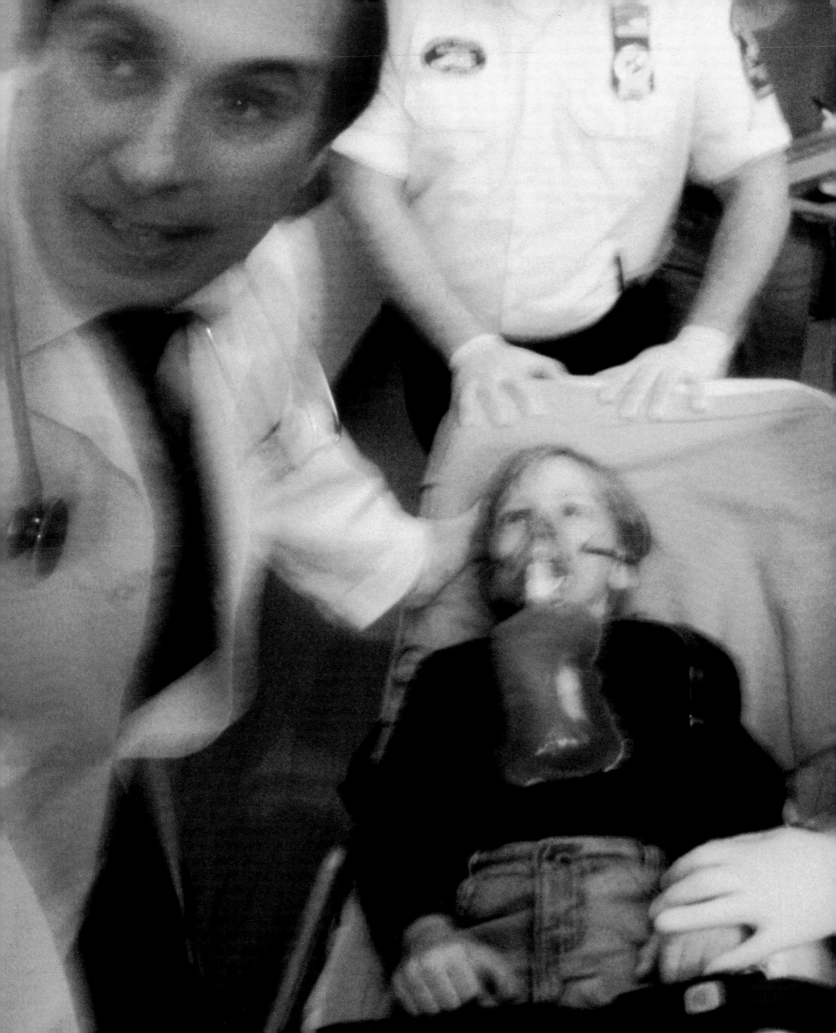

substances constrict the bronchial tubes much more easily than they do in nonasthmatics. When exposed to an allergen, the asthmatic's immune system manufactures histamines and other chemicals that irritate the twitchy airways, causing an attack.

The diversity of asthma's causes and effects—there is huge variability in the patterns and severity of the disease, the multitude of contributing allergens, and the way symptoms change over time—is a great mystery. Asthma can be the result of an allergic reaction to pollen, smoke, animal dander, mold spores, or dust. But asthma is not always caused by allergies; it also can arise from cold air, exercise, strong emotions, and other sources.

Asthma affects 4 to 5 percent of American adults and 7 percent of children—and it is on the rise in cities and suburbs alike. In the inner cities asthma is particularly prevalent, with rates often double those found elsewhere, according to Peyton Eggleston of Johns Hopkins University. Research has shown that the ubiquitous cockroach is the leading cause of severe childhood asthma in the country's poorest city neighborhoods. Cockroaches give off proteins, mainly in their saliva and droppings, that can trigger strong allergic reactions. From 1980 to 1990 the prevalence of asthma increased a whopping 40 percent among children under 18 years of age; no one knows why. Some scientists believe it is partly due to the fact that energy-efficient homes and offices are more airtight than ever, reducing fresh ventilation and creating more concentrated exposure to allergens.

Ironically, allergies actually may have helped us survive during our evolutionary history. This theory is ventured by scientists who have noted that the production of large amounts of IgE antibodies—the defining feature of allergy—occurs in only one other case: when the immune system tries to eliminate parasites. Perhaps the allergic response initially evolved to help the body cope with parasites. If so, people with allergies enjoyed a survival advantage, which they passed on to their offspring. But now, perhaps because we rarely encounter those early parasites, our immune system can focus—inappropriately, at times—on other things. At any rate, allergies have gained firm foothold in our gene pool.

The fundamental cause of allergies is still not known. Why do allergic people generate abnormally high levels of IgE antibodies? Why does someone become allergic to some substances and not to others? Also unknown is why some allergy attacks occur immediately after exposure to the offending substance, while others take minutes, hours, or even weeks.

It is clear that allergies run in families: If one parent has allergies and the other does not, chances are that one in three of their children will have them. If both parents suffer from allergies, all their offspring will inherit the condition. However, the inheritance pattern of allergies demonstrates that they are complex genetic disorders; they do not follow simple dominant, recessive, or sex-linked patterns of transmission from generation to generation.

Moreover, nonhereditary factors also can play a part. According to doctors at Johns Hopkins University, infections of certain common viruses can trigger the development of allergies and asthma. Test tube studies on human cells shows that weak viral infections can cause immune system cells to produce immunoglobulin E. Researchers have found that people who had more viral infections as youngsters were more likely to have allergies and asthma later in life.

Scientists have classified allergic reactions into four categories, based on the type of tissue damage that occurs. Type I, or anaphylactic, reactions harm because the antigen-antibody combination triggers the release of substances called histamines, which cause the blood vessels to dilate and the airways to narrow. Examples are hay fever or an asthma attack. Type II, cytotoxic, reactions are created when antibodies turn deadly, killing healthy cells. In Goodpasture's syndrome, for instance, antibodies are formed that attack capillaries in the lungs and kidneys, eventually causing breathing difficulties and kidney failure. When groups of interlocking antigens and antibodies build up, they cause Type III, or immune complex, reactions. Immune complexes are quickly eliminated from the bloodstream normally, but sometimes they continue to circulate. Eventually the groups get stuck in blood vessels or tissues of the kidneys, lungs, skin, and joints, where they initiate reactions with complement that cause inflammation and tissue damage, particularly in blood vessel walls. Immune complexes are at work in many

----

Struggling for air, a young asthmatic requires emergency room care to keep his airways open and to ensure that enough oxygen gets to his lungs. Asthma varies greatly in severity, affecting some only occasionally but others on a daily basis.

diseases, among them viral hepatitis, malaria, and many auto-immune diseases such as lupus. Type IV, delayed or cell-mediated, reactions occur when an antigen interacts with lymphocytes that release inflammatory and toxic substances, then attract other white blood cells and injure normal tissues. This type of reaction takes more than 12 hours to develop. The skin test for tuberculosis is an example of this type of reaction.

....

AUTO-IMMUNE DISEASES: CIVIL WAR BETWEEN CELLS

....

As we've seen, the immune system normally can distinguish self from nonself, friend from foe. And when it finishes its attack on an invader, it simply shuts off. Peace reigns. However, in auto-immune diseases, the attack is chronic. This is because the body fights not invaders but itself—damaging normal tissues, causing illness and even death. An ugly and endless war within the body, auto-immune disease strikes 5 percent of the adult populations of North America and Europe; many of those people have more than one illness at a time, and two thirds of those stricken are women. There are more than 80 known auto-immune diseases; the three most prevalent are Graves' disease, rheumatoid arthritis, and Hashimoto's thyroiditis.

Nobody yet knows for sure why the immune system turns against the body. Our immune T and B cells are propagated by the bucketful—but the vast majority, perhaps 95 to 98 percent, are juvenile delinquents configured in shapes that would cause damage at home if allowed to survive. Normally, the body gets rid of these self-reactive cells as quickly as it makes them.

What happens in auto-immune disease, however, is that the immune system's delicate surveillance system breaks down, and the body begins to overlook these self-reactive T cells and antibodies (in this scenario, called autoantibodies) that are directed against the body's own cells and organs. Normally such cells are destroyed or quieted in the thymus during development. In this instance, however, they made their way to the rest of the body, where they release chemicals that damage tissues.

There is emerging evidence that auto-immune disease cannot be blamed solely on the ability of these self-reactive T cells to slip past the body's careless screening mechanism. Provocative new research suggests that we all seem to have some self-reactive T cells circulating in our bodies—but that they are generally dormant. The solution to the riddle of auto-immunity hangs on what activates those cells.

To complicate things still further, researchers are beginning to suspect that not all auto-immune diseases have the same mechanisms. They suspect that the immune system breaks down in various ways, or that different conditions trigger self-reactive cell activation. Numerous factors are likely to be involved. For example, something in the environment, perhaps a virus or a drug, may change the surface proteins of human cells so that the immune system interprets them as nonself. The immune system may respond to a foreign substance that is similar in appearance to a natural body substance—and target them both. Perhaps these diseases are the price we pay for having such a complex and nuanced system—a system that protects most of the time, but because it is so complicated, occasionally breaks down.

Virtually any organ can be attacked by misguided immune cells, resulting in many diseases. Different organs and tissues become inflamed in different people, and the severity of the disease ranges from mild to debilitating, depending on the number and variety of antibodies that appear. Each auto-immune disease is diagnosed by its particular symptom pattern.

For instance, in hemolytic anemia, autoantibodies attack red blood cells. In myasthenia gravis, a neurologic disorder characterized by muscle weakness, they attack a vital protein on muscle cells that receives signals from nerves; once these proteins are missing, the muscle can't respond to nerve impulses. Another autoantibody, known as rheumatoid factor, commonly occurs in persons with rheumatoid arthritis. It attacks the tissue that lines and cushions joints, causing huge numbers of lymphocytes to rush to the joint tissues, resulting in pain, swelling, and redness. Eventually, the cartilage, bone, and ligaments of the joint erode. In multiple sclerosis, T cells target the protein covering, or myelin sheath, that coats nerve cells. Like wires stripped of insulation, demyelinated nerves fail to properly conduct impulses, causing a bizarre and debilitating array of complaints. Other auto-immune diseases include pemphigus, scleroderma, Sjögren's syndrome, and pernicious anemia.

One of the most devastating auto-immune diseases is insulin-dependent diabetes mellitus. This disease stems from a self-directed attack on the pancreas, causing that organ to no longer produce the hormone that is critical to maintaining blood sugar levels. The resulting insulin deficiency is so severe that a person with this type of diabetes, called Type 1, must regularly inject insulin to survive. Diabetes is a highly selective disease. Rather than affecting the majority of pancreatic cells, which secrete digestive enzymes, it restricts itself to the hormone-producing cells, which are clustered in spherical groups called the islets of Langerhans, scattered throughout the pancreas.

Researchers Mark A. Atkinson and Noel K. Maclaren of the University of Florida College of Medicine in Gainesville, Florida, discovered that three of the four cell types in the islets of Langerhans are spared; only the insulin-producing beta cells, which make up the core of an islet, are targeted. As more and more of these insulin producers succumb, the remaining healthy cells work overtime to supply the needed hormone. This stresses the cells, perhaps leading to an intensified self attack. The surviving beta cells are killed with increasing speed.

Eventually, too few cells remain to supply insulin to the body, and the symptoms of diabetes suddenly appear. But there is nothing sudden about the disease, which typically has brewed silently for several years, as the immune system slowly destroyed beta cells. The classic symptoms appear only when at least 80 percent of those cells are gone; the remainder are eliminated over the next two to three years. Diabetes can lead to complications such as blindness, kidney failure, and infection.

Because researchers have found clusters of auto-immune diseases in families, heredity is suspected to play a role in these disorders. Genetics may explain, for instance, why some people who have been infected by an adenovirus—a type of virus that causes respiratory infections such as the common cold—develop debilitating multiple sclerosis while the vast number of us do not. There are genetic differences in individuals' HLA (Human Leukocyte Antigen) types, a group of molecules crucial to the regulation of white blood cells in immune functions. It is their job to determine exactly which fragments of a pathogen are displayed on the cell surface for presentation to T cells. While most individuals' HLA structure may bind a fragment unique to the pathogen, not mimicking self, some people may have an HLA structure that binds a self-mimicking pathogen fragment and presents it to the immune system, triggering it to inappropriately attack the body's own tissues. This

theory has been preliminarily confirmed by recent research linking certain auto-immune diseases with specific HLA types. For instance, people who carry the gene for DR4, a particular kind of HLA molecule, are six times more likely than others to acquire rheumatoid arthritis. Those with HLA-DR2 are four times more likely to inherit multiple sclerosis. Juvenile diabetes is 20 times more common in people who have HLA-DR3 and HLA-DR4 genes.

Some HLA types seem predisposed to more than one auto-immune disease, explaining why people with myasthenia gravis, for instance, have a 30 percent chance of acquiring Graves' disease as well. Frank Arnett of the University of Texas Health Science Center at Houston studied 32 patients with auto-immune disease—and found that more than half of them had family members who also had an auto-immune disease. Totally healthy individuals can also have these HLA types, too, indicating that other factors must be involved.

The genetic link is not perfect. Auto-immune disease can be selective even among identical twins. Although both individuals share identical genetic codes, one twin may have rheumatoid arthritis while the other is affected less than half the time.

Injury, infection, or some other environmental factor may trigger some forms of auto-immune disease in people with the "correct" susceptibility genes. Perhaps such stimuli send false signals, prompting the immune system to activate cells in a tissue where it should not. New research suggests that a strain of reactivated herpes virus (HHV-6) may be linked to multiple sclerosis in genetically susceptible individuals.

One might expect that auto-immune disease, once started, would proceed without interruption. Yet in many cases—rheumatoid arthritis, multiple sclerosis, lupus, and myasthenia gravis—periods of deterioration are interspersed with remission. Why? Scientists have found that periods of decline correlate with three factors: stress, infection, and female hormones. Stress is thought to worsen auto-immune disease by affecting two glands in the brain—the hypothalamus and the pituitary—which secrete hormones that promote inflammation. The female hormone estrogen stimulates a DNA sequence which promotes the production of the cytokine called interferon, which triggers the process.

Infectious disease has been variously described as an inconclusive negotiation between microbe and human, as an overstepping of the line by one side or the other,

and as a biologic misinterpretation of borders. When, despite great odds, microbes do succeed in entering and causing massive illness, it is due to either a traumatic injury, the introduction of a particularly aggressive germ, or a defect in the immune system. This defect—when the immune system lacks one or more of its parts or the parts malfunction—results in an immunodeficiency disease. More than 70 disorders are included in this broad group. Immunodeficiency diseases are present from birth or are developed later in life—involving many different cells, tissues, and products of the immune system. Without a properly functioning immune system, the gates to the body are flung wide; invaders storm the sanctum.

Disorders of the immune system caused by defects in specific genes are heartbreaking but, fortunately, rare. Much more common, worldwide, are immunodeficiencies acquired due to malnutrition, stress, age, cancer, or immunosuppression due to infection with the AIDS virus. When a person's body weight drops to less than 80 percent of ideal, the immune system becomes slightly impaired; when it falls below 70 percent, impairment is severe. Unfortunately, infections due to immune problems both depress the appetite and increase metabolism—resulting in a vicious cycle of worsening malnutrition.

The spleen helps trap and destroy bacteria in the bloodstream and is one of the places in the body where infection-fighting antibodies are produced. People with certain spleen problems, or who lose that organ due to trauma, are more susceptible to infection and must receive antibiotics and vaccines.

Some immunodeficiencies arise simply as a consequence of age. The thymus, vital to the production of T cells, reaches its peak size at puberty and steadily shrinks thereafter. This results in a decline in the number and activity of infection-fighting cells. Age-related immune decline is also blamed on changes in the types of T cells in the body. As we age, new T cells, freshly emigrated from the thymus, are replaced by memory T cells with experience—but also with many defects. There is a third reason why our immunity lessens as we grow older: Our B cells lose the ability to produce antibodies in response to both new and previously encountered antigens.

Scourge of AIDS claims yet another victim in Uganda: Eighty-year-old Christina Mukakibibi mourns the death of a daughter, her eighth child to succumb to the dread disease. One in five Ugandans carries the AIDS virus.

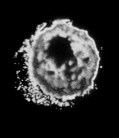

HUMAN IMMUNODEFICIENCY VIRUS

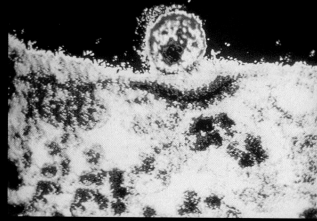

HIV ADHERES TO THE CELL MEMBRANE OF A LYMPHOCYTE

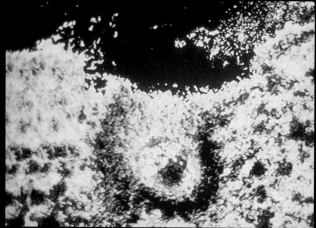

HIV ENTERS THE CELL

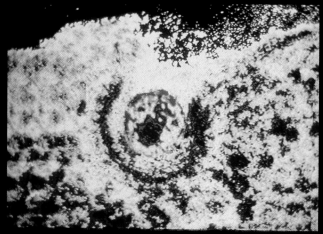

HIV BURROWS FARTHER IN, SEEKING THE NUCLEUS

Commandeering the body's own defenses to further its existence, a single human immunodeficiency virus (HIV) invades a type of white blood cell known as a lymphocyte. Drawn within the cell's nucleus, the virus integrates its genetic material into the DNA of its host. It then uses the cell's machinery and nutrients to replicate itself over and over. In time, the new virus particles push through the cell's membrane, the cell dies, and they go off in search of other lymphocytes to infect. The strategy is a cunning one: Not only does the virus disable the very cells that should be fighting it, but it also evades the rest of the immune system by hiding within "friendly" cells. Like a successful spy, it can't be caught if it can't be seen.

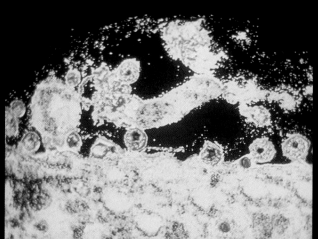

HIV DIRECTS THE CELL TO MAKE COPIES OF ITSELF

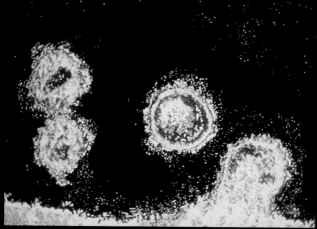

NEWLY FORMED HIV PARTICLES BREAK OUT OF THE CELL

Temporary and minor immune deficiencies can develop due to common viral infections, such as influenza, infectious mononucleosis, and measles. But some viruses cause permanent and devastating damage to the immune system, destroying the body's ability to fight off infection.

AIDS

The most notorious example of permanent and devastating damage is the human immunodeficiency virus (HIV), which causes AIDS (acquired immunodeficiency syndrome), one of the principal threats to human life and health around the globe. Every day in 1997, HIV infected 8,500 adults and 1,000 children throughout the world. About 25 percent of adults in Botswana are infected, 22 percent in Zimbabwe, and 19 percent in Namibia, according to Elhadjas Sy, head of the Joint U.N. Program on HIV/AIDS. A highly lethal viral infection, HIV not only destroys T cells but also colonizes and infects macrophages, the scavenger white cells that are another one of the indispensable components of the immune system. The damage caused by HIV accumulates slowly and is not immediately apparent, so people may live for many years in good health. But without treatment, it is almost inevitably fatal.

"To be certain, other diseases…kill more people than AIDS does," said Warner C. Greene, director of the Gladstone Institute of Virology and Immunology in San Francisco. HIV's attack strategy is far more sophisticated than that of any other human virus: It debilitates the immune system, not by overwhelming it from the outside but by infiltrating from within. Once it has infected a cell, its set of viral genes integrate themselves into the human DNA. HIV has mastered what no other human microbe has ever learned: to systematically target a person's defense system against itself.

Think of the movie *Alien*. When the virus enters a T cell, it takes over the genetic machinery of that cell, directing it to crank out new HIV genes and proteins that become the next generation of virus. The infected person's immune system becomes the means for the virus to reproduce itself. And, ironically, the very thing the human depends on for defense has turned against him. Once infiltrated, the immune system still has resources to fight the virus—if it could only find it. But HIV doesn't just disable the immune system, it hides from it, invisible to surveillance. The idea is as old and as effective as espionage: If you can't find me, you can't stop me. Finally, after turning the T cell into its personal factory, the virus kills it. HIV progeny have been spawned; the invader's mission is complete.

It was long thought that the virus lies dormant after infection. It is now known that HIV reproduces itself from the moment of infection. Even when the virus does not evoke symptoms, it is quietly disassembling the immune system. Of course, the immune system tries to fight the virus. But the virus replicates too quickly and destroys too many T cells, leading to the entire system's eventual collapse. In destroying the immune system, the virus destroys the scaffolding of the self.

TRANSPLANTATION

Every year, transplanted organs—hearts, lungs, kidneys, livers, and pancreases—extend the lives of thousands of Americans. But because the immune system "sees" an implanted organ as foreign, it attacks it. Despite major improvements in surgical techniques, tissue preservation methods, and immunosuppressive drugs, 10 to 50 percent of transplanted organs and tissues undergo a rejection episode in the first year following transplantation. For a transplant to take, doctors must successfully overcome the body's natural tendency to attack and eliminate any foreign tissue. Immunosuppressant drugs work by suppressing the production and activity of lymphocytes, enabling the nonself to grow and function in peace.

Transplanting an organ from one identical twin to the other will often preclude rejection because identical twins

FOLLOWING PAGES: United in anger, AIDS activists protest what they see as government inaction in the face of a modern plague. Although controversial, their tactics have helped accelerate research, awareness, and the drug-approval process.

have the same basic genetic material, and the immune system identifies protein from the twin as self. Often a brother, sister, mother, or father may have enough proteins so similar that the patient's immune system becomes confused about whether the transplanted organ is self or not. Rejection can be minimized by ensuring that the donor's tissue matches, as closely as possible, that of the recipient.

Another approach is to quiet the recipient's immune response by administering cyclosporines—laboratory-grown antibodies designed to attack mature T cells—or other powerful immunosuppressive drugs.

There is a major downside to this path, however: these drugs must be taken for a lifetime. As soon as treatment is stopped, the immune system swings back into action against the tissue. And while the drugs keep the immune system from rejecting the transplant, they also impair the ability to fight off dangerous infections, boosting the risk of infection and of the development of certain cancers.

Organ transplantation has long been stymied by antirejection medicines so broad in their approach that they suppress not just the immune cells that go after the new organ, but also everything else. The patient is left defenseless. A new generation of medicines is kinder, gentler—and much more specific. The magic of these therapies is that they suppress only the immune cells that attack new organs, without shutting down the entire immune system.

SCID AND OTHER IMMUNE DISEASES

Of all the babies born each year, a few lack all of the body's major immune defenses. Missing even one of these links can have widespread repercussions; for instance, when B lymphocytes aren't present, neither are its protective antibodies. This causes a disease called agammaglobulinemia, a rare but grave condition that leads to recurrent bacterial infections of the lungs, sinuses, or bone. When the thymus or its T lymphocytes are missing, the body can't fight off viruses or fungi. One result might be a persistent yeast infection called candidiasis, affecting the skin, mouth, or throat.

The deficiency of both prongs of the immune system—the B cell and T cell systems—is called severe combined immunodeficiency (SCID). People with SCID are born with the immunological equivalent of an AIDS patient in the final stage of illness, and have a similar life expectancy. Exposure to any bacterium, virus, or parasite can cause fatal infection. Living in an artificial, antiseptic world has enabled some children with SCID to survive for a number of years. Nothing from the world outside is allowed to come into contact with them without first being carefully treated to make it germfree, and they are not allowed to touch other human beings except through gloves. As they grow, children with SCID have larger and larger plastic "houses" in which to live. They go outside the germfree cage by dressing in a plastic space suit, like moon-walking astronauts. If they leave this germfree plastic house, they become sick with a cold, a sore throat, or an earache, going from one infection to the next in spite of treatment, until finally one infection turns deadly.

Perhaps the most well-known, but ultimately tragic, case of SCID occurred in the 1970s with a boy named David. A bright-eyed, intelligent, and mischievous boy, he lived in a plastic bubble for years, his life tracked on television. Then, in an attempt to build him an immune system and a normal life, doctors took a chance: They performed a bone-marrow transplant with tissue donated by his sister. David's physicians, however, did not know that Epstein-Barr virus lay hidden within his sister's cells. Although the virus normally causes only mononucleosis in most people, in David it ran wild, triggering widespread cancer. Within four months tumors riddled his intestine, liver, lungs, and brain, and he died.

A hundred years ago, a girl born in the United States could expect to live to be 44 years old. A girl born at the end of this century has a life expectancy of 78 years—a 77 percent gain. The increase in life expectancy for boys is almost as great. While public health measures such as improved hygiene, diet, and water have aided this historic progress, a key factor is our ability to ward off infectious disease by boosting, manipulating, and reconstructing the immune system.

From its modest beginnings just a little more than a century ago, immune science is finally revealing many of the delicate checks and balances that control human defenses, and is opening the door to new therapies. Using this new knowledge, we now can quiet a self-directed attack, as in auto-immune disease. We can forestall organ rejection. And we can strengthen the body's response to an infectious invader.

The goal of some auto-immune therapies is to suppress immunity; the immune system needs to be scaled down and redirected. Research is proceeding, albeit slowly, as scientists try to devise ways to thwart these attacks on the self. For instance, multiple sclerosis, long an incurable auto-immune disease, may soon surrender to treatment.

One promising strategy is to selectively eliminate the overzealous T cells that cause auto-immunity. Leroy Hood, of the California Institute of Technology, has used antibodies—directed not at external microbes but at the body's own destructive T cells—to prevent multiple sclerosis in mice. Another new strategy under study is to identify the antigens that trigger the aggressive T cells, then find a way to block their ugly behavior.

Rheumatoid arthritis, which results when the immune system is provoked to release inflammatory chemicals called cytokines, may someday yield to new drugs that block the release of cytokines, thereby protecting future generations of joints and bones.

Researchers suspect that lupus is caused by a genetic defect in the cellular process of apoptosis, also known as "programmed cell death." In apoptosis, the body's damaged or potentially harmful cells commit suicide. If they don't, other bodily tissues are at risk because harmful cells linger and may cause damage. Gene therapy offers potential here: Studies on mice show that when the gene that controls apoptosis is defective, it can be replaced with a normal gene—and the illness recedes.

Genetic engineering enables scientists to extract genes—segments of DNA—from one organism and insert them elsewhere. These transplanted genes then go about their jobs normally. In this manner, missing genes can be replaced; defective ones can theoretically be fixed; the scaffolding of the immune system can be constructed—or reconstructed. One example of this is a pioneering treatment for SCID.

The first attempt to treat SCID with gene therapy shows promise. A missing enzyme caused by the lack of a single gene is what causes this immune deficiency. Doctors at Children's Hospital in Los Angeles took T cells from two girls with SCIDS, successfully introduced the missing gene, then reinjected the bioengineered T cells into the patients' bloodstreams. The gene signaled the cells to produce the needed enzyme, and the immune system began to function in both girls.

CANCER

Normally, the immune system becomes alerted to trouble when healthy cells turn cancerous, because some of the antigens on their surface change. These altered antigens trigger immune attack by killer T cells, macrophages, and others. The immune cells that patrol the body for intruders are thought to ferret out and destroy most cancerous cells; those that develop into clinically detectable cancers are ones that have managed to escape the body's surveillance system.

When that system fails to detect cancer, it is because the immune defenses are too late, too weak, or unable to recognize the dangerous cell, says William Clark of the University of California at Los Angeles. Scientists now believe that it is crucial to augment this early-detection system. Tumors develop when the system has broken down or is overwhelmed—or if the tumor cell escapes recognition. The chance of eliminating cancer would be vastly improved if the immune system could recognize and attack tumor-associated antigens before a tumor reaches a critical size.

The goal of cancer immunotherapy is to use the body's own natural defenses. In the 1800s physicians noticed that the tumors of cancer patients who contracted bacterial infections sometimes regressed. This discovery led to the conclusion that boosting the overall activity of the immune system might restrain the development of cancer. The approach is nonspecific. Zap the entire immune system, as you might whack a stubborn vending machine to get it to deliver your snack, and inspire it to increase its capacity to rid the body of cancer cells.

William Coley, a surgeon at Memorial Hospital in New York City from 1892 to 1936, pioneered efforts to harness the immune system to restrain cancer. First, he deliberately attempted to infect cancer patients with bacteria. Later, he created a vaccine made of killed bacteria in an effort to promote a tumor-killing response in patients. In some individuals Coley's treatments brought about complete tumor regressions. The results were inconsistent, however, and the scientific community did not embrace his methods.

Early attempts at cancer vaccines also failed because they weren't specific enough. Doctors injected patients with a "chicken soup" of molecules, including mashed bits of the tumor and other biochemical boosters. The

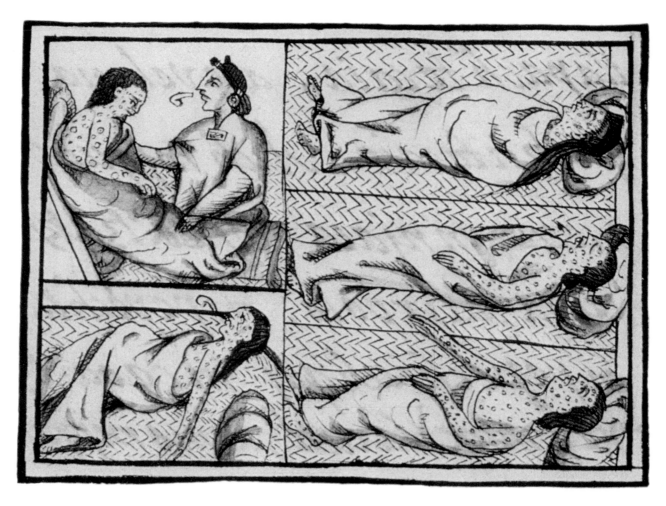

immune response, when it occurred, was often violent. But the fury was rarely unleashed directly on the tumors. Doctors also tried a different tack, "vaccinating" patients with introduced cancer cells, killed so as to be safe, in the hope of producing a more vigorous response to their own tumors. A significant shortcoming of this approach was that researchers could not track the vaccine's effect on the immune system. At that time no information existed about cancer antigens and the immune response they provoked.

A basic discovery of tumor immunology launched modern efforts for a cancer vaccine in the 1940s and 1950s: When chemicals or viruses stimulated the growth of tumors in mice, those tumors developed antigens that triggered cancer-fighting antibodies that protected other mice against transplants of the tumors. It was found that not only antibodies but also T lymphocyte immune cells taken from vaccinated animals could confer a degree of protection against cancer to healthy animals. Since then, scientists have been trying to make existing tumor cells more "visible" to the immune system, forcing them to display their antigens more clearly in the hope they will provoke a quick and angry immune response.

With each experiment, excitement has grown and died over the hope that scientists might harness the disease-fighting powers of the immune system to destroy cancers. But waves of optimism and pessimism have been part of tumor immunology throughout its history.

. . . .

"A man could not set his foot down, unless on the corpse of an Indian," wrote conquistador Hernán Cortés in 1522. This drawing of smallpox-infected Indians (above)—who lacked natural immunity to the disease—recalls the "gift" Spaniards unwittingly brought to the New World in 1518; within four years, the population of Mexico's chief city had shrunk by half. Today, other once isolated diseases can traverse the globe overnight, sparking medical concerns. Inside a containment chamber during a drill (opposite), a U.S. Army medical evacuation team member plays the role of an infected patient.

. . . .

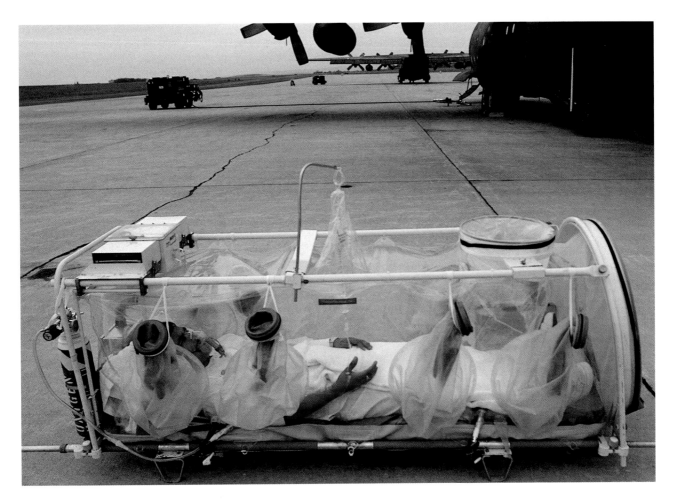

As one scientist said, "Cancer cells are masters of deceit and disguise—veritable Houdinis that can readily alter themselves to evade immunologic recognition and attack."

Today, cancer immunology is in a resurgent phase. The key has been identification of a number of natural and man-made substances that can boost, direct, or restore the body's immune system. As researchers have learned more about the very cells that guard the body against invaders, they have acquired a deeper understanding of the hows and whys that may someday help them win the war against cancer.

Scientists are also defining previously elusive tumor-associated antigens and testing a significant number of immunotherapies. Unfortunately, turning their findings into workable weapons has been frustratingly slow.

A major challenge has been for an immunotherapy to identify and distinguish cancer antigens from normal cells so that the immune system knows what to target. In 1975 researchers from the University of Cambridge made the search for cancer antigens easier, earning a Nobel Prize for their work. They created a technique that allowed scientists to manufacture unlimited numbers of identical antibodies produced by the descendants of one plasma cell. These so-called monoclonal antibodies can be directed to search for cancer antigens. Their discovery enabled the Cambridge scientists to create enough antibodies to test antibody-based therapies.

Antibodies designed to recognize cancer cells can be loaded with killer drugs and radioactive materials, then launched like smart bombs to home in on their targets. In addition to destroying cancer cells, antibodies can slow a tumor's growth by attacking connective tissue between the cells or neutralizing the chemicals they need for growth.

The goal—still years away—is to create a vaccine that offers general and permanent protection against many different cancers. The hope is that this vaccine would contain cancer antigens that would trick the immune system into triggering a potent anticancer response, protective in the event a real cancer cell comes along. But to create that, doctors first need to better identify specific cancer antigens present on tumor cells of all the victims of the disease. They have created a list of tumor antigens to be tested in a vaccine for humans, but there is still much work to be done.

Meanwhile, scientists have sought other approaches to battling cancer. One of these, called adoptive immunotherapy, aims to press into service those elements of the immune system most able to fight cancer, rather than to boost the system's overall response. In this type of immunotherapy, the patient donates his own T cells, which are then stimulated by being exposed to tumor cells in the laboratory. The treated T cells are subsequently injected back into the patient. Because the patient serves as both donor and recipient of his own T cells, there is no danger of rejection of the treated cells. Continuing efforts focus on making this therapy more effective and less expensive than others.

Adoptive immunotherapy may also be helpful in people with immune systems already weakened by disease. Vulnerable to infection, these patients can be injected with T cells that have been treated to fight the specific infections they suffer. Virologists have gotten into the act: Any cancer caused by a virus, such as cervical cancer, is a prime target for vaccine-based therapies.

Researchers have come to appreciate the potential of cancer immunotherapy, and many are trying to transform the concept into real and viable therapies. Another approach to controlling cancer is to attempt to change its progressive and deadly nature—to disable it so that it can be managed throughout a person's life.

. . . .

VACCINES

. . . .

Immunotherapy has established a much longer and more successful track record in combating infectious

. . . .

Netting ticks with a blanket drag, researcher Sam Telford works the wilds of Nantucket Island. Similar fieldwork helped identify the organism responsible for Lyme disease, transmitted to humans by tiny deer ticks.
FOLLOWING PAGES: Spread by rats, the viral infection known as Lassa fever was first identified in 1969 in Lassa, Nigeria. Rodent control and new drug therapies have since reduced deaths, but the virus still claims many lives.

. . . .

disease than in fighting cancer. Before the advent of modern vaccines, some Americans wore malodorous bags around their necks to ward off germs; Russians preferred a clove of garlic. Still, *Hemophilus influenzae* bacteria—which can cause pneumonia—and polio claimed the lives of thousands of people worldwide and left many others permanently disabled. In the U.S. alone, 20,000 children a year contracted these diseases. Vaccination has made many once deadly diseases a distant memory.

But vaccinations also can have drawbacks. They occasionally create adverse reactions. And vaccines are available only for some, not for all, diseases. Scientists have been unable to create an immunogenic form of cholera, for instance. The major limitation of vaccines is that they elicit only an antibody response—and antibodies are rarely effective against viruses and other non-bacterial bugs. Effective preventive agents against many fungi, parasites, and viruses have not yet been developed.

Efforts are currently under way to make existing vaccines both safer and more specific. Again, biotechnology has come to the rescue. Using molecular techniques, scientists can now rapidly locate the genetic component of a microorganism that gives rise to an illness. They can

isolate the protein, or series of proteins, arising from these genes, then manufacture them in pure form and in great quantity. Thus people can be vaccinated with the specific element rather than the entire organism—a much safer approach. In fact, researchers have found that it is not even necessary to administer the protein; simple injection of the gene alone (so-called naked DNA) also works well, according to immunologist William R. Clark of the University of California at Los Angeles. Further-

more, if a gene or its proteins are found to be harmful, they now can be easily deleted or modified.

Also on the horizon are vaccines that may elicit an immune system response that will protect against viruses and other nonbacterial microbes—goals not possible for conventional vaccines.

Advances in vaccine design have been accompanied by more effective ways to administer immunizations. Adjuvants, substances that strengthen the effectiveness of a treatment or enhance the immune response to an antigen, are now used in the preparation of many vaccines. Adjuvants ensure that an antigen will be "seen" by the correct immune system cells. In the future, timed-release applications and innovative combinations of today's vaccines could reduce the total number of immunizations.

All these efforts are helping make the world a safer place. A single vaccine has virtually eliminated measles and mumps and rubella in the U.S.—in the process saving the lives of an estimated 25,000 American children. Vaccines against *Hemophilus* pneumonia and hepatitis B are helping make those diseases disappear. In the past several years, chicken pox and rotavirus vaccines have become available. A large number of new vaccines are in trials, including pneumococcal and meningococcal vaccines and vaccines for Group B streptococci, respiratory syncytial virus, *Helicobacter*, herpes, and various human papillomaviruses.

But despite these improvements in vaccines, many microorganisms maintain their capacity to outwit the immune system. They have an intrinsic advantage: New microbial generations take only a few minutes to mature, compared to decades for human offspring. A virus such

Delicacy or disease-carrier? Ducks carry influenza viruses, which normally infect pigs before changing to a strain capable of striking humans. A mysterious avian flu virus in 1997 bypassed the porcine pathway and proved lethal to several Hong Kong residents—prompting fears that a rare exchange of key genes between ducks and humans could cause a pandemic.

as influenza can be an elusive quarry because it alters its appearance with each rescrambling of its genetic makeup. The flu virus undergoes spontaneous mutations from one year to the next. When there is a small change in the virus, there are minor epidemics; when there is a large change, antibodies in millions of people no longer recognize this virus. A massive pandemic—even more widespread than an epidemic—ensues.

Many diseases have not yet been tamed. Normal medications are powerless against some parasites, viruses, and deadly fungi. And many chronic infectious diseases, such as tuberculosis, leprosy, and leishmaniasis, do not always respond to conventional treatments.

ANTIBIOTICS

Antibiotics can help the natural defenses provided by the body's immune system eradicate an invading organism. Antibiotics work in different ways. Some pull water into the invading cell, causing it to expand, burst, and die. Others halt bacterial reproduction, giving the immune system a chance to cope with the infection.

Since the introduction of antibiotics in the early 1940s, most common bacterial diseases occurring in the developed world have been treatable. But they continue to be major health problems in developing countries. And even the industrialized world has been sobered by the increase of antibiotic-resistant bacteria.

It was quite by accident that the British scientist Alexander Fleming first observed, in 1928, that the mold *Penicillium notatum* had a damaging effect on certain bacteria. He noticed that colonies of staphylococcus bacteria, growing on a discarded culture plate in the sink of his lab, had been affected by some mystery substance diffusing through the jelly from the mold that had by chance contaminated the culture. Though he made this observation and identified the mold as *Penicillium*, Fleming never succeeded in purifying it and converting the accident into a useful drug.

By 1940, scientists had isolated and purified the substance produced by *Penicillium*—and recognized that it could be used to attack other microbial invaders. It was first tested in a desperate moment on February 12, 1941. A 43-year-old British police officer, Albert Alexander, had been scratched on the corner of his mouth by a rosebush. Infection by two types of lethal bacteria, streptococci and staphylocci, set in. The infection progressed to the rest of his face and scalp, right shoulder, and both lungs, and he was hospitalized. Two months later his temperature soared to 105°F. Death seemed imminent. A young researcher, Charles Fletcher, injected 200 milligrams of purified penicillin into Alexander. Within 24 hours the policeman began to improve. His temperature dropped. The scratch began to heal. His appetite returned. Then the world's supply of penicillin ran out. On March 15, while doctors stood by helplessly, Albert Alexander died.

Although there was no happy ending for the patient, that single experiment launched the antibiotic revolution. Together with the other antibiotics that came in its wake, penicillin has been credited with adding an entire decade to human life expectancy. Antibiotics were so successful that, in 1967, the surgeon general of the United States, William H. Stewart, announced that victory over infectious diseases was clear at hand—a victory that would close the book on modern plagues.

Sadly, we now know differently. There is a chink in the armor: resistance. Today, some 5,000 different antibiotics are known. Of them, about 100 are currently used to treat infections. Some are broad-spectrum weapons, while others have more specialized applications. Earlier versions, once hailed as wonder drugs, don't help as they used to. The overuse of antibiotics—using them to treat diarrhea, colds, and other nonbacterial infections—has created drug-resistant strains that can sidestep treatment, leading to serious illness, even death. Many researchers now fear that the antibiotic spree in the decades after World War II will turn out to have been a terrible mistake, creating superbugs that have since mutated out of reach. Consider the arithmetic: In 1945, penicillin cut fatalities by staph to only one-fourth of patients infected. By 1953, only half could be saved. Just two years later, 80 percent of those infected with penicillin-resistant staph died. In other words, less than two decades after Albert Alexander's treatment, penicillin was rendered virtually ineffective against the most dangerous microbial organism of the day.

Many doctors consider vancomycin, an antibiotic developed in 1958, to be the best weapon against bacteria now invulnerable to other drugs. As resistance to other antibiotics has increased, doctors have turned increasingly to vancomycin. But this drug, the last line of defense, is losing its punch against two serious bugs:

Enterococcus and *Staphylococcus aureus*. "Super Staph" is particularly scary: If it gains a secure foothold, it could claim thousands of lives.

These new strains of superbacteria could bring back the therapeutic impotence of pre-antibiotic days. Many human infections—malaria, gonorrhea, pneumonia, and leprosy among them—are now resistant to at least one major class of antibiotics.

Increased resistance has underscored the need for new therapies, giving us an incentive in the arms race between scientists and microbes. There now is a huge push to find new versions of old, once-effective medicines. Because antibiotics are difficult to synthesize and tinker with, researchers largely have had to accept what nature gives them: compounds made by bacteria and fungi. So they've gone to the lab to figure out a way to reproduce the antibiotic-making chemical pathways of these microbes, hoping to exploit them to produce a wide variety of new antibiotic compounds.

To combat resistance, scientists also are seeking ways to turn off the genes that make bacteria so hardy, a discovery that could help head off the growing medical crisis. For instance, a Yale University team led by Nobel Prize laureate Sidney Altman has found a way to insert into bacteria artificial genes that make the germs highly sensitive to two antibiotics currently in wide use, chloramphenicol and ampicillin. But so far the work has been demonstrated only in laboratory cultures, not in the human body, and no one knows yet how to deliver the artificial genes into live cells.

....

VIRUSES/HIV

....

A single virus is incredibly tiny—only .02 to .3 microns (millionths of a meter) in diameter—so small that thousands of them could fit in the period at the end of this sentence. But they cause more sickness than anything else on earth: They are blamed for almost 80 percent of all acute illnesses, from deadly Lassa fever to the familiar common cold. On the other hand, many are harmless. Some may even be beneficial. There is an emerging theory that, over time, viruses have accelerated the relatively slow process of heredity, transmitting genes from one species to another, thus introducing new genetic material and speeding the course of evolution. Our intimate companions, they are also our co-travelers through time. Not quite alive yet not dead, viruses survive on chance—they float, directionless, on wind and water; they hitch rides on animals and birds; they rest on doorknobs and computer keyboards. They become animate only when they encounter a living cell. Then they attack and take over, commandeering the host's genetic machinery to crank out replicas of themselves.

Antibiotics, which have proved such a boon in fighting bacterial infections, are helpless against viral infections. And viruses are relatively safe from the body's defenses. Because they climb within cells, they can be hidden from the immune system. Only if the infected cell is destroyed does the virus perish also.

Of all viruses, human immunodeficiency virus (HIV), which causes AIDS, is one of the most intimidating. Within the first decade of entering our contemporary life, it penetrated almost every nation on the globe. It is almost invariably lethal. HIV is like a ticking time bomb. It doesn't kill quickly, but bides its time, slipping into the genes of a cell and slowly but steadily overwhelming the immune system. All the while, unseen, it is transmitted from person to person via blood and semen.

If AIDS is viral, why is there no AIDS vaccine? One problem is that scientists disagree over which type of immune response is needed to protect against infection by HIV. Also, they have been daunted by the virus's ability to mutate quickly and elude the immune system. And although some trials have shown that vaccines can improve the body's immune response to this virus, that is a far cry from demonstrating useful protection against HIV infection. So far, no vaccine has provided protective immunity. But there is an encouraging sign that the virus can be defeated: HIV produces no symptoms in humans for years, which suggests that immune processes hold the virus in check for long periods, just not long enough.

Recent work at Massachusetts General Hospital offers one explanation for how one very small group of people—healthy despite their long-term HIV infection—have been able to avoid progressing to AIDS. Researchers found that some individuals, dubbed "nonprogressors," produce large numbers of helper T cells specifically targeted against HIV. These HIV-specific helper T cells seem to help control the virus. People with the weakest HIV-specific T cell response had the highest levels of virus in their blood; those with lower viral levels showed a stronger T cell response. This suggests that the body

might be able to prevent AIDS if these important helper T cells were somehow protected. There may be a window of opportunity following infection when prompt antiviral treatment can boost the helper T cell response to HIV.

While some people with HIV never progress to AIDS, an even smaller number somehow avoid HIV infection entirely—although repeatedly exposed to the virus.

There is evidence that helper T cells may not be alone in their battle to suppress or extinguish HIV. Jay A. Levy, of the University of California at San Francisco, believes that people who remain uninfected despite repeated HIV exposure have a unique immunologic ability to marshal a different kind of T cell—a killer T cell—that prevents the virus from reproducing. In the

laboratory Levy found that, when he attempted to infect blood that had been fortified with these killer T cells, the virus died. But when the killer T cells were removed, it grew again.

Perhaps these HIV-resistant people were once exposed to very low levels of the virus, enough to get the body's protective immune response going, but not enough to overwhelm the immune system, Levy speculates. The virus was quashed yet its memory lingered, offering a lifetime of protection.

There is still much to be learned about how the body fights HIV. Until then, we must assist the beleaguered immune system with antiviral drugs. In response to the AIDS epidemic, federal and private labs around the

Invisible dangers—along with nourishment—await Alaskan hunters carving bowhead whale blubber. Such traditional foods harbor high levels of pesticides, which may kill immune cells and weaken the body's resistance to bacteria and viruses. Despite their isolation, native peoples of the far north show marked immune system deficiencies. Pollutants concentrate in the milk of nursing mothers, and their children suffer frequent infections.

country have launched the most aggressive medical research effort in recent history, in order to create tools with which they can fight HIV. Happily, there has been a huge payoff: New combinations of antiviral drugs have helped people who were very sick get well and return to leading productive lives. For the first time in the AIDS epidemic, the number of people dying from the disease has decreased significantly.

Almost miraculously, some new drugs reduce the level of HIV in the body so dramatically that it cannot be measured in the bloodstreams of many patients. This is no cure, however: The virus may still lurk in some cells, undetectable by current tests. Moreover, the drugs don't work for everyone. Some people show resistance to the agents and do not respond. Others lose their sensitivity to the drugs over time. But for now, these innovative therapies are prolonging lives—and buying time to design even better, more potent drugs.

Even if such innovative therapies send the virus into hiding, will the ravaged immune system of an HIV patient ever recover? Doctors are unsure whether the current antiviral drugs will allow the immune system to heal completely. Without immune cells, patients cannot even fight off the most ordinary of infections.

There are promising early reports that after much therapy, people with advanced HIV infection show partial, although not complete, restoration of normal immune activity. It is still unknown, however, whether these restored cells will act like normal, functioning, pre-HIV cells—and thus whether they will offer adequate protection. Scientists are hard at work inventing new strategies to give the immune system a hand. In one, called cell transfer, immune cells from a healthy person are given to someone who is infected. Another technique uses cytokines, such as interleukin, to stimulate healthy T cells to reproduce.

Also on the horizon are ways to protect new cells from infection. One novel approach involves the use of genes to find and destroy HIV-infected cells. Another plan seeks to insert HIV-resistant genes into immune cells, thus protecting them from future infection. Researchers are also exploring the possibility of equipping immune cells with genes that would block HIV replication.

Encouraging though these discoveries are, an enormous amount remains to be done in the realm of immunotherapy. Much has been revealed about the immune system; much more remains uncharted territory.

Can humans retain the upper hand in their ongoing battles with microbes? The two have a long history together. Throughout this co-evolution our immune system has helped us prevail over parasites, bacteria, viruses, and other pathogens.

One reason, of course, is the extraordinary variety of our immune system. Immune system molecules—more than those of any other part of the body—vary widely from person to person, increasing the odds that some individuals will survive even the most devastating epidemic. Moreover, these various genes recombine from one generation to the next, allowing for quick evolution of the fittest. Also, a live-and-let-live truce—with a few notable exceptions—has been struck between people and microbes. Microbes can kill us, but they seldom do so quickly. They and we have a shared agenda: survival.

As George J. Armelagos, an anthropologist at Emory University, has observed, "Logic would suggest that the best interests of an organism are not served if it kills its host; doing so would be like picking a fight with the person who signs your paycheck."

Already we have shown great cleverness in creating drugs and vaccines to outwit many invaders. But there is heightening concern that these historic protections won't last forever. The world is changing, and our environments with it. International travel, global commerce, and rapid population growth increasingly expose us to microbial threats that even our highly adaptive immune system can't handle.

For instance, construction of the Aswan Dam in 1971 resulted in vast quantities of stagnant water; mosquito-transmitted Rift Valley fever later killed 600 Egyptians. Similarly, the expansion of American suburbs into sparsely populated areas has been blamed for outbreaks of both Lyme disease and *Hantavirus*. Ship travel has

Deadly to bacteria yet safe to human cells, antibiotics can turn microbial invaders such as *Staphylococcus aureus* (opposite, top) into a lifeless husk (bottom). Despite the successes of such "wonder drugs," more and more pathogens today are growing resistant to these medicines.

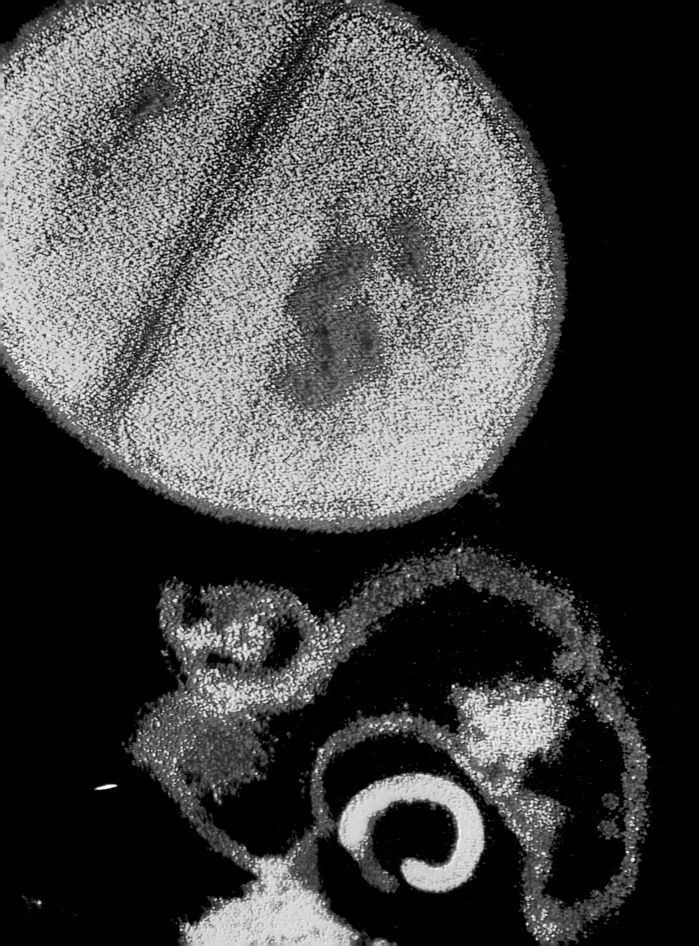

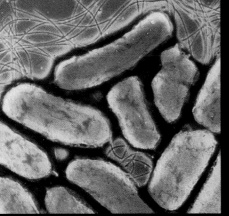

SALMONELLA BACTERIA

People and pathogens share a long history together. An estimated 5,000 types of viruses and 300,000 species of bacteria cohabit the globe with us. Some are ancient and familiar killers: Infections have been detected in human bones more than a million years old. Others, new or old, have been identified only recently.

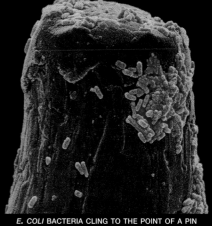

E. COLI BACTERIA CLING TO THE POINT OF A PIN

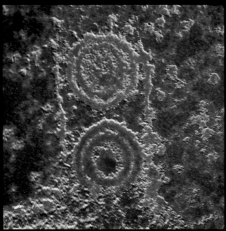

HERPES SIMPLEX VIRUS

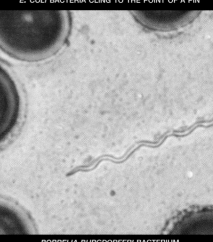

BORRELIA BURGDORFERI BACTERIUM

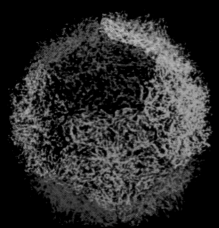

RHINOVIRUS

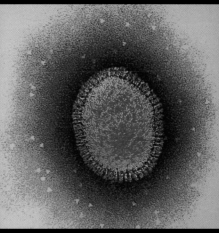

INFLUENZA VIRUS

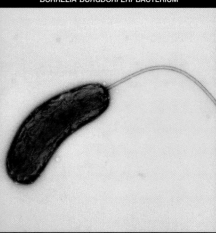

CHOLERA BACTERIUM

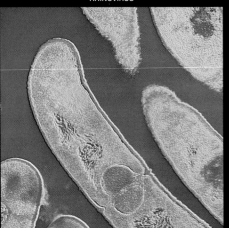

HELICOBACTER PYLORI BACTERIA

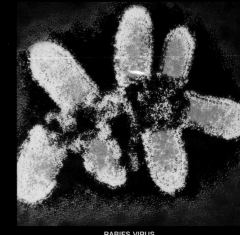

RABIES VIRUS

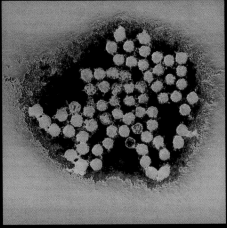

RHINOVIRUSES, SOURCE OF COMMON COLD

Often existing in remote corners of the globe for millennia, some deadly microbes have only lately arrived on our urban doorstep. Global travel, overpopulation, sexual promiscuity, ecological destruction, and misguided science all contribute to this growing problem. Overuse of antibiotics has helped create drug-resistant variants.

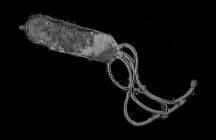

ULCER-CAUSING *HELICOBACTER PYLORI* BACTERIUM

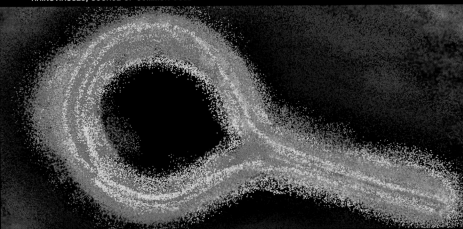

EBOLA VIRUS

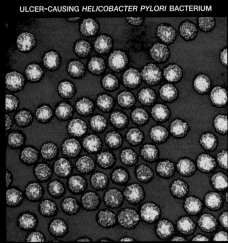

PICORNAVIRUS

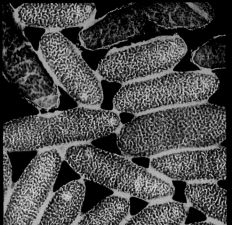

LEGIONELLA BACTERIA

Earlier in this century, victory over infection seemed imminent—but not now. In the time it takes us to produce a new generation, bacteria and other microbes can reproduce half a million times, evolving innumerable virulent modifications that can quickly make even the newest drugs obsolete.

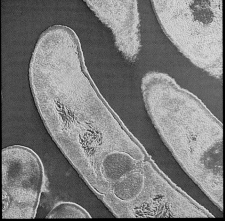

LISTERIA BACTERIA, CAUSE OF MENINGITIS

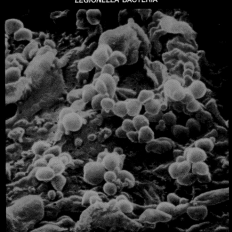

MEASLES VIRUS

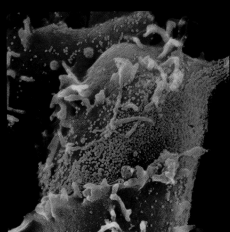

RHINOVIRUS

prompted the deadly cholera microbe to move from Asia to Latin America, killing thousands.

And diseases that just a few decades ago seemed to have been tamed are returning in virulent, drug-resistant varieties. Tuberculosis, the ancient lung disease that haunted 19th-century Europe, is aggressively mutating into strains that defy standard medicines. The bacterium that causes meningitis was once routinely controlled with ampicillin; now, about 20 percent of such infections are resistant to the drug.

Many doctors fear we are living in the twilight of the antibiotic era: Within our lifetimes, even minor wounds such as scraped knees and cut fingers may result in infections that could become life threatening.

Moreover, deadly and previously unimagined novel illnesses have recently emerged. On the international scene the newly recognized strain of Creutzfeldt-Jakob disease, linked to bovine spongiform encephalopathy in England, attacks the brains of victims. Marburg and Ebola are two exotic African viruses with high fatality rates and grotesque symptoms, as well as a mysterious past and an unpredictable future. They are so deadly that they roar through populations quickly, killing too fast to cause large and prolonged epidemics.

In the United States an outbreak of *Hantavirus* in 1993 caused victims' capillaries to leak blood, leading to organ shutdown before the immune system had time to react. The just-discovered human herpes virus type 8 (HHV-8) is the agent behind the purple lesions of Kaposi's sarcoma, a once rare disease now common in AIDS patients. Hepatitis C virus, also newly defined, has emerged as possibly the most common cause of chronic liver disease and cirrhosis in the U.S.

New forms of infection, such as viruses that have recently crossed to humans from an animal species, bring with them their own agendas: Rarely do they honor the polite "live and let live" philosophy of microbes that have co-evolved with humans.

Microbes that long ago reached a truce with their host animal species can be highly dangerous when they manage to invade human beings. Milk from infected cattle can transmit tuberculosis, for instance, a slow killer that destroys the lungs. Wool from sheep can carry fatal anthrax. And when a certain monkey virus apparently jumped to humans, it led to the global epidemic of AIDS.

THE SUCCESS OF THE SYSTEM

Despite the terrifying variety of microbes on our planet, however, most scientists trust in the power of the human body and mind to survive. Our immune system has not reached its apogee, they say. We have a huge storehouse of genetic variability and still undiscovered protective genes that remain at the ready, like missiles in their silos. And if conventional wisdom is correct, it is not in a germ's best long-term interest to completely annihilate its host. Our destinies are bound together.

Finally, *Homo sapiens* is no ordinary host—we are sophisticated, resourceful, and well organized, creating massive disease-fighting organizations like the National Institutes of Health, the Centers for Disease Control and Prevention, and the World Health Organization. In the last few decades we have begun to appreciate just how complex the immune system is and how elegantly it strives to protect us. As our molecular understanding of the immune system races forward, therapeutic advances seem certain to follow.

But to stay safe, we will grow increasingly reliant on vaccines, improved drugs, and more stringent measures in hospitals to limit contact between people and microbes. Better infection-control practices, such as increased use of condoms and clean hypodermic syringes, must be implemented. And the affluent, industrialized nations must export their public health measures to the poorer, developing countries that still lack clean drinking water, sanitation, and other essentials.

Increasingly, we find ourselves ruled by what English zoologist Richard Dawkins of the University of Oxford calls "The Red Queen Principle": Like Alice in Lewis Carroll's book *Through the Looking Glass*, we need to run ever faster just to stay in place.

Handful of germs commonly found on human skin—*E. coli, Staphylococcus aureus, Proteus mirabilis,* and others— blossoms into a full-blown handprint on a culture medium. Even with today's space-age medicines, the best infection-fighter is prevention: in this case, simply washing the hands.

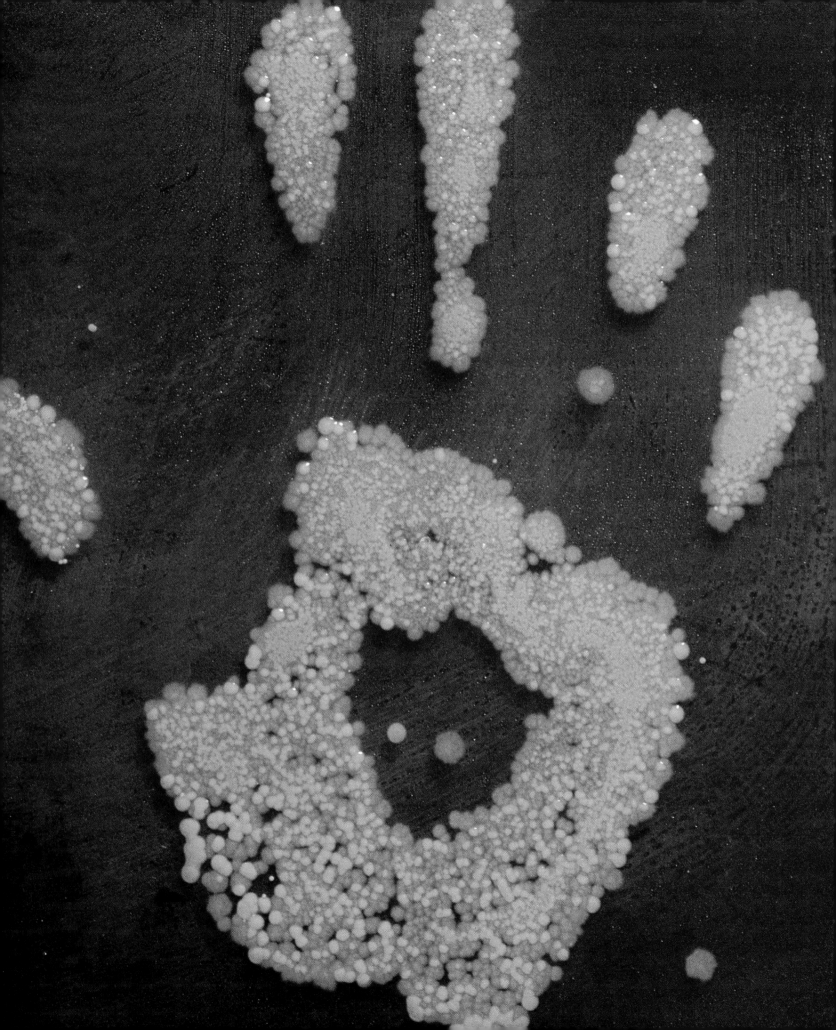

Brain

ROBERTA CONLAN

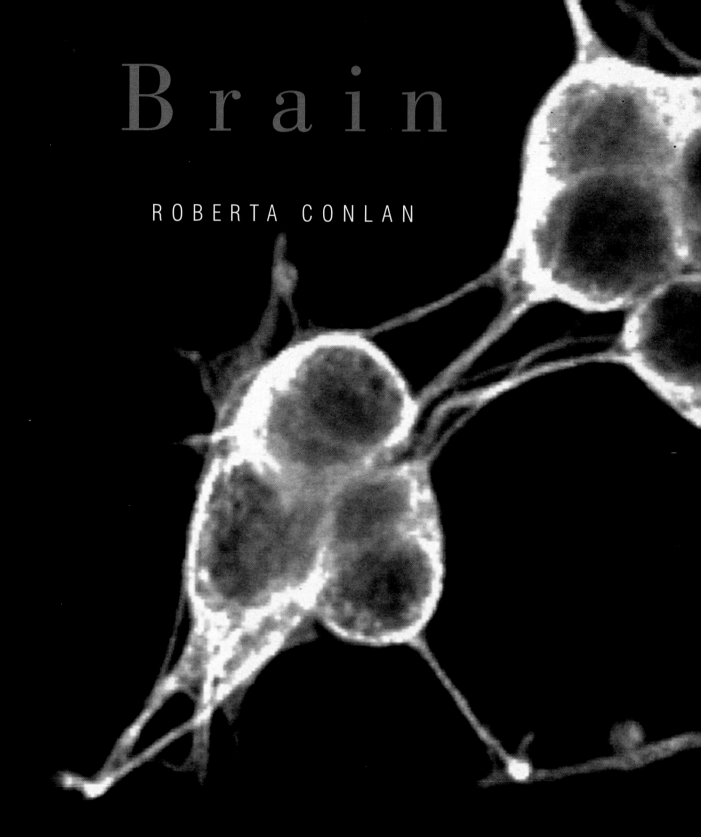

Agents of all that we do, see, feel, think, and remember, the brain's
neurons—nerve cells—transmit messages among themselves
and throughout the body.

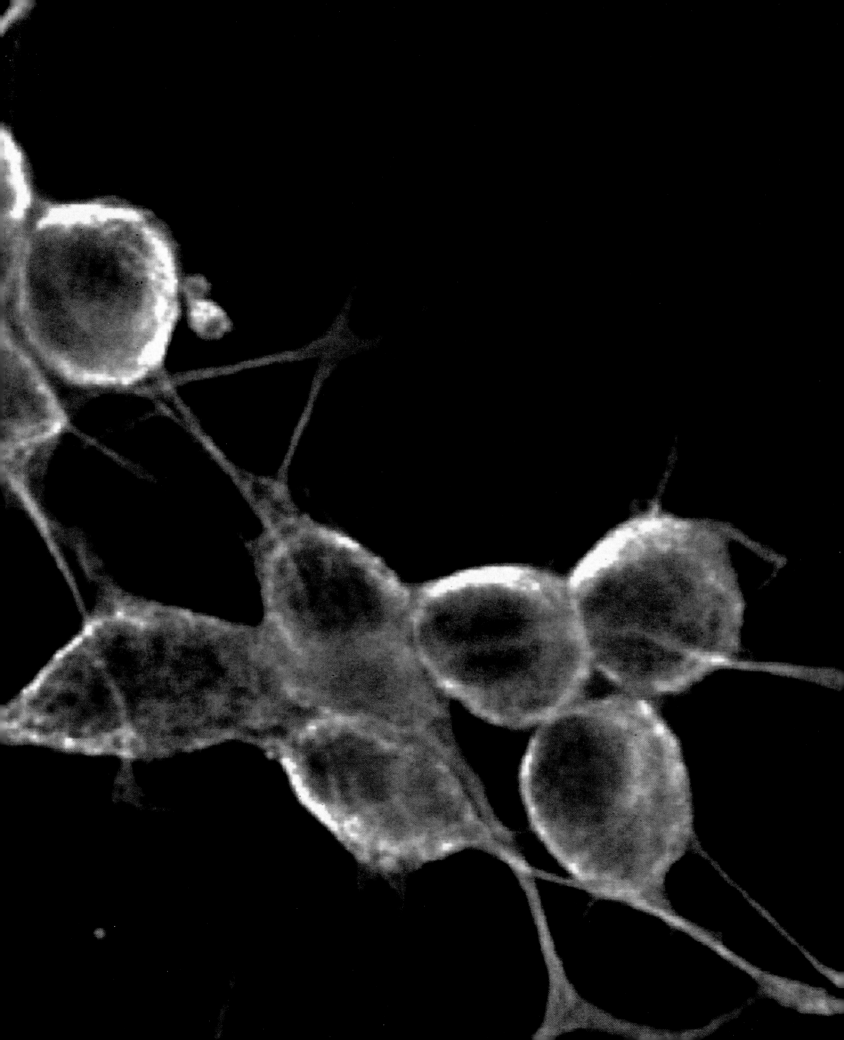

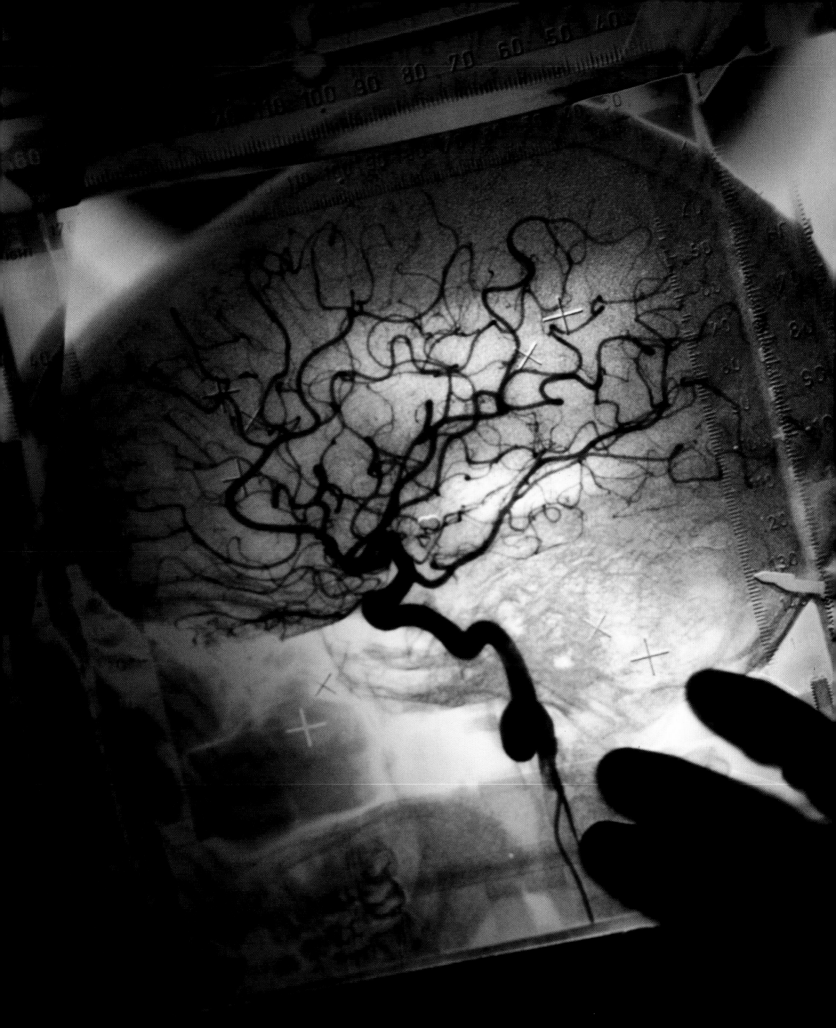

Monday morning. The alarm goes off, dragging us more or less easily from dreamland. Some of us turn the alarm off, stretch, and have our feet on the floor within a minute or two. Others of us wake long enough to punch the snooze button, roll over, and go back to sleep. Sooner or later we all get out of bed and start our morning routines: Brushing teeth, making breakfast, taking a shower, getting dressed, reading the paper, getting the kids organized for school, leaving for work.

As we carry on this completely ordinary activity, we pay no attention to the extraordinary symphony taking place inside the hard casing of our skulls. We go to sleep, we wake up. Our lungs pull in air and let it out, our eyes blink, our hearts beat. We sit, stand, and walk. We talk and listen. We hold a cup, pet the cat, read the paper, stare out the window, think. All involve different parts of the brain acting together, each chiming in at just the right time like the different instruments of an orchestra.

This orchestra seems to have no conductor, however. When the brain is functioning normally, its workings are clear to us, for which (if we think about it) we should be profoundly grateful. Indeed, as Lewis Thomas noted, were we foolish enough to try to take control of our own brains, we'd be in deep trouble in no time flat. "It might be something of a temptation to take over my brain, on paper," he wrote in his classic *The Lives of a Cell*, "but I cannot imagine doing so in real life. I would lose track, get things mixed up, turn on wrong cells at wrong times, drop things. I doubt if I would ever be able to think up my own thoughts. My cells were born, or differentiated anyway, knowing how to do this kind of thing together. If I moved in to organize them they would resent it, perhaps become frightened, perhaps swarm out into my ventricles like bees."

Only when the brain is not functioning normally, or when we become aware of something "not quite right"— when Dad forgets, again, how to get home from an errand to the store, or a friend confesses that for several months she's been so exhausted and depressed she can scarcely drag herself out of bed—might we wish its workings were more amenable to scrutiny and repair. For a long time, most of what we knew about how the brain worked came from observing people who had something wrong with theirs—brain injuries resulting from accident or illness.

Nineteenth-century surgeons and pathologists would note the problems and difficulties these unfortunates had, and then, after they died, would do autopsies to see where the brain was damaged. In this way we began to home in on which areas of the brain seemed to be involved with language, for example, or the inhibition or expression of emotion. In the 1930s we learned more from people about to undergo brain surgery for such disorders as severe epilepsy. Doctors inserted fine-gauge electric probes into the exposed brain of conscious patients and stimulated parts of the brain to identify healthy tissue to avoid during surgery. (Despite its hundred billion nerve cells, the brain has no pain sensors, so once the scalp was anesthetized, the patient would feel no pain.) Sometimes the stimulation would elicit a finger twitch, or the patient would report hearing a phantom noise. But sometimes the patients also reported what seemed to be quite vivid memories.

As modes of research these methods left a lot to be desired. Since only severe or life-threatening conditions were justifiable reasons for opening up the skull, progress was painfully slow, and sometimes just plain painful. Today we have machines that, in effect, let us see through the skull and watch the living brain in action, bringing tremendous advances in understanding which parts of the brain do what. These techniques also promise to unite two disciplines that traditionally have been poles apart: neurobiology, the science of the brain, and cognitive psychology, the science of the mind. Already we are gaining fascinating insights into a variety of mental functions such as language, perception, learning, and memory.

One such technique, functional magnetic resonance imaging (fMRI), takes advantage of the fact that active areas of the brain require fresh supplies of oxygen-rich blood. Through fMRI we've recently learned, for example, that people who suffer from dyslexia and have great difficulty reading seem to use a different pathway in the brain from the one used by those who read easily. Other fMRI research indicates that people who learn a second language later in life use a different part of the brain for that language than the area they use for their native language, offering clues not only to how the brain handles language but also perhaps to the brain's remarkable ability to adapt its own systems to new uses. Still

Angiogram reveals the branching blood vessels that nourish the brain, source of our humanity. Though this three-pound organ has yielded numerous secrets to science, many of its workings continue to puzzle.

other fMRI efforts have captured the moment-by-moment activity involved in seeing faces or a series of letters, holding them briefly in so-called working memory, the brain's temporary blackboard, and later recalling them.

Today researchers are probing into the brain's secrets at both the molecular and genetic levels, addressing in the process the age-old nature versus nurture debate. Results suggest a partnership between the two so integrated that its components cannot readily be teased apart. As neuroscientists often describe it, genes give a kind of best-guess blueprint for building a brain, sketching out arrangements of neurons—nerve cells—for vision, language, and movement. But the environment is the architect that refines the blueprint to build a particular brain. For example, genes have instructions to do the basic wiring that links the retina to the various parts of the brain involved in the process of vision, but timely stimulation from the environment—in this case light impinging on the retina—is needed to fine-tune and stabilize the connections. Without such stimulation, potential connections wither; an otherwise normal eye will be permanently blind, as can occur when a child born with a congenital cataract does not have the cloudy lens removed in time.

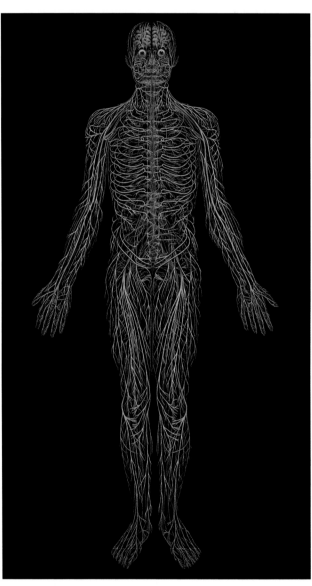

Stimulation—that is, input from the environment—is essential to normal brain development. Without it, brain cells atrophy, and neural circuits fail to wire themselves properly. Whether the context is vision, language, mood disorders, learning disabilities, or addiction, research affirms that the brain is above all else a learning machine, designed by evolution to adapt to experience and to reorganize its systems as much as possible on the basis of external stimuli. This constant change, or plasticity, as it is called, is how the brain builds itself, even in the womb, and it survives in the world by learning and adapting. Sometimes, though, what the brain learns proves deleterious in the long run, as when it adapts to an addictive drug or makes adjustments to survive a violent environment. Research shows the pernicious effects of harsh environments on the malleable brains and nervous systems of infants and children. Research also demonstrates that given proper stimulation the brain learns to see and hear, to understand language and speak, to walk and ride a bicycle, to feel affection or distrust, to make a soufflé and savor its lightness—to be, in short, human.

As human beings, however, we have to remember we're the newest kids on the block. Many of our brain functions are "primitive," in the sense that evolution has conserved them over

Weblike peripheral nerves connect the brain and spinal cord to the rest of the body, providing the wiring that enables all sensation and action. Slice by slice, multiple MRI images (opposite) build a composite view of the brain in cross section.

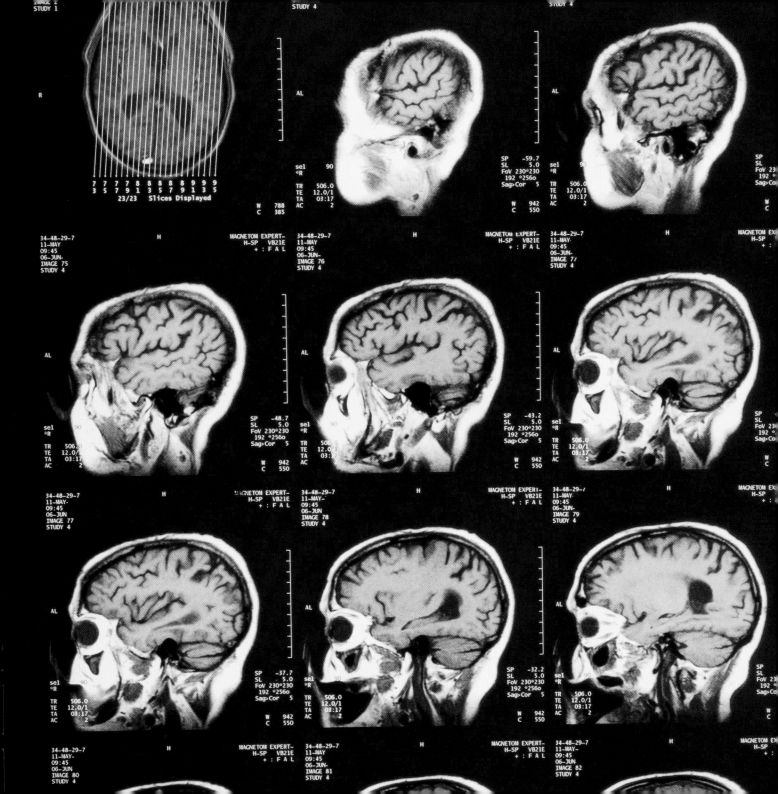

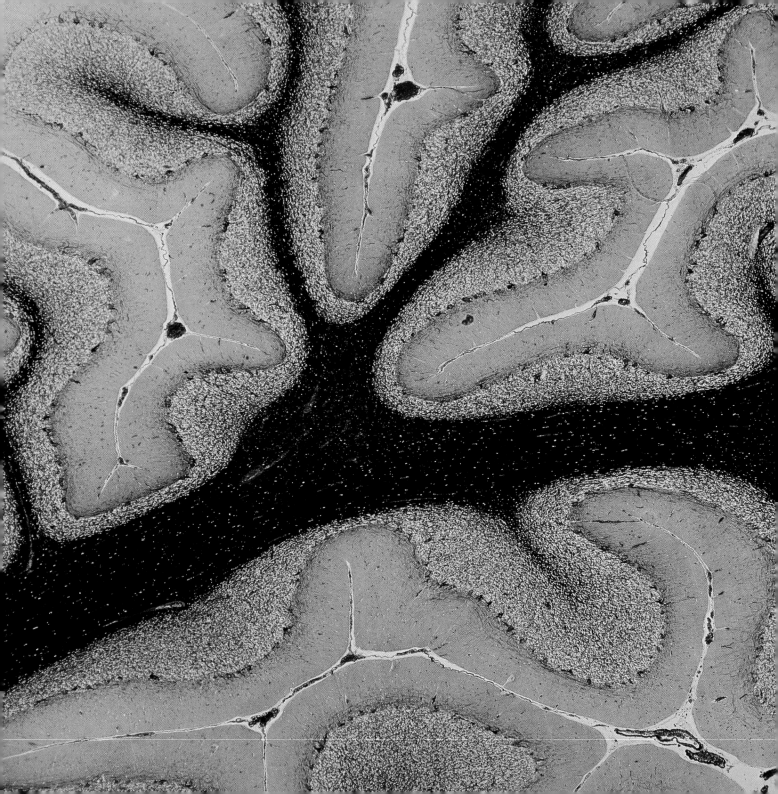

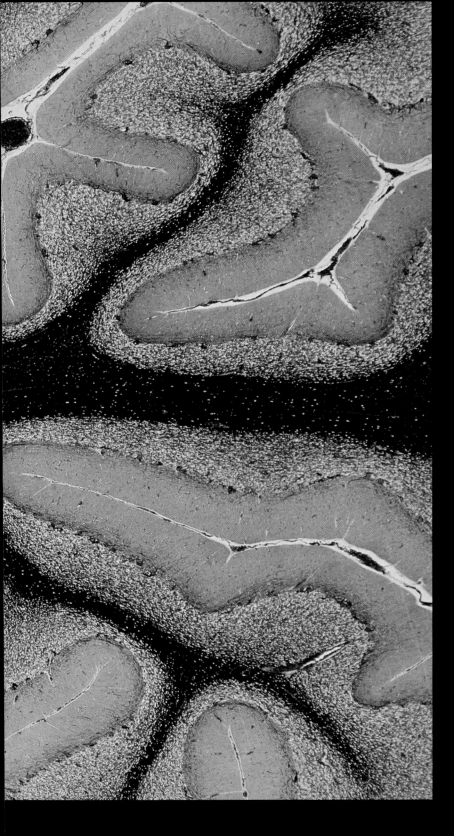

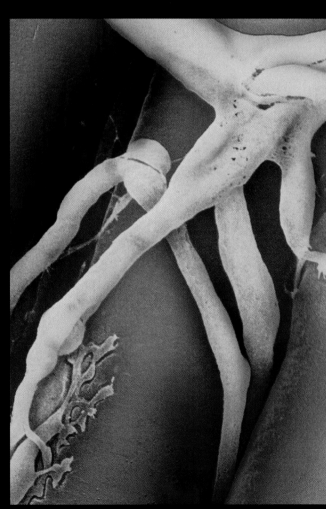

millennia and across many species. The central nervous system, for example—the brain and spinal cord and the bundles of nerves that relay messages to every muscle and organ in the body—is represented, albeit sometimes in very rudimentary form, not only in our fellow vertebrates but also in invertebrates.

THE CENTRAL NERVOUS SYSTEM

The central nervous system is itself made up of two more or less independent systems. The autonomic nervous system handles all the involuntary acts such as reflexes, heartbeat, and breathing. The peripheral nervous system enables us to perform voluntary movements such as picking up a pen or walking out the door. Even though our voluntary movements are, by definition, things we do at will, we aren't aware of the myriad contractions and relaxations of all the different muscles involved, nor of the barrage of data and feedback coming in from our senses. This sensory feedback enables us to exert just the right amount of pressure on the pen or raise our leg just enough to climb a step. The brain handles all the sensory input and all the motor instructions at lightning speed and usually below our level of consciousness.

Indeed, the brain operates as a kind of committee of the whole, with each member responsible for certain tasks but usually relying on input from other members as well. No part of the brain operates in total isolation, including

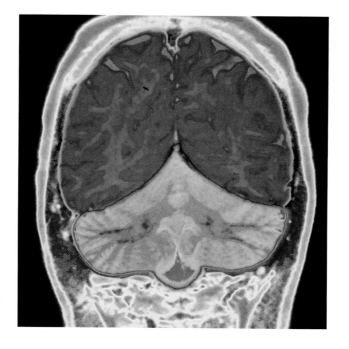

our much vaunted cerebrum. The two hemispheres of the cerebrum constitute the largest part of the brain, and its eighth-inch thick outer layer, the cerebral cortex, is the newest brain feature in evolutionary terms. The so-called higher cognitive functions of the cerebral cortex in which we humans take such pride—"rational" thought and decision-making, for instance—are permeated by nonconscious contributions from parts of the brain that we share with all other vertebrates, from chimpanzees to cats and iguanas. These parts include the cerebellum, or little brain, and the brain stem. Together the brain stem and cerebellum function as our autopilot for vital life-support functions, such as heart rate and breathing, that go on without our conscious effort. The cerebellum—two mounds of folded tissue at the base of the brain—is responsible for balance, muscle tone, posture, and voluntary movement. Thanks to the cerebellum we can get out of bed and walk down the hall without crashing into walls. It is also the structure that helps us learn to ride a bike or dance—and remember how to do these things years later.

The brain stem, a three-inch-long segment that meets the spinal cord, is the oldest and most primitive region of the brain. It controls our basic life functions and is also responsible for the processes that let us drift off to sleep. It regulates the cycles of dreaming and nondreaming sleep, then pulls us toward waking in the morning. The brain stem is made up of the medulla, the pons, and the midbrain. The medulla, the lowest part of the brain stem, sits just above the spinal cord. It regulates breathing, blood pressure, and such responses as sneezing and swallowing. The pons, above the medulla, is a key player in regulating the changes in brain

Tucked at the rear of the skull under the cerebrum's twin hemispheres (green), the cerebellum (red) and brain stem together govern involuntary activities that require no conscious effort, such as breathing, heart rate, and blood pressure.

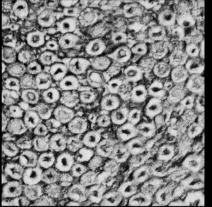
CROSS SECTION OF NERVE FIBERS

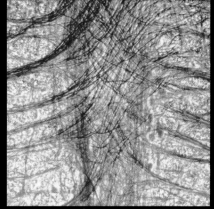
LEFT BRAIN / RIGHT BRAIN CROSSOVER

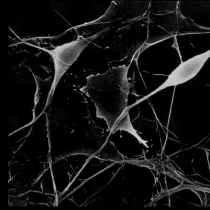
NEURONS AND GLIAL CELLS

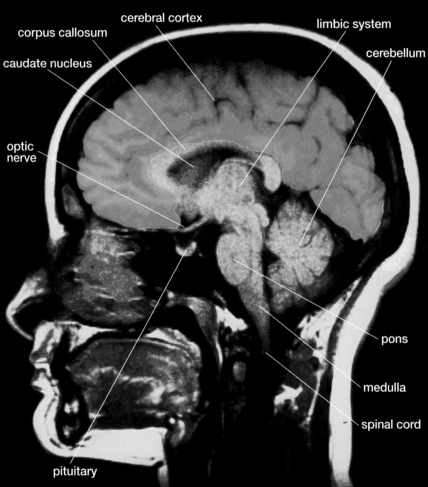

corpus callosum

cerebral cortex

limbic system

caudate nucleus

cerebellum

optic nerve

pons

medulla

spinal cord

pituitary

CELLS OF THE CEREBELLUM

FETAL NEURON UNDERGOES MITOSIS

Specialized into myriad shapes and regions, the brain displays structural as well as functional diversity. Billions of neurons, arrayed on threadlike glial cells, communicate through chemical messengers called neurotransmitters. Due to a neural crossover at the base of the brain, each cerebral hemisphere controls the body's opposite side.

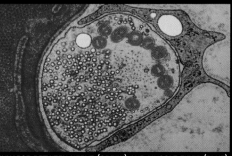
SYNAPSE BETWEEN NERVE (GREEN) AND MUSCLE FIBER (BLUE)

chemistry of sleep and waking, relaying signals between the two hemispheres of the cerebral cortex and the rest of the central nervous system. The pons, together with the midbrain that sits above it, activates the muscles of the eye, whether the movement is involuntary, as during the rapid eye movement that characterizes dreaming sleep, or voluntary, as when we open our eyes to look at the clock. Even that seemingly simple act has a number of components, requiring the simultaneous control of 12 muscles, 6 for each eye. The midbrain, along with the rest of the brain stem, is one of the relay stations for pain signals coming from the body to the brain, be it a stubbed toe or the ache of arthritis. A portion of the midbrain called the periaqueductal gray area initiates signals that trigger the release of natural painkilling biochemicals to block the upward passage of pain messages from injured tissue and dampen the sensation of pain.

All vertebrates possess similar structures above the brain stem and deep inside the brain—the thalamus, pituitary, hypothalamus, hippocampus, and amygdala. Together with several other elements, these structures make up what is sometimes called the limbic system, which encircles the top of the brain stem. The amygdala, hippocampus, and thalamus are paired organs—half of each resides in the left hemisphere, the other in the right. The limbic system manages many basic brain activities, including processing sensory information, expressing emotions, and initiating the so-called fight-or-flight response, responsible for what we modern creatures experience as "stress."

The thalamus, often considered the threshold of consciousness, is the first relay station for sensory input from the eyes, ears, and body, passing a crude translation of those signals on to the separate regions of the cortex that process information from each of the senses. At the same time, though, the thalamus is signaling the amygdala, which plays a key role in emotional memory. This route, shorter and quicker than the path through the cortex, produces an emotional reaction to sights, sounds, tastes, and touches well before the cortex has had time to form a conscious, rational opinion. An even more potent emotional response is evoked through the sense of smell: Signals from the nose bypass the thalamus and go directly to the amygdala, a short cut that explains why the aroma of frying bacon can take you right back to Sunday morning breakfast from your childhood.

The hypothalamus is the commander, as it were, of the body's autonomic functions, producing the physical reactions of anger, sexual arousal, or fear in response to signals that come from other parts of the limbic system and the cerebral cortex. It also controls the endocrine system through the workings of the pituitary gland, regulating growth, sexual development, and a host of functions that, together with the autonomic nervous system, help maintain homeostasis, the body's physiological equilibrium. The autonomic system is itself divided into two counterbalancing branches. One, the sympathetic branch, mobilizes the body's energy and resources in times of danger or for high performance; the parasympathetic branch helps turn off the fight-or-flight response, restoring the body's equilibrium after a threat has passed. When the body is in the throes of arousal, whether sexual or fearful, the hypothalamus triggers a chain reaction that produces racing pulse, flushed face, and a constricted or labored feeling in the chest. As these sensations feed back to the rest of the brain, we feel physical, emotional, and rational interplay.

In recent years, considerable research has shown that prolonged activation of the body's fight-or-flight response, with its increases in blood levels of the stress hormone cortisol, can lower the body's resistance to disease. The so-called HPA axis (hypothalamus-pituitary-adrenal) is the main pathway for the stress response. When the brain perceives stress, the hypothalamus secretes a chemical called corticotropin-releasing hormone, or CRH, that triggers a chain-reaction release of other hormones. CRH stimulates the pituitary gland to send its chemical messenger ACTH (adrenocorticotropin hormone) into the bloodstream, causing the adrenal glands to release cortisol. Cortisol does several things: It increases the supply of blood glucose for energy, especially to the heart and brain; it helps turn fat into energy; and it depresses the reproductive system while activating the immune system in case the body has to cope with injured tissue. All of these are the brain's way of preparing the body to confront danger or run from it. Normally, once the danger has passed, cortisol itself acts in a feedback loop to turn off production of CRH by the hypothalamus, and the body's fight-or-flight response winds down. Blood pressure and breathing return to normal, and the immune system relaxes its vigilance.

But stress in modern life tends to be chronic. We don't have intermittent encounters with lions and bears. We have daily clashes with commuter traffic and demanding jobs. Since the stress response acts within the autonomic and endocrine systems, chronic exposure to cortisol has an effect throughout the body—in the form of increased risk for coronary heart disease, hypertension, diabetes, gastric ulcers, colitis, and asthma. Doctors can't say that stress causes these diseases, but it certainly accelerates the disease process. Recent research also shows that chronic exposure to cortisol can damage the hippocampus, a key component of the brain's memory system, responsible for factual memory. Luckily, research also shows that this damage is reversible; if the stress is not overly prolonged, hippocampal neurons can recover.

On top of and nearly surrounding the brain, the two symmetrical-looking cerebral hemispheres are linked by a band of fibers called the corpus callosum. The symmetry is deceptive. One of the most consistent findings of early brain research was the division of labor between the left and right cerebral hemispheres, with each half controlling the opposite side of the body. This reversal is brought about by a crisscrossing of nerve fibers in the medulla that relay information between the brain and the spinal cord. Handedness is the most obvious proof: About 90 percent of the human population is right-handed, and they're usually right-footed and right-eyed as well, which means their left hemisphere is dominant. Why the brain is not more evenhanded, so to speak, about left- or right-dominance has long been a puzzle to brain scientists.

Many scientists see a possible relationship between the uniquely human tendency toward right-handedness and our equally unique capacity for language—which is also largely a left-hemisphere function. In 1861 Frenchman Paul Broca studied patients who could understand language but could not speak. Thirteen years later Carl Wernicke in Germany studied patients with the opposite problem: They could speak coherently but could not understand speech, not even their own. Autopsies revealed that these individuals all had lesions in one or the other of two separate regions in the left hemisphere, regions now called Broca's area and Wernicke's area. Today, brain imaging has revealed that language is considerably more nuanced than either Broca or Wernicke could guess. Positron emission tomography (PET) scans, for example, have shown that the precise location of brain activity shifts markedly depending on whether the person is hearing words, or reading them, or speaking them. Such findings help explain the patchwork of language difficulties sometimes experienced by people who have suffered strokes or head injuries.

Over many decades, often through studies of patients with head injuries or other brain damage, investigators have found that language and a host of other functions are dispersed to different regions of the cerebral cortex, the deeply fissured outer layer of the cerebrum that not only controls our sensations and voluntary movements but also stores our memories, makes our decisions, and conceives our every thought. Conventional brain mapping divides the cortex into four lobes in each hemisphere. The frontal lobes, just behind the forehead, handle complex mental activities such as language and decision making. They are considered home to our personality, the source of spontaneity as well as impulse control, especially in areas of social and sexual behavior.

The temporal lobes, near the temples, deal with memory, expression, and hearing. Injuries to the left temporal lobe impair our ability to recognize words, while right temporal lobe injuries interfere with the ability to talk. Patients with epilepsy who suffer seizures in the temporal lobes often become paranoid and erupt in aggressive rages. The parietal lobes, at the sides of the brain, are responsible for sensory perception from different parts of the body. When the parietal lobes are damaged, we lose a number of abilities we take for granted. We can't integrate visual and tactile input, for example, to reach for and pick up a glass on the table in front of us. Some parietal lobe injuries cause patients to stop sensing part of their body, such as the right arm. The occipital lobes, at the rear of the brain, are largely devoted to vision. Due to their location, they are less vulnerable to head injury than other parts of the brain, but damage here can result in hallucinations as well as the illusion that objects seem larger or smaller than they are.

NEURONS: THE GREAT COMMUNICATORS

Communication between and among cortical areas and the rest of the brain is complex, even when we're just staring out the window. Sensory information forwarded by the thalamus (or, in the case of smell, by the olfactory bulb) goes first to three primary cortical areas—the visual, auditory, and sensory cortices. So, for example, when we hear the alarm clock in the morning, hair cells—sensory cells in the inner ear with bundles of hairlike projections that bend in response to sound waves—convert the mechanical force of the alarm into electrical impulses that are carried by the auditory nerve to the neurons, or nerve cells, in the primary auditory cortex. These neurons relay the signals to their cousins in higher-order cortical areas where the sound is interpreted (yes, that *is* the alarm) and then passed on to the association areas of the cortex. These areas rope in information from parts of the brain such as the limbic system that store emotions and memories, bringing to bear a lifetime's habits and experiences with getting up in the morning. Meanwhile, the neurons in the association cortices are integrating sensory inputs from other parts of the body—hunger, perhaps, or a subtle ache from yesterday's workout—and the wispy remnant of a dream. After some back and forth with the decision-making frontal lobes (get up? Or snooze ten more minutes?), signals are sent to the premotor cortex, which in turn relays orders to neurons in the motor cortex for execution. The eyes open, the arm reaches out, the hand locates the alarm, and the fingers find the proper button—each action in turn requiring its own neuronal concert.

The cells that make this possible, from the smallest finger twitch to the most profound philosophical insight, are unique to the brain and central nervous system. Nerve cells communicate with one another, ceaselessly sending and receiving electrochemical messages. Out of this incessant chatter arises not only our ability to see and hear, walk and talk but also the quintessentially human attributes of self-awareness, abstract thinking, and creativity. Scientists believe that the first neuron appeared in animals some 500 million years ago, about three billion years after the initial appearance of the DNA molecule. Both events represent quantum leaps in evolution. Researchers have found what seems to be a direct correlation between the cognitive power of a brain and its quantity of neurons and the number of connections between them. A sea snail, for instance, has a mere 20 thousand brain cells; a fruit fly has 100 thousand, and a mouse has 5 million. In comparison, the brain of a monkey has roughly 10 billion neurons. This may sound like a lot of brain power, but it pales beside the potential housed in the brain of a newborn human, which has about 100 billion nerve cells. A baby's billions of

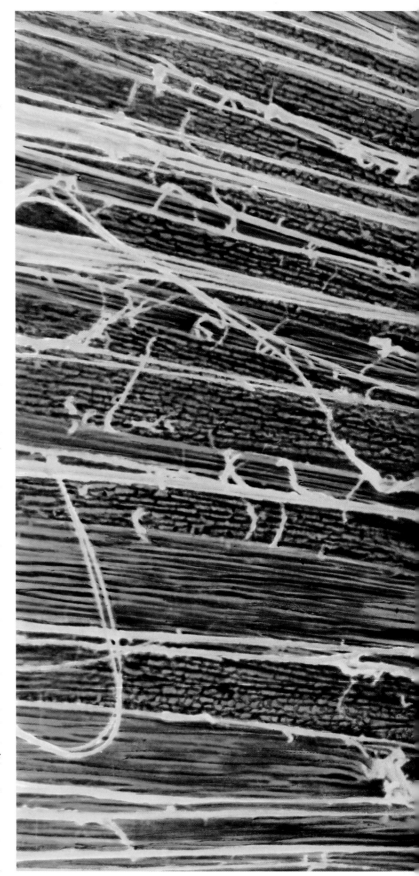

....

Microstructure of the eye includes a muscular iris (purple) ringed by ciliary cells (red) that produce nourishing fluids. Filaments suspend the unseen lens. Like all sensory organs, the eye sends information to the brain for interpretation.

....

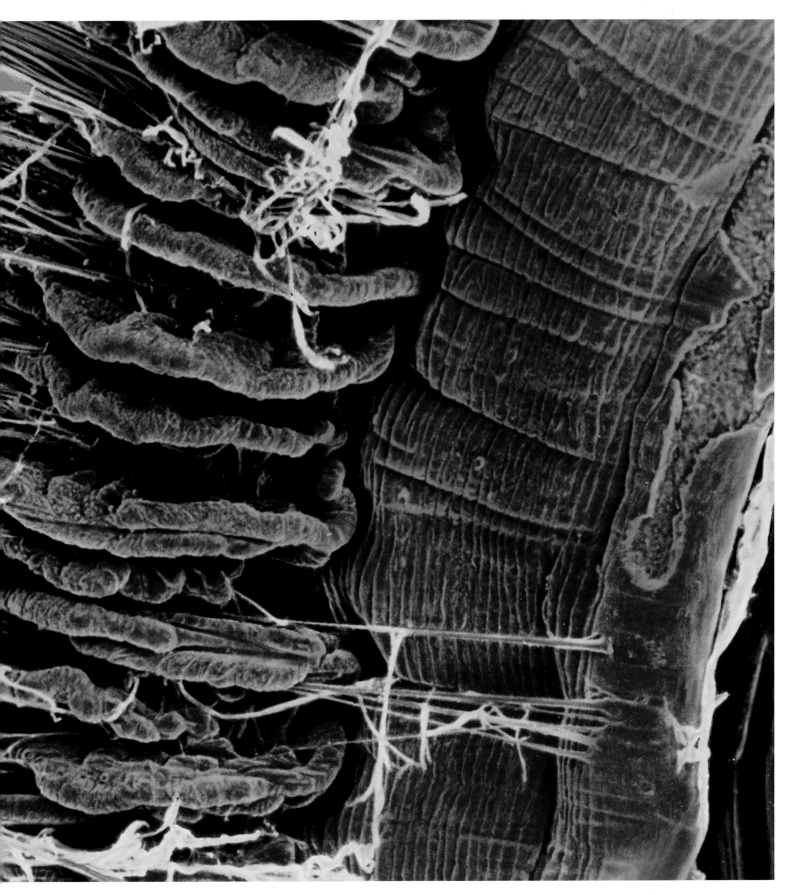

neurons are supported by about a trillion cells known as glial (for glue) cells. Until recently scientists assigned a relatively passive role to glial cells, which account for about 90 percent of the cells in the human brain. Now scientists know that glial cells may be vital for the correct wiring of the brain. They form the highway along which nerve cells migrate in the developing brain as they move into their final positions. In addition, research suggests that glial cells produce a substance, called glial factor, that strengthens the lines of communication between neurons. Without glial cells or glial factor, the transmission of messages between nerve cells can short-circuit or become inefficient.

Efficiency of transmission is essential. Each of the brain's billions of neurons makes an average of about 10,000 to 15,000 contacts with other cells; some of the cerebellum's neurons may communicate with as many as 200,000 others. Neurons release chemical messages to each other, triggered by electrical impulses. The impulses travel from each cell body down a single, specialized fiber called the axon, which may have multiple branches if the neuron is linked to many target neurons. In some cases glial cells wrap around the axon, insulating it with myelin, a protein that helps speed message transmission. Axons vary considerably in length; some are a tenth of an inch long, but those that activate arm and leg muscles may travel more than three feet. Given the billions of neurons in the brain, there are probably something on the order of three million miles of axons in the body. On the receiving end of the message are fibers called dendrites, which form a kind of fringe around the cell body. Each dendrite in turn may have a few branches or many branches, with multiple receptor sites, allowing it to receive signals from numerous other neurons.

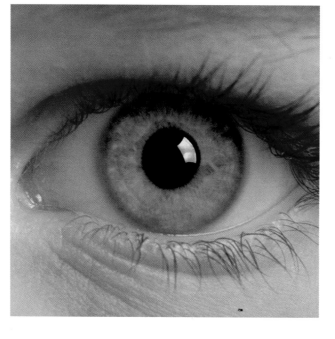

NEUROTRANSMITTERS

The actual communication between cells takes place across a small gap called the synapse, or synaptic cleft. The terminal end of the sending axon releases chemical molecules known as neurotransmitters into this synaptic cleft to be picked up by receptors that are each tuned to a specific neurotransmitter. Glial cells located at synapses help this exchange take place, and they absorb excess neurotransmitter molecules. Neurotransmitters can either excite or inhibit, and a given target neuron might receive a host of these conflicting messages at once. The target neuron fires, initiating a nerve impulse of its own, only when the sum of the excitatory messages exceeds a threshold.

When neurotransmitters bind to receptors on the outside of a cell, a cascade of information flows into the cell body and on into the cell nucleus, with its cache of DNA subdivided into 60,000 to 80,000 genes. This is where the work of the cell takes place. Each gene contains the information to make one protein, although some make more than one. Proteins are the crucial building blocks of cells: The cells are shaped by structural proteins; receptors on the outside of cells are proteins; some neurotransmitters are themselves small proteins; and enzymes, which control the chemical reactions in cells, are also proteins. Given that every cell in the body

The eye's colorful iris adjusts the size of the pupil to admit more or less light, which the lens focuses on the retina; retinal rod cells (opposite) react to light and, with color-detecting cone cells, forward information to the brain via the optic nerve.

contains the same 60,000 to 80,000 genes, each cell needs to turn on just the right genes in the right cell. Bone-marrow cells should make hemoglobin, for instance, and fingernail cells should make keratin, and not vice versa.

When the proper genes in brain cells are activated, they direct the synthesis of proteins whose work, in the long run, is to prune or build synapses. In this way the connections between neurons are strengthened or weakened, determining the distinct pathways in the brain that are responsible for everything from vision, hearing, and movement to language and memory. This modifying of synaptic connections is what enables people who have suffered the loss of an arm to learn to use the other one and allows stroke victims to regain control over movements and recover the ability to speak and understand.

The plasticity of the brain not only makes all forms of learning and relearning possible but also holds the key to unlearning behaviors and strategies that are deleterious to the body and brain, such as drug addiction. A deeper understanding of what goes on at the molecular level as the brain makes these changes can help neuroscientists fine-tune their treatments and therapies for various

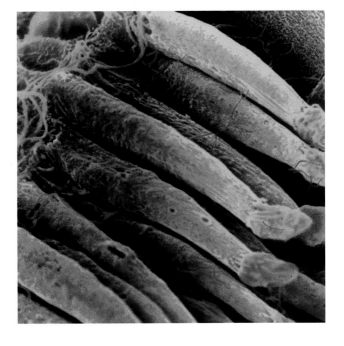

illnesses and disorders such as manic-depressive illness, post-traumatic stress disorder, and Parkinson's and Alzheimer's diseases.

These are just a few reasons researchers have been excited by two relatively recent discoveries. One is that some neurons communicate using several different neurotransmitters. The other is that a given neurotransmitter, rather than being matched in strict lock-and-key fashion to only one or two receptor proteins, may work with anywhere from five to several dozen or more. This chemical diversity, unknown only a few years ago, means that brain cells are capable of a much greater subtlety of response than was previously suspected. We knew, of course, that our senses, for example, do not report predetermined, calibrated readings of the world. Rather, our eyes, ears, taste buds, nose, and fingertips are capable of distinguishing exquisitely nuanced differences in color, sound, flavor, odor, and texture.

The same nuances are reflected in our day-to-day emotional responses—which might explain why sometimes we feel both happy and sad at the same time—and in the differences in basic temperament between any two individuals. So far, researchers have found five different receptors for the neurotransmitter dopamine, which, among other things, seems to play a role in the debilitating disorder known as schizophrenia. They have found six different receptors for norepinephrine and dozens for GABA (gamma-amino butyric acid), both of which are players in anxiety disorders and depression. Some 15 receptors have been found for serotonin, which is involved in depression. Scientists estimate that receptors for glutamate, the primary excitatory messenger in the brain and critical to learning and memory, may number more than a hundred, with each receptor controlling functions as divergent as digestion and sexual interest.

The diversity of our biochemical makeup also explains some of the well known but unintended effects of the current crop of psychiatric medications. Most of these drugs act indiscriminately, lessening or increasing activity at all the receptors for a given neurotransmitter. Side effects such as stomach problems often result. As more receptors are identified, the fervent hope is that the next generation of pharmaceuticals will be able to target the specific receptors that control the symptom in question, making side effects a thing of the past. Much basic research is necessary, however. As Steven Hyman, director of the National Institute of Mental Health has cautioned, "We have to integrate everything from molecular biology and genetics to the study of people's behavior."

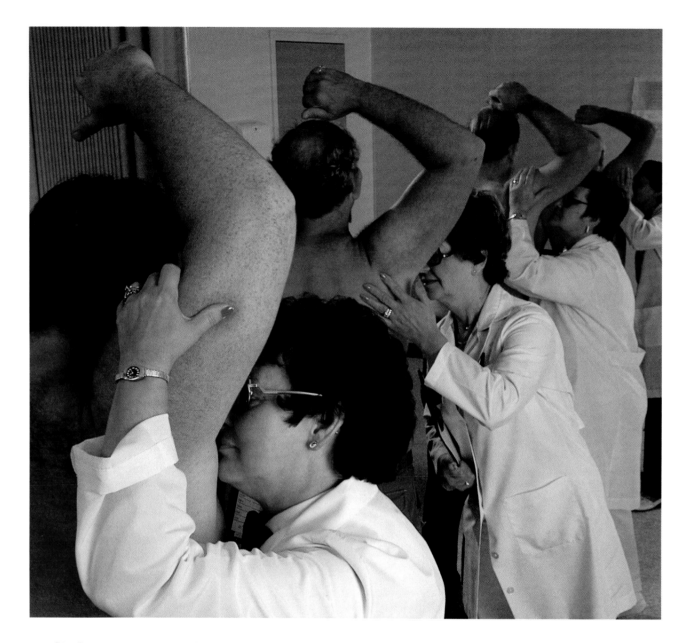

Similar cautions are applicable to another recent finding in the effort to understand how neurons work. Research reported in early 1998 strongly suggests that, contrary to long-held belief, the human brain is capable of growing new brain cells after adulthood. If this proves true, and if ways can be found to stimulate new cell growth in specific parts of the brain, therapeutic vistas open up for victims of many neurodegenerative diseases, such as Alzheimer's, as well as for victims of brain injury due to trauma or stroke. It also means, however, that neurobiologists must reevaluate conventional notions of how the brain changes in response to learning and life experiences, and must rethink where—and how—memory is stored in the brain.

Studies have demonstrated that neurogenesis, as new nerve cell growth is called, occurs in parts of the brains of birds, mice, and rats. These instances of nerve regrowth were largely regarded as anomalies, however, and not applicable to the human condition. Now scientists have found evidence that neurogenesis occurs in the brains of monkeys, specifically in the hippocampus, a structure integral to many functions of memory and emotions. The assumption is that similar growth is likely to be found in human brains. The question closest to the hearts (and minds) of aging baby boomers concerned about ordinary memory loss is: How do you stop it? The simplest answer, for the moment, is advice that neuroscientists have been giving for years: Use it or lose it. Adult mice kept in a stimulating environment replete with toys, exercise wheels, and tunnels to explore develop up to 15 percent more neurons than those in boring environments. And everyone knows of men and women who remain mentally sharp into their 80s and 90s, most of whom keep active both mentally and physically.

NEURAL PLASTICITY

The importance of environmental stimulation cannot be overstated. But the brain has to start somewhere, and neural plasticity goes to work in the womb. The brain starts working months before it is finished. During gestation, fetal brain cells multiply at the rate of about 250,000 per minute as the brain and spinal cord assemble themselves. Research suggests that these cells generate pulsing waves of electrical activity that physically shapes the structure of the brain even as it is creating itself. Programmed by some 30,000 genes, neural cells lay the brain's foundations, traveling great distances to put in place the wiring that will connect one part to another. For example, the billions of cells destined to form the cerebral cortex, a structure that comes late in the development of the human brain, must shove their way through dense clumps of cells already formed—a migratory mass journey that is staggering to contemplate.

Although nature plays the dominant role at this point, the environment is—as always—a vital factor. Maternal malnutrition or drug abuse, for instance, can derail cell

assembly lines, resulting in forms of epilepsy, mental retardation, autism, and schizophrenia. Even small variations in the timing of genetic activity can have huge repercussions. For example, the human cortex, the thin layer responsible for higher thought processes, has ten times more neurons than a monkey's cortex has—a difference that accounts for the fundamental intellectual capacity of the human species. That tenfold difference is made possible by a crucial three-day delay in the activation of one gene, whose task is to trigger the movement of neural cells from the tissue where they are created to their intended position. The delay allows more cells to be born before the migration occurs. Despite all that can go wrong, everything goes right most of the time, and an infant comes into the world with its full complement of a hundred billion neurons. At that point, however, the brain's work has barely begun.

These billions of cells are not yet completely functional or properly positioned. During the first several years of life, the infant brain quadruples in size as it generates tens of thousands of connections to each neuron, which in turn develop axons and dendrites tying those neurons together with others in a tangle of interconnections that number on the order of a hundred quadrillion (10^{17}). This remarkable process occurs as the brain uses the electricity generated by environmental stimulation—sights, sounds, smells, tastes—to activate the neural cells and organize them into the necessary systems. Shapes and colors, the emotional and physical give-and-take of every mealtime and every playful game of patty-cake—all contribute to the business of brain construction.

In one layer of the visual cortex alone, a study at the University of Chicago shows, the number of synaptic

Hate your job? Try sniffing armpits to test deodorants, as these women are doing (opposite). Smell-sensing cilia adorn millions of receptor cells in the upper nose; rat cells (above) contain more than 20 cilia each, at least 8 more than human ones.

connections rises from about 2,500 per neuron when the baby is born to perhaps as many as 18,000 per neuron by the time the child is six months old. Not surprisingly, another study has shown that children who are rarely touched, or who don't have opportunities to play, develop brains 20 to 30 percent smaller than normal for their age.

Supporting evidence from work with laboratory animals demonstrates that early stimulation in an enriched environment results in dramatic alterations to brain structure—greater branching of nerve cells, increased numbers of glial cells, thicker capillaries, and a thicker cortex. The key to sustaining such development is novelty. When one researcher didn't change the toys in her experiments soon enough, new synaptic connections that had formed in her lab animals withered as quickly as they had taken shape. Of particular interest for human development, these same experiments showed that nurturing social relationships were critical, too. Animals that received enrichment training but didn't have a rich social environment showed no increased brain growth.

Although the microscopic connections of the human brain continue to form throughout life, they reach their highest average densities, roughly 15,000 synapses per neuron, when a child is about two years old. For the next six years or so a child's brain vastly outpaces an adult's, with twice as many neurons and twice as many connections between them. This is evolution's way of making sure that the developing brain will be able to adapt to whatever eventualities life may throw its way.

When a person reaches the age 10 or 11, however, the brain begins ruthlessly pruning its most underused and therefore its weakest synapses. These will die off at the rate of thousands per second. Only connections that have been repeatedly exercised will have turned on the genes that direct the synthesis of powerful synapse-preserving growth factors.

Thus, by the time we become old enough to vote, our interests, habits, and experiences have sculpted patterns in the brain—patterns of talents, tendencies, emotions, and thoughts so unique that no two individuals, not even identical twins raised together, are likely to perceive the world in precisely the same way.

Cacophony of kindergartners can be painful to the ears—even theirs. Sensitive both to faint whispers and to the clash of heavy-metal bands, these organs transform sound waves into electric signals for processing by the brain.

THE BRAIN AND EMOTIONS

Genetic inheritance and life experiences ensure that each person perceives the world in a different way. The unique wiring of our brains—the particular patterns of strong connections and weak ones—also influences our responses to people and events. Nowhere is this more evident than in our emotional reactions. On the one hand, emotions are at the core of who we are as individuals. On the other hand, emotions link us to the rest of the human species. They are what make it possible for a theater full of people to cry during one scene in a movie and cheer during another. Indeed, someone who seems to be unfeeling is regarded as "inhuman," a veritable Mr. Spock of *Star Trek*. At the same time, though, emotions

such as choosing a mate to making mundane selections at the supermarket—they draw on something beyond rational thought. But getting at that "something" has been a difficult challenge. Part of the problem is that, in effect, "everybody knows" what emotion is—until they're asked to define it. One productive approach has been to let the brain speak for itself, as it were, rather than trying to define emotions in the abstract. Advances in animal research and pharmacology along with high-tech brain scanning techniques are allowing scientists to begin decoding what goes on when we hug someone we love—or when we think we see a snake in the path.

Presumably, different emotions are activated by neural systems that evolved for different purposes. Joseph LeDoux of New York University's Center for Neural Science, has chosen to study the fear system, in part because it is at the root of the so-called anxiety

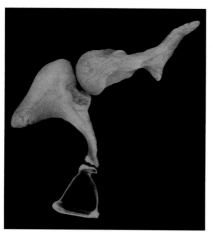
BONES OF THE EAR

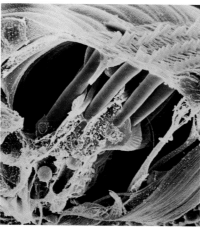
ORGAN OF CORTI, INNER EAR

INNER EAR'S VESTIBULAR APPARATUS

are often regarded as "irrational," as Spock would say, something to avoid when required to make a "rational" decision. Ancient philosophers categorized emotional passions and desires as "of the body," part of our more primitive animal nature, which ought to be controlled by the higher-order, presumably more human, "mind," with its capacity for thinking, planning, evaluating.

Clearly, emotions are normal and must be a crucial element in any effort to understand how the brain works. Psychologists have long known, for instance, that when people make decisions—from enormously important ones

disorders—panic attacks, phobias, obsessive-compulsive and other disorders—that afflict more than 23 million Americans. These often debilitating illnesses make up about half of all the psychiatric conditions treated every year, not including substance abuse problems. Increased understanding of how the fear system works has given researchers insights into the specific brain circuits that are malfunctioning and offers the possibility of finding better therapies.

Species ranging from our primate cousins to invertebrate sea slugs have some rudimentary fear system. The

Smallest bones of the human body, the malleus (hammer), incus (anvil), and stapes (stirrup) transmit vibrations from the eardrum to the fluid-filled cochlea. There, hairs in the organ of Corti convert motion to electrical nerve pulses. The cochlea also harbors the vestibular apparatus, which provides our sense of balance (opposite) by working in concert with the brain.

actions humans take and the way we respond when we're afraid are very similar to the behaviors and physiological responses of other animals, so we can study animal behavior as well as the neurological processes in animal brains to understand how the human fear system works.

"Fear is a relatively traceable emotion, unlike love or hope which are difficult to pin down," LeDoux says. For example, when we hear an unexpected bang, we jump. Researchers call it the startle reflex, which is immediately followed by a momentary freeze. Freezing is nature's way of buying time: Predators generally respond to movement, so if we stay still, we might live long enough to figure out whether to run, to stand and fight, or to realize the noise was only a car backfiring. To study fear, says LeDoux, it is easy to create experiments in which a simple stimulus that warns of impending danger triggers a set of responses in a laboratory animal, such as a rat, that are similar to the kinds of responses a human facing danger would make. By tracing the flow "from the stimulus processing pathways to the response control networks," LeDoux says, "it's possible to identify the basic neural circuits involved."

In addition to the startle and freeze reflexes, other things are going on automatically in the body that are part of the fight-or-flight response. The hypothalamus activates the alarm phase, causing the adrenal glands to secrete epinephrine, which accelerates heart rate, and norepinephrine, which raises blood pressure. Breathing quickens; digestion shuts down. If the threat continues beyond a few seconds, the hypothalamus prepares the body for the stress response by producing corticotropin releasing hormone (CRH) to trigger the chemical cascade that culminates in the stress hormone cortisol. In short order, the autonomic system has readied the body to move quickly and to suppress its reaction to pain. All these fear responses, from the freezing response to changes in heart rate and blood pressure, are evident when a rat sees a cat, and all are easily measured—an essential component of research.

Over several years, LeDoux and his colleagues ran painstaking experiments to pinpoint which parts of the brain are involved in producing the fear response. First, rats were conditioned to associate a particular sound with a mild foot shock. Within very few tries, the sound alone produced the fear response. Then, to determine which parts of the brain are essential to producing the fear response, the scientists made a series of lesions, small holes in the brain, to interrupt the flow of information

between one neuron and another, or between one set of neurons and another. By systematically blocking various brain pathways, scientists were able to determine whether damage to a given structure did or did not interfere with fear conditioning.

The normal pathway for processing sound in the brain is well known. Sound comes into the ear, sparking nerve impulses that enter the brain, make their way up to a region called the auditory midbrain, then to the auditory thalamus, and ultimately to the auditory cortex, the highest level of processing. What scientists wished to learn was whether a signal must travel all the way to the auditory cortex in order for an animal to know that the sound meant danger. Does an unconditioned animal need its cortex to learn to associate the sound with the shock? Secondly, is the cortex required for a conditioned animal to react to the sound alone? Surprisingly, researchers found that damaging the cortex did not interfere with learning the conditioned response to the sound. No higher processing is required.

Since sound gets to the brain by way of the auditory system, and the fear response is enacted by way of the autonomic system, where do the two link up? Eventually, the key player emerged: the amygdala. Researchers had known from animal studies that this small, almond-shaped structure is important in emotional responses but beyond that, not much was known.

Now it appears that the amygdala receives information about the outside world directly from the thalamus and immediately sets in motion a variety of bodily responses. Indeed, if the amygdala of a rat conditioned to pair the sound and foot shock is damaged, the animal's fear responses to hearing the sound alone are greatly reduced. Its blood pressure and freezing response, for instance, are virtually no different from those of rats that have not been conditioned to fear the sound. The conditioned animal has been made to forget what it has learned. Presumably similar emotional remembering (and, potentially, forgetting) goes on in the human brain.

LeDoux has nicknamed the thalamo-amygdala pathway the "low road" because it acts without waiting for information to be processed by the cortex, which also communicates with the amygdala. Studies of human patients with damaged amygdalas by neuroscientist Antonio Damasio of the University of Iowa corroborate the importance of this structure in emotional learning.

Researchers have used fMRI imaging to determine whether the amygdala is activated in response to emotional stimulation even when the subjects are not consciously aware of that stimulation. In one experiment pictures of different facial expressions were shown to the subjects. Fearful and happy faces were masked by longer viewings of neutral faces; the subjects reported seeing only the neutral expressions. But fMRI scans revealed that even though the subjects weren't aware that they had seen any strongly emotional faces, their amygdalas had heightened activity whenever fearful faces were presented and decreased activity when happy ones were shown.

This discovery explains how we can begin to respond to an emotional stimulus even before we are conscious of it. Say you have a big argument with a coworker one day that leaves you shaken and upset. The coworker happens to be wearing a red-and-white checked shirt. Several days later you walk into a new restaurant and suddenly find yourself back in the emotional state you were in during the argument. It makes no sense, so you attribute it to having to wait too long. Actually, however, you are reacting on a nonconscious level to the restaurant tablecloths—those red-and-white checks are a visual stimulus that your brain has paired with the highly charged emotion of the argument with your coworker.

. . . .

KINDS OF MEMORY

. . . .

This is an example of an emotional memory, an enduring effect of neurological and physiological responses to an experience. Such memories are stored in the brain as a result of that experience, then reactivated—unconsciously—in the future. As such, they are the likely sources of "gut feelings," hunches, and intuitions—reminders from the unconscious that can serve us well or ill, depending on the nature of the original event in the past and of the triggering stimulus in the present.

People often advise one another to "go with your gut," but we might do well to take that advice with a grain of salt in terms of day-to-day reactions. When we avoid a restaurant because of unpleasant associations with red-and-white checked fabric, our only loss may be a potentially fine meal. When we take an instant dislike to someone because of unpleasant unconscious associations with, say, mustaches, we may be writing off a potential good friend. If the mustached stranger is our new boss, we're setting ourselves up for a difficult relationship.

Still, acting on the advice of our emotional memories is probably the rule rather than the exception. As Damasio has put it, "Emotions and feelings are not a luxury. They are a means of communicating our states of mind to others. But they are also a way of guiding our own judgments and decisions. Emotions bring the body into the loop of reason." Damasio has studied brain-damaged patients who seem to have lost the ability to make good decisions. With colleagues he set up an experiment using four decks of cards—decks A and B were "bad"; decks C and D were "good." The players had to turn over cards in any order with the aim of winning as much money as possible and keeping their losses to a minimum. Turning over most of the cards carried a reward—$100 in decks A and B and $50 in decks C and D—but turning over some cards carried a penalty. Penalties in high-dollar decks A and B were larger than in decks C and D. The "good" decks, C and D, were rigged to yield a higher gain in the end but smaller rewards along the way; the "bad" decks were rigged to give big immediate earnings but an overall loss. The players could not predict when penalties would arise or determine with any precision what the net gain or loss would be from each deck.

During the game players' palms were wired to a device that detected changes in electrical conductance of the skin. Ten normal subjects reached what Damasio calls a "pre-hunch" stage very quickly. By about card ten the normal players were getting sweaty palms whenever they considered drawing a card from decks A and B. By the time they had turned over 50 cards, all the normal subjects were expressing a hunch that two decks were good and two were bad. The six other subjects, patients with damage in both hemispheres to a region of the cerebral cortex just behind and above the eyes, never reached the stage of being able to express a hunch, nor did they generate anticipatory sweaty palms before turning over cards from the bad decks. However, half of them did, at some point, realize that there were "good" decks and "bad" decks—even though they continued to choose cards from the bad decks. Such counterproductive behavior is analogous to the kind of decisions that many brain-damaged patients make in real life, choosing poorly in marriage or in financial matters, for instance.

Damasio and his team suggest that the missing decision-making element among cortically-damaged patients is an inability to draw on memories based on past emotional learning. Damasio believes the region studied— the ventromedial prefrontal cortex—is part of the system

that triggers nonconscious emotional memories. Normal people draw on these nonconscious memories to make decisions they think are based on intuition or hunches.

Viewed through the long lens of evolution, it is good that we have developed the sort of neural circuits that prepare us to pick up on the briefest of environmental cues; quick reactions keep us alive. A stick on the path might be a snake: The amygdala says, better to react to the stick as if it were a snake than to a snake as if it were a stick; if the alarm is false, we can turn it off. But the amygdala, important as it is, does not have total control of what we learn and remember. The hippocampus, which lies deep within the temporal lobes, also communicates with the amygdala and recognizes, for example, that you needn't fear a snake in the zoo as much as you might dread coming upon one loose on your path in the woods.

The hippocampus in particular, as well as other parts of the brain, are critically involved in the formation of long-term memory, the stable memory that lasts for days, weeks, and even a lifetime. By contrast, short-term memory, also known as working memory, lasts only minutes. Moreover, as much research has shown, there are many different kinds of memory. The two major systems seem to be "knowing how," or knowledge of motor skills, and "knowing that," knowledge of facts and events. Remembering how to ride a bicycle or play golf are examples of knowing how. This type of memory, also called implicit or non-declarative, does not require conscious awareness. Rather, it is expressed through performance: We can pedal a bike without having to consciously recollect how to move our legs. By contrast, remembering the year of the Norman Conquest or a

From feminine nurturing to boyish bravado, gender differences exist—and the brain plays a role. While psychologists long have argued nature versus nurture, neurologists are beginning to find structural and functional differences between the brains of both sexes. In other words, as parents have known for a long time, boys and girls ARE different.

name from the past—examples of explicit or declarative memory, of "knowing that"—require focusing attention to bring the relevant fact or idea to conscious recollection.

Each of these major memory systems can be further subdivided. Declarative memory has two subsystems. Semantic memory deals with general knowledge of facts and vocabulary (the capitals of all 50 states and how to recite them in proper English). Episodic memory is what lets you tell stories about your life ("I remember the year I lived in Oregon"). Non-declarative memory includes such things as classical conditioning of the sort Pavlov did with dogs, as well as skills, like playing the piano, and habits, such as knowing which way to turn a key.

People who have suffered brain damage through accident or illness often retain some kinds of memory while losing others. One of the most dramatic cases

After the operation, H.M. had reasonably good long-term memory—that is, he could remember much of his childhood and recall other events prior to the operation. He could speak English coherently and without difficulty, and his overall intelligence was essentially untouched. H.M.'s immediate memory also seemed fine in that he could repeat a new telephone number or the name of someone he was introduced to directly after being told, and he could carry on a normal conversation, provided it didn't last too long or switch topics too often.

What H.M. lacked, to an overwhelming degree, was the ability to put new information into long-term memory storage. He appeared to forget, less than an hour after a meal, that he had eaten. He would read the same magazine over and over, each time as if it were new. He

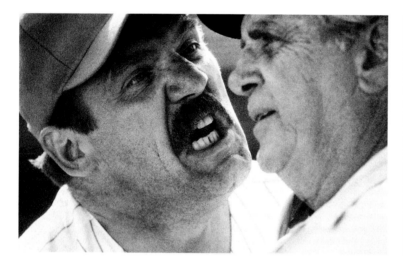 

involved a patient known as H.M., who in 1953, at the age of 27, underwent an experimental surgical procedure to relieve incapacitating epileptic seizures. The surgery removed parts of both sides of the medial temporal lobe including the anterior hippocampus. His seizures were strikingly reduced, but H.M. was left with a devastating memory loss. Still alive today, he cannot convert a new short-term memory into long-term memory.

Psychologist Brenda Milner, of the Montreal Neurological Institute, has followed H.M. for more than 40 years, carefully recording which aspects of memory function were lost and which retained.

did not recognize anyone he was introduced to after his operation, even when he met them repeatedly. These findings indicated that the hippocampus, although it is not where long-term memories are stored and is not where short-term memory takes place, seems crucial to the ability to create one out of the other.

However, Milner discovered that even this inability to convert new information from short-term memory to long-term memory was not absolute. H.M. could learn new motor skills (the skills and habits of nondeclarative memory) that were not dependent on awareness or cognitive processes. For example, he could trace the

All states of mind—from in-your-face rage to pensive reflection to unadulterated joy—link the thinking centers of the cerebral cortex with the more primitive limbic system, center of emotions and of the basic drives for food, sex, and survival.

outline of a star while looking at a reflection of his hand in a mirror rather than looking directly at what he was doing. From one day to the next, H.M. never remembered doing this exercise before. But his performance improved daily, as would that of a normal subject.

Milner's painstaking work with H.M. provided the first evidence that the brain has more than one memory system. A breakthrough in its own right, the finding also underscores the extraordinary challenge researchers face in trying to map the neural circuitry that enables a child to learn to read or an octogenarian to remember details of her wedding day 60 years before, or that triggers a nightmarish flashback in a war veteran or survivor of child abuse. Understanding that the amygdala is key to emotional memory and the hippocampus crucial to making the switch from short- to long-term storage are essential starting points, certainly, but how to discern, among the brain's 70 trillion synaptic connections, which are strengthened and which weakened to encode a given event or experience? And what determines whether a given memory is permanent or fades?

Studies have shown that an animal conditioned to respond fearfully to a sound eventually will not respond if it hears the sound enough times without the unpleasant event (the foot shock) occurring. Its fear response will be extinguished because the prefrontal cortex is providing the trigger-happy amygdala with current information that the unpleasant stimulus isn't happening anymore. But extinction isn't necessarily permanent. Weeks later the animal might suddenly respond fearfully again. Human phobias are similarly resistant to extinction, and stress can trigger a relapse even if some improvement has occurred. This suggests that therapy and the extinction process don't eliminate emotional memory but simply allow the prefrontal cortex to inhibit the amygdala's output, holding that memory in check. Indeed, animal studies show that if the prefrontal cortex is damaged, the fear response is difficult, if not impossible, to extinguish even temporarily.

The fact that the emotional residue of events that can elicit a fear response is nearly impossible to extinguish would seem to give credence to the controversial notion of victims of child sexual abuse somehow repressing and later recovering memories of their traumatic experience. Memories often can be altered, intentionally or not, by suggestion, particularly in the case of young children. Furthermore, some researchers argue that very young children, in whom the hippocampus is not yet fully developed, cannot form factual and contextual memories of traumatic events even though the emotional aspect of the trauma may be registered by the amygdala. Research also has shown that during intense stress, the flood of the stress hormone cortisol released into the bloodstream has opposite effects on the hippocampus and the amygdala. Cortisol affects the hippocampus adversely, interfering with its job of converting information from short-term to long-term memory. In fact, if the stress continues long enough, some studies suggest, neurons in the hippocampus begin to die off, and the structure itself begins to shrink. At the same time, however, the cortisol overdose seems to help the amygdala register emotional memories even more efficiently than usual.

Thus, Joseph LeDoux, for one, argues that if the hippocampus is unable to record an event (whether because of stress or because the person was too young to have a fully formed hippocampus), even if an emotional memory is seared into the amygdala, it is not possible to "recover" any kind of factual information. "It's not possible to take a memory that was not coded through the hippocampus and turn it into a hippocampal memory," LeDoux says. "The amygdala does its business, the hippocampus does its business. They communicate with each other, but their coding and representation is different." If the hippocampus is able to make a weak, partial record of the event, LeDoux cautions, the memory might be retrievable under certain circumstances, if, for example, the person were placed in a similar emotional state. As Daniel L. Schacter, author of *Searching for Memory*, notes, the issue has "many intermingled parts that need to be disentangled. Although it is likely some therapists have helped to create illusory memories of abuse, it also seems clear that some recovered memories are accurate."

MEMORY'S WORKHORSE

New insights into these and other thorny problems of memory and cognition are likely to emerge as scientists gain a better understanding of working memory. This short-term storage is what we draw on when we look up a phone number and remember it long enough to dial, or read and understand a sentence, or follow through on a previously decided plan of action. It is also where we hold information temporarily copied from long-term memory, as when we bring to mind something we previously learned—the capital of Indiana, for instance, or the name of the movie we saw last week.

Without working memory we lose the ability to reason and make judgments that require remembering information and its context. Scientists believe, for example, that defects in working memory are at least in part the cause of schizophrenia; people with this mental illness suffer from hallucinations and delusions, loose associations, inattention, and often wildly disordered thought processes. Understanding which parts of the brain are involved in working memory and how the whole system operates may offer avenues for treatment of patients with schizophrenia as well as for victims of Alzheimer's disease. Scientists still don't know, for example, if the memory problems associated with Alzheimer's patients stem from disruptions of the brain's ability to retrieve memories or of its ability to encode and store memories in the first place. Both functions involve what now appear to be different aspects of working memory, one for retrieving memories from long-term storage, the other for encoding new information.

Efforts to unravel the puzzle of working memory began some 40 years ago when researchers working with monkeys used electrodes to record the activity of individual neurons in the animals' brains. Not until the recent advent of positron emission tomography (PET) and functional magnetic resonance imaging (fMRI), however, have investigators had a tool that could capture the human brain's fleet signals in anything approaching real time. With fMRI scientists have not only demonstrated that the prefrontal cortex is the home of working memory but also that working memory itself has subdivisions: That is, different circuits handle working memory for different types of information.

Not everyone agrees on the organizing principle of such differentiation, however, and the evidence is far from clear-cut. Some studies seem to suggest that these subregions of the prefrontal cortex are organized to temporarily store information about different sensory domains—the appearance of an object, for instance, as opposed to its location. In a series of experiments led by Leslie Ungerleider, chief of the Laboratory of Brain and Cognition at the National Institute of Mental Health, subjects were scanned while three faces flashed briefly in different spots on a screen. The subjects were asked to remember either the locations of the faces or their

identities. After a nine-second pause a face appeared somewhere on the screen for a few seconds. In one test, for spatial working memory, subjects pushed a button if the face was in one of the three locations they had seen previously. In the other test, for non-spatial working memory, they pushed a button if the face matched one of the previous three. Subjects performed a simple eye movement as a control test, allowing researchers to confirm the location of an area known as the frontal eye field, located in the prefrontal cortex. In the test for spatial working memory, the region that seemed to remain active during the nine-second pause was located just in front of the frontal eye field; this region had never before been distinguished in the human brain. Clearly differentiated from this newly identified area for spatial working memory—in this study—was a region in the lower left frontal cortex that showed sustained activity during the pause to remember the identity of the faces.

Other researchers, however, believe a two-tier hierarchy exists and that a region known as area 46 is a kind of central processor for all working memory, regardless of the type of task. According to this view, an area in the lower prefrontal cortex retrieves information from long-term storage elsewhere. Area 46 monitors the brain's processes and enables it to keep track of several items or events. If you had various projects on your plate at work, area 46 would let you remember which ones you had completed and which were left to do.

This facility at juggling mental "balls" may be a key ingredient of intelligence. Studies over the last two decades have found a high correlation between working memory capacity—how many different things a person can keep in mind at one time—and skill at solving both verbal and nonverbal problems. In a recent imaging study by John Gabrieli of Stanford University, subjects' brains were scanned while they completed tests that required them to figure out rules governing a series of geometric designs. Investigators found that this task activated most of the same regions seen in the studies of spatial and non-spatial working memory, strongly suggesting that, as one researcher put it, "Reasoning seems to be the sum of working-memory abilities." Leslie Ungerleider has put it poetically. "Working memory," she said, "what we're aware of from one moment to the next—bridges time and is the content of our consciousness."

Some sleuths on the memory trail have bypassed the macro scale of neural circuitry and zoomed in for a look at the neurochemicals that float in the synaptic gaps.

Putting a face on body chemistry, color-enhanced photomicrographs portray serotonin (above) and dopamine (opposite), two of dozens of neurotransmitters—powerful chemicals that bridge synapses, the gaps between neurons. Imbalances can cause mental and physical troubles: Too little serotonin can lead to debilitating depression, while excessive dopamine accompanies some forms of schizophrenia.

What they've learned could lead to therapies that manipulate memory in ways that sound like science fiction, both boosting our powers of retention and recall and erasing recollections of past trauma. Optimism stems from mounting evidence that a particular molecule—CREB—is one of the cornerstones of lasting memory. CREB is short for a jawbreaker of a name: cyclic adenosine monophosphate-response element binding protein.

The first hints of CREB's importance came from studies by pioneering learning and memory researcher Eric Kandel. Kandel focused on sea snails, creatures whose nervous systems are manageably small—a mere 20,000 neurons—compared with millions or billions of nerve cells in a mammalian brain. Moreover, *Aplysia*, as Kandel's snail is named, has among the largest neurons in the animal kingdom, some large enough to see with the naked eye. Also, these giant cells are distributed to different regions of the animal's brain: As few as a hundred cells might be designated for the performance of a certain behavior. It would thus be possible, in the same way that other researchers used fear conditioning with rats, to train *Aplysia* to perform a simple behavior, track down the cells involved, and then look for physical evidence of the learning process in the form of changes in the synapses between cells. Over more than two decades of research, Kandel and his associates did just that, demonstrating that *Aplysia's* synapses grew more connections when the animal learned.

Meanwhile, other researchers discovered that if animals were given a drug that blocked their ability to make new proteins, they lost the power to form long-term memories. They could still learn, but they couldn't convert the learning to permanent storage. Since the only

way to make a new protein is to switch on a gene, it followed that genes were involved in long-term memory.

Then the hunt began for the cellular mechanism that turns genes on and off. This time, rather than working with the entire animal, Kandel's team snipped critical memory cells from *Aplysia* and maintained them in a dish to study their biochemical communication. In short-term learning a single stimulus, such as a mild electric shock, activates systems within the nerve cells involving the neurotransmitter serotonin. A single pulse of serotonin unleashes a molecular chain reaction mainly in the neuron's cytoplasm, a signaling system known as the cyclic adenosine monophosphate (AMP) pathway, that results in enhancing the release of another neurotransmitter, glutamate; glutamate strengthens communication among different neurons for a period of minutes.

In 1990 Kandel's group discovered that repeated stimulation—the essence of training and long-term learning—caused a component of the cyclic AMP pathway to move into the neuron's nucleus, its bastion of DNA. The last molecule in the chain of the pathway was CREB, whose job, it was discovered, is to bind to genes and switch them on. The genes then produce the proteins that promote the growth of synapses. Kandel's team showed that blocking CREB stopped all the events that characterized long-term memory, such as protein synthesis and synaptic growth. CREB, in other words, was the molecular gateway to memory.

This news launched a series of studies that helped decipher the CREB process in sea snails and other organisms. A few years later Tim Tully and Jerry Yin at Cold Spring Harbor, a lab in New York, discovered a substance that represses CREB-1. They called it CREB-2. Consider the times in life when we seem to learn and remember things easily—new material sinks in at a glance, or we trounce the opposition in Trivial Pursuit, coming up with answers we didn't even know we knew. Not only has CREB-1 been activated, but also CREB-2 has been somehow vanquished. Then consider all the other times when we read a sentence over and over and it won't stick, or we find a name or fact "on the tip of the tongue" but can't pull it all the way out of memory storage. In this instance the repressor is interfering—big time.

Tully and Yin demonstrated the stunning implications of this arrangement for human memory when they genetically engineered fruit flies to produce an excess of CREB-1. What they got were, in essence, flies with photographic memories, flies that could learn the required exercise after one training session instead of the usual ten. Kandel's lab later achieved the same result in *Aplysia* by developing a specific antibody that removed the CREB repressor. Conversely, Cold Spring fruit flies engineered to possess an excess of repressor CREB couldn't form the strong connections needed for long-term memories.

The commercial possibilities for improving human memory are breathtaking, not only for victims of Alzheimer's and other illnesses and disorders that disable memory, but also for the legions of baby boomers who want to stave off age-related memory loss. The ability to manipulate the CREB switch also holds out hope to people with the opposite problem—victims of trauma who cannot forget. None of this is around the corner, of course. As Kandel has pointed out, "The reason you're hearing so much enthusiasm for CREB is that it hasn't been tested yet!"

Drugs already exist that attempt to strengthen memory. Tacrine, for instance, acts by enhancing brain levels of the neurotransmitter acetylcholine. The major neurotransmitter at the junction between nerves and muscles, acetylcholine is also used at many synapses throughout the brain. Most neurons that synthesize acetylcholine and whose far-reaching axons transmit it to much of the cerebral cortex are concentrated in an area called the nucleus basalis of Meynert after its discoverer. Located in the lower part of the basal ganglia (a

structure adjacent to limbic system structures such as the hippocampus), the nucleus basalis degenerates in Alzheimer's disease, resulting in a cortex that contains well below normal levels of acetylcholine. By artificially boosting levels of that neurotransmitter in the synapses between neurons, drugs such as tacrine attempt to strengthen the synaptic connections necessary to memory formation. Other research efforts have found memory-enhancing qualities in the neurotransmitter dopamine, long associated with the brain's reward circuits. Despite the promise inherent in these research findings, much work remains to be done before physicians can start prescribing memory-enhancing drugs. As for genetically engineered super memories: So far they remain in the province of fruit flies and sea snails.

THE BIOLOGY OF TEMPERAMENT

The brain chemicals that make memory possible also shape the kinds of things we remember. Those of us who see the glass as half full, who tend to view most of life's experiences in a positive way, prefer to let go of the bad or unpleasant things we experience. Those of us for whom the glass is half empty put a darker spin on many of the same events.

It turns out that these ways of experiencing, interpreting, and dealing with the world—our personality or temperament—are also a function of brain chemistry. Ancient scholars as well as contemporary psychologists have recognized that people tend to fall into two broad categories of temperament. One category tends to be wary and restrained around strangers, hesitating when they encounter something unexpected. These people are especially cautious when faced with choosing a course of action that holds the possibility of failure. People of the second category are sociable and exuberant, bold in the face of the unknown and the unexpected. Greek physician Galen called the wary personality melancholic and the bold one sanguine. Carl Jung called the personalities

Shooting up and shorting out: Many addictive drugs such as heroin (opposite) interfere with normal neurotransmitters. Cocaine's euphoria, for example, results when clogged receptors create a buildup of dopamine. FOLLOWING PAGES: Blood vessels that feed the brain may also cause pain when dilation of external vessels produces a throbbing migraine.

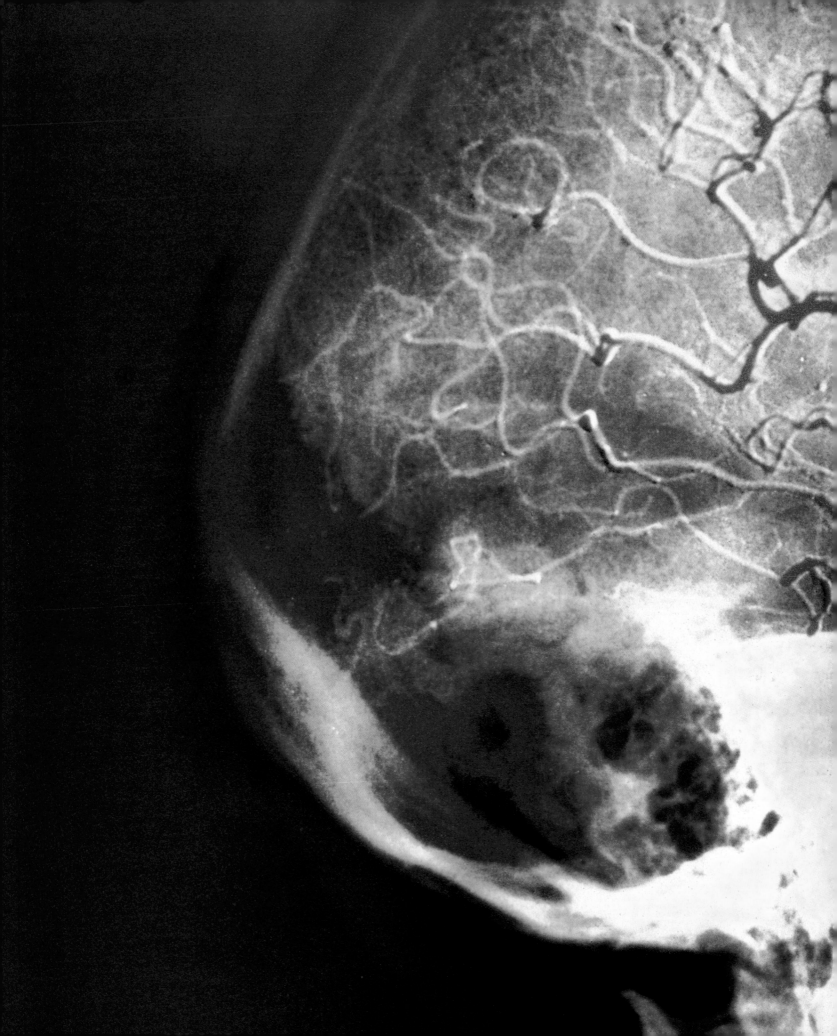

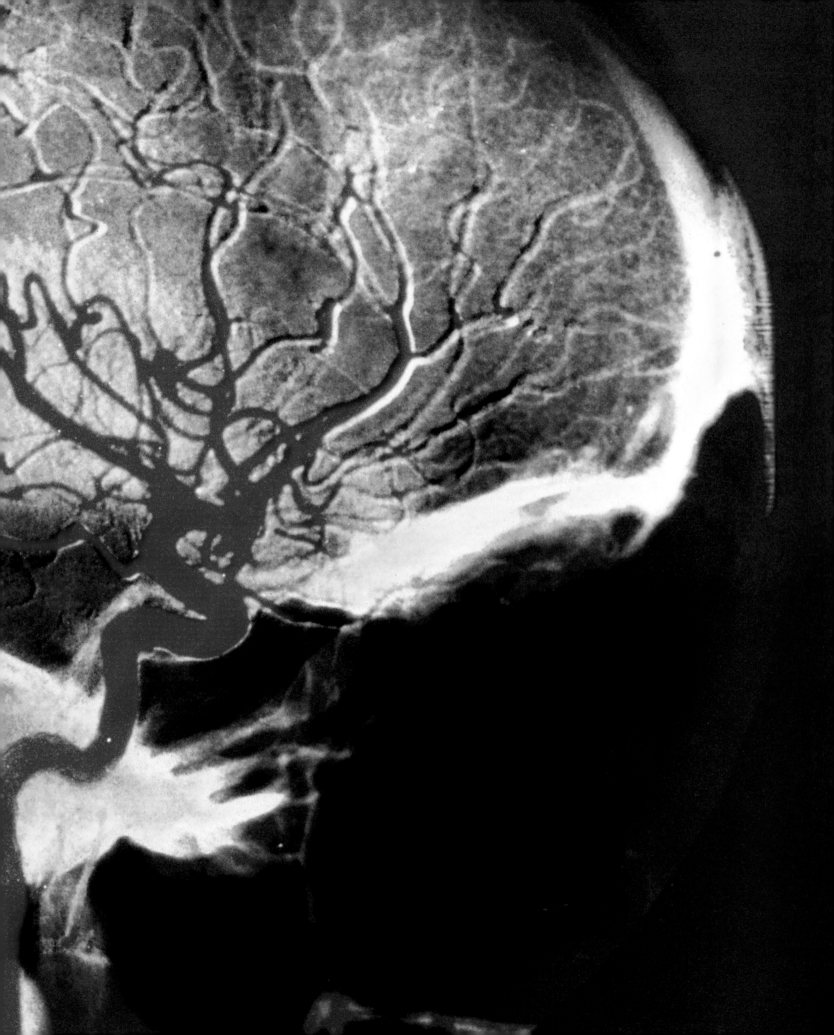

introverts and extroverts. Current theories of personality can be extremely complex, but all involve these two categories, even if by different terms.

In recent years neuroscientists and psychologists have come to believe that a person's temperament is influenced by the specific blend of neurotransmitters that bathe the brain and affect the body's stress response. Specific brain chemicals, among them norepinephrine, acetylcholine, dopamine, and serotonin, influence behavior and subjective mood as well as the integrity of the immune system and vulnerability to mental illness. Indeed, drugs that act to modify blood levels of these brain chemicals are used to treat disorders ranging from manic-depressive illness and depression to schizophrenia and obsessive-compulsive disorder.

But we're a far cry from knowing enough to say precisely which chemical profiles will produce an extroverted personality or an introverted one. Indeed, given the numbers of known neurochemicals and the numbers of receptors, the possible outcomes are many. Some temperaments will be quite rare—a saintly Mother Teresa, on one hand, a diabolical serial killer like Jeffrey Dahmer on the other.

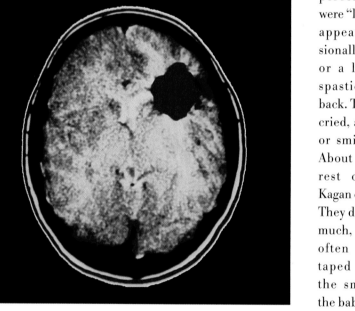

But even though the underlying biochemical recipe remains a mystery, researchers have built a strong case for a biological basis of the temperament types from behavioral and physiological studies. In work that has spanned some two decades, Harvard professor Jerome Kagan and others have tracked children's temperaments both forward and backward to learn both how early a stable personality type may be detected and how much influence the environment has in modifying it over time. Kagan began working with four-month-old infants and their mothers. In a 40-minute battery of tests the babies' reactions were monitored and videotaped as they were presented with a variety of harmless but unfamiliar or unexpected stimuli—a series of tape-recorded sentences read by female voices, colorful mobiles, a balloon popping. Of the more than 500 infants studied, all were born of healthy, well-educated mothers who had taken care of themselves during pregnancy.

When the videotapes were analyzed, researchers found that 20 percent of the infants showed what Kagan labeled "high-reactive" behavior. That is, they arched their backs, moved their arms and legs a lot, and cried in response to many of the events—all behaviors that animal research has shown occurs when the amygdala is stimulated. About 40 percent of the babies were "low-reactive": They appeared calm, occasionally moving an arm or a leg, but with no spastic arching of the back. These babies rarely cried, and some babbled or smiled occasionally. About 25 percent of the rest of the children Kagan called "distressed": They didn't move around much, but they did cry, often in reaction to taped voices. Finally, the smallest group of the babies moved around a lot but didn't cry. This category, representing about 10 percent of the sample, was called "aroused."

Focusing on the high- and low-reactive infants, Kagan followed up on both groups when they were 14-months and 21-months old. His working hypothesis was that high-reactive infants would grow into what he called "inhibited" children, while the laid-back low-reactives would be "uninhibited" toddlers, or extroverts. The tykes, always with their mothers, underwent a battery of procedures during which they encountered

Once considered untreatable, many brain tumors such as this ganglioma (above, purple mass) can be successfully removed by surgery (opposite). Research continues to explore and develop less invasive procedures and techniques.

a variety of unfamiliar situations. This time researchers were especially interested in "fear" reactions, which they defined quite precisely: Simply becoming quiet when something new happened, widening the eyes, or freezing did not count as fear. But bursting into tears or wails or running back to Mom was considered a fear response. So was a failure to approach a designated person or object within a reasonable amount of time after being asked. In one exercise, for instance, the examiner showed the child a toy, smiled and said a nonsense word in a friendly voice, and then spoke the child's name. Next, the examiner held up a different toy, frowned, and said the same nonsense word and the child's name, this time in a stern voice. Most toddlers simply looked back at the examiner or wrinkled their brows, but a small number of them cried; that reaction was coded as fearful.

Kagan's predictions held up: Children who had been high-reactive infants were more likely to have fearful responses at 14 and 21 months of age than those who had been low-reactive infants. At 4½ years, the same children were tested again. The criterion this time was how frequently they spontaneously smiled or commented to the examiner. (When people are tense or

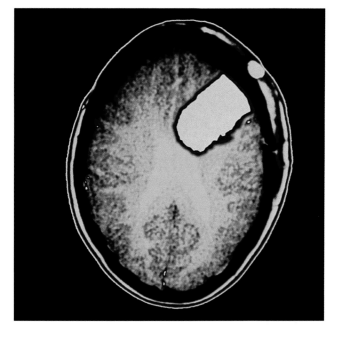

apprehensive, one of the first reactions is to stop smiling or talking.) Simply answering a question didn't count as spontaneous, but any elaboration on the answer did. True to expectations, the children who had been high-reactive infants tended to be low on spontaneous smiling and talking. Some high reactives never smiled or said anything beyond answering when asked a question. Most of the low reactives, of course, were happy chatterboxes by comparison. In another test a few weeks later, the subjects returned to meet two unfamiliar children of the same age and sex as their own. Along with their mothers, the children were in a large room full of toys. More of the children who had been high-reactive infants were shy compared with the low-reactive children and

tended to stay close to their mothers for a large part of the session. The low-reactives were sociable souls, initiating play with the other children and rarely going to their mothers.

Even so, all children didn't always exhibit the defining characteristics of an inhibited or uninhibited profile. Among high-reactive infants, only 13 percent were extremely fearful at 14 and 21 months as well as subdued and shy at age 4½. Environment eventually had a moderating influence on the initial temperamental bias of the remaining 84 percent of the children originally classified as high-reactive. Presumably interactions with their parents and others had helped them overcome their timidity, and such efforts helped produce four-year-olds who appeared to be average in sociability and confidence. Kagan, however, is quick to point out that it was rare for a high-reactive infant to show at age four the exuberance, boldness, and emotional spontaneity that characterizes a typical uninhibited child. "The environment has power," Kagan says, "but that power has its limits"—at least in the context of turning high-reactive babies into outgoing glad-handers.

That easily aroused babies grow to be shy children and introverted adults seems reasonable. But what makes a high-reactive baby? What makes one person a born worrier and another always ready for whatever comes next? Temperament researchers have done a remarkable sleuthing job, akin to that performed by the memory investigators who sought a molecular gateway to remembering and forgetting. They have measured heart rates of fetuses three weeks before birth and when the babies are two weeks old, and they have correlated high heart rates at these stages with the propensity to develop high-reactive personalities by the age of four months. (Heart rate, one function of the autonomic nervous system, is influenced by the amygdala.) Researchers have looked at the differences

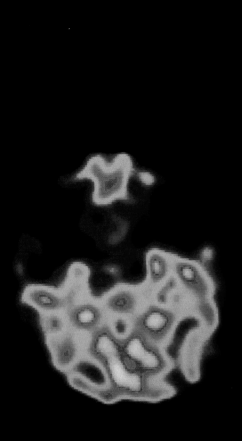

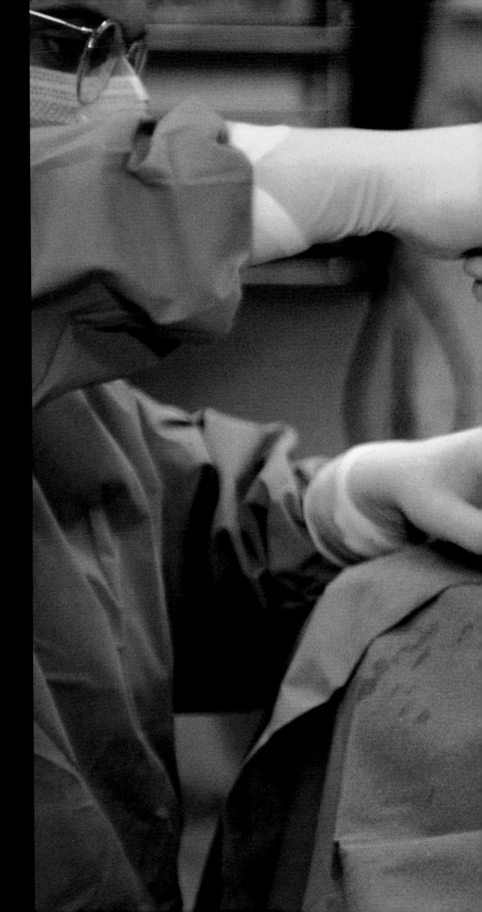

Desperate measures for debilitating diseases: A lifetime of nearly continuous brain seizures led this epilepsy sufferer to seek a surgical solution. Doctors screw a calibrated frame in place to help pinpoint malfunctioning areas of her brain (right); the frame supports a probe that admits diagnostic electrodes. Brain scans (above) show areas of intense neuronal activity that indicate a seizure in progress.
FOLLOWING PAGES: Surgeons ready the brain of another epileptic for radical hemispherectomy, in which half of the cerebrum is removed. Though such surgery can impair visual and motor control on one side of the body, the remaining hemisphere often compensates by taking on many duties of the missing portion.

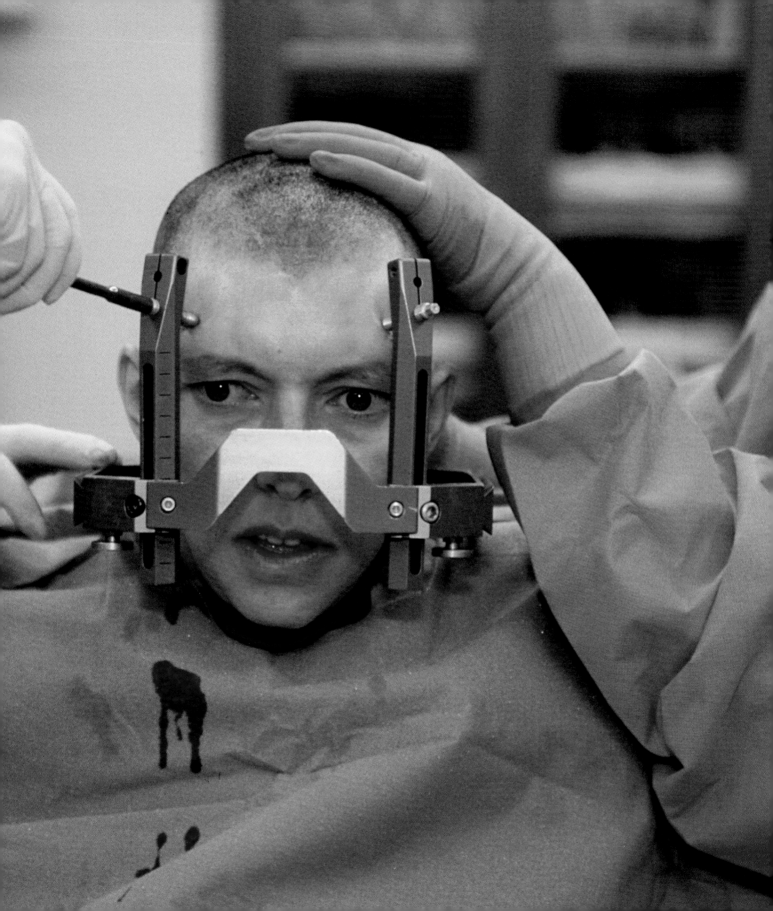

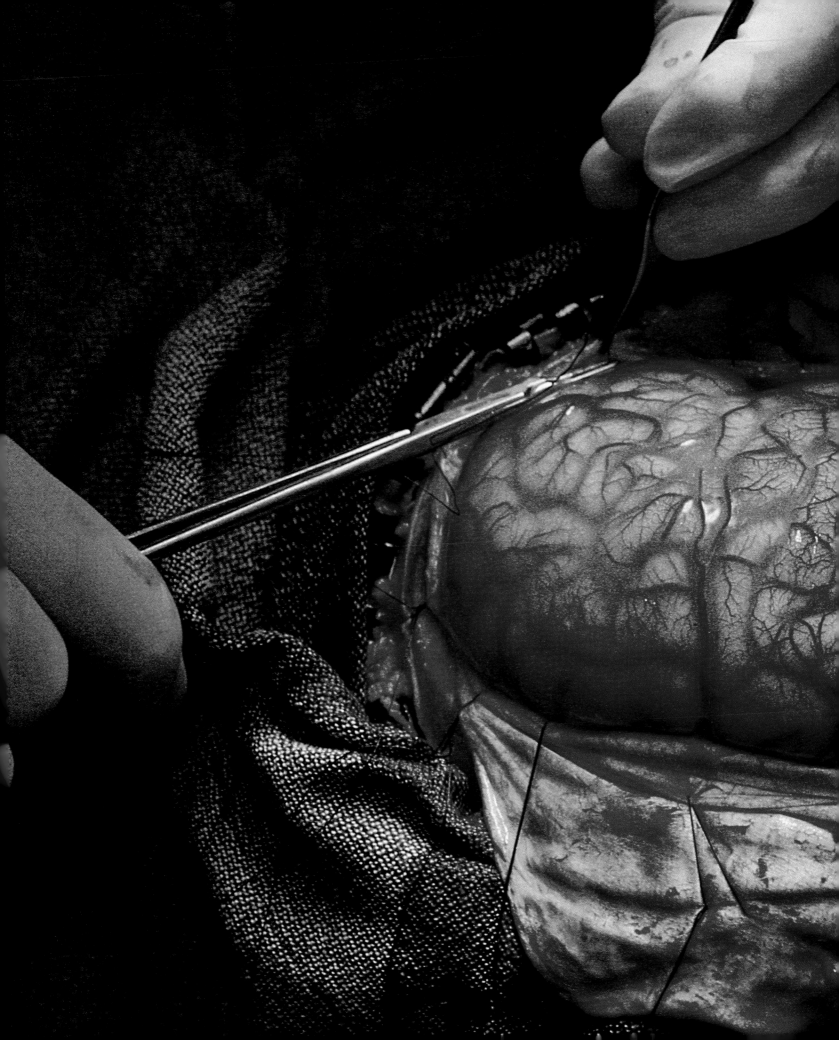

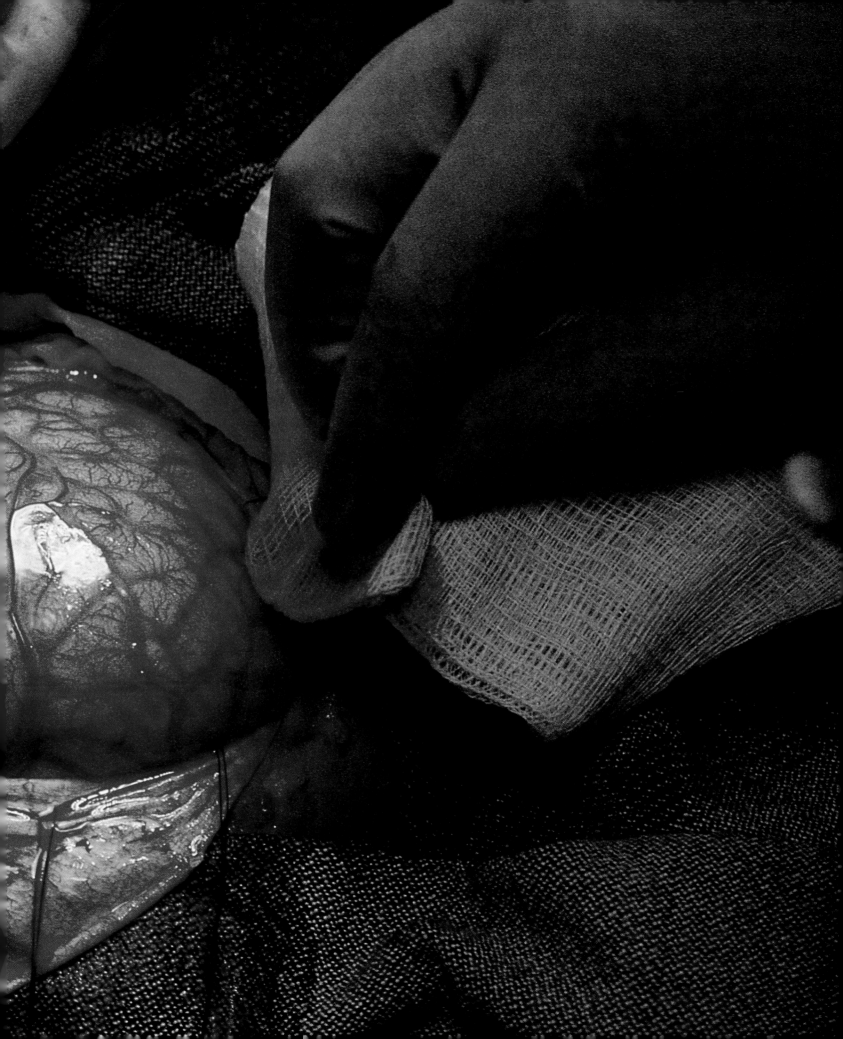

in response between the two sides of the brain. A considerable body of literature, some based on stroke patients with damage to one hemisphere, indicates that the right side is usually dominant in states of fear or anxiety, while the left hemisphere participates more fully in states of joy or relaxation. High-reactive, inhibited children have more activity in their right hemispheres, especially frontal areas, where functions such as decision making and stream-of-consciousness working memory seem to reside; low-reactive, uninhibited children have more active left frontal areas.

Other researchers have looked into levels of gamma-amino butyric acid (GABA), the brain's most prevalent inhibitory neurotransmitter. When a neuron releases GABA into a synapse, it sends an inhibiting signal to the receiving neuron; GABA is thus a natural modulator of stress. Other studies have monitored dopamine levels for clues to explain why some people are upbeat extroverts. Cornell University clinical psychologist Richard Depue has found evidence that dopamine is strongly associated with what some researchers call extroversion but that Depue refers to as "positive emotionality." The more responsive a particular brain happens to be to dopamine or the higher its level, the more likely that person will be to respond to various incentives and rewards and to be elated and excited when achieving a goal. People who are highly reactive to dopamine usually score high on personality tests for positive emotionality.

The connection between temperament and developing behavior is of enormous interest not only to parents but also to educators, mental-health professionals, and social workers, all of whom are concerned with ensuring that children reach their potential. Research shows that even though most high-reactive, inhibited children will never see a psychiatrist because of excessive anxiety, they are probably at higher risk than others for developing anxiety symptoms. Kagan's studies suggest that by the time they are adolescents, about 10 percent of inhibited children will show signs of social phobia. They won't like parties and will be uncomfortable in large groups of strangers. Inhibited teens will also be less likely to disobey or become delinquent.

The same cannot be said of low-reactive, uninhibited children, youngsters who don't faze easily. "I do not believe there are genes for crime," says Kagan, "but there are genes for the development of a fearless profile." Given a home and an environment that tempers and socializes aggressive behavior, low-reactive children are likely to

grow up to be CEOs, surgeons, jet pilots, investment bankers—all vocations that involve taking risks. But if the youthful environment permits aggression and offers illegal or antisocial temptations, these risk takers are themselves at higher risk.

Research suggests that low serotonin levels may make people more responsive to dopamine, the neurotransmitter associated with positive feedback. Low levels of serotonin, Depue points out, seem to cause depression, irritability, and volatile emotions. Individuals with this particular neurochemical profile thus may be more susceptible to drugs that activate the dopamine system, such as alcohol, cocaine, and nicotine. One theory is that different dopamine receptors in the brain may be related to different types of drug abuse. Given recent findings that neurotransmitters mix and match with scores of different receptors, the prospect of arriving soon at some kind of neurochemical recipe for happiness or social responsibility would seem unrealistic at best.

MINDS SUSCEPTIBLE TO DISRUPTION

The ebb and flow of a number of biochemicals throughout the brain helps determine susceptibility to various neurological disorders. A wide range of conditions—from Lou Gehrig's disease to drug and alcohol addiction to chronic pain—are the result of something gone wrong in the brain. This includes many disorders long labeled—and often stigmatized—as "emotional" or "psychologically based." Schizophrenia, for instance, was once blamed on upbringing; incapacitating depression was seen as merely a failure of willpower. Children who had difficulty paying attention in school were chastised for being lazy or stubborn. One of the most important changes in the practice of medicine in recent decades has been the recognition that these and other debilitating, sometimes fatal disorders have a biological base, that mood and personality can be influenced by internal chemistry.

At the same time, however, neuroscientists also recognize that environmental factors play a role in neurological disorders. Interactions with other people, for good or ill, as well as such practices as meditation and psychotherapy can have profound effects on brain

function, modifying synaptic connections and the changing flow of brain chemicals.

As scientists reach a better understanding of how our neurochemistry works, they will be able to offer more options for treatment of a wide variety of ills. New research on dopamine, for instance, has far-reaching implications for the treatment of addiction, learning disorders, and certain forms of schizophrenia. Since the late 1970s scientists have known that dopamine is involved in the brain's reward system, which operates between certain areas of the midbrain and the limbic system. As Steven Hyman, Director of the National Institute of Mental Health likes to put it, this reward circuit tells the brain, in effect, "That was good, let's do it again, and let's remember exactly how we did it." Research has shown that production of dopamine in the brain is a signal that a recent experience was either emotionally valuable or just pleasurable—eating or sex, for example. Dopamine production also seems to explain why people become addicted to drugs such as cocaine, which is a dopamine mimic: Such drugs block dopamine receptors in the brain, clogging normal removal mechanisms and leaving excessive amounts of dopamine in the synapse—which makes the cocaine user feel euphoric and alert.

Early on, however, the pleasure principle made little sense in the context of what was known about schizophrenia. Research shows that people with this disorder have abnormally high brain levels of dopamine, but this excess doesn't seem to confer much happiness or enjoyment in life.

Clearly there was more to the story, and in the 1990s clues began to emerge. In Switzerland a group of researchers led by Wolfram Schultz, a neurophysiologist at the University of Fribourg, was studying the action of dopamine-producing cells in a region of the midbrain called the substantia nigra ("black substance" for its normally dark hue). Scientists had long known that dopamine from the substantia nigra helps the brain stem,

Charred remains of a cow infected with bovine spongiform encephalopathy—mad cow disease—burn in an English field. Scientists debate whether a virus or an abnormal protein called a prion causes spongelike holes to develop in the brain.

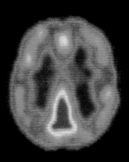
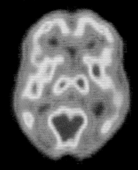
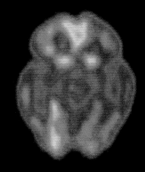

PET SCANS OF BRAIN SHOWING OBSESSIVE-COMPULSIVE DISORDER

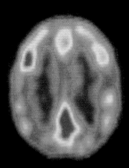
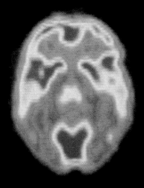
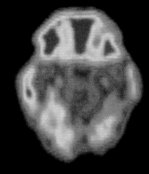

SCANS OF NORMAL BRAIN

Beyond x-rays: Positron emission tomography (PET) not only reveals soft tissues but also records activity within various parts of the brain. A patient with obsessive-compulsive disorder displays overactivity in the brain's central area, which acts as a gatekeeper for incoming information. Stuck in the "on" mode, it fosters repetitive behavior. In schizophrenics, overly active lower regions—areas that deal with sensations—interfere with thinking and language in the frontal lobes, leading to delusional thought.

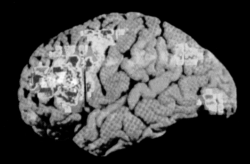
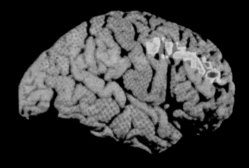

PET SCANS SHOWING EVIDENCE OF SCHIZOPHRENIA

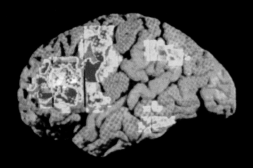
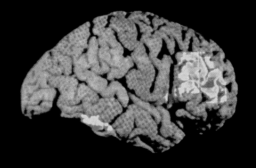

SCANS OF NORMAL BRAIN

cerebellum, and basal ganglia carry out their muscle-control duties. In the absence of sufficient dopamine, those nerves short out and the person develops Parkinson's disease, suffering, among other things, tremors of the extremities and rigid arms and legs that translate into slow movement. The Swiss researchers, specifically interested in learning how the dopamine cells in the substantia nigra related to movement, recorded the electrical activity of individual dopamine-producing cells in the substantia nigra of monkeys that had been trained to do a movement task. The researchers hypothesized that those neurons would fire when the animals moved or were getting ready to move. To their surprise, however, the dopamine-producing cells remained quiet. But when they gave the monkeys a reward, Schultz reported, "the neurons started going crazy."

The scientists first thought the firing might be preparing the animal to move toward the reward—a piece of apple—but they found that the cells fired in response to a reward even when the monkeys were not required to move. It began to look like a straightforward case of dopamine-as-signal of reward. A series of follow-up experiments, however, gradually led the team to conclude that dopamine-producing cells respond to reward, as Schulz put it, "only when it occurs unpredictably—for example, during learning." The cells fired when an animal unexpectedly happened on a response that brought a reward, but ceased firing once it knew that a given response would bring a reward. This behavior of the dopamine-producing cells, the researchers reported, "which reveals the difference between what is predicted and what occurs, looks like a perfect teaching signal."

More recent evidence suggests a much more general role for dopamine in learning. Research at Cold Spring Harbor suggests that dopamine helps retain memories by blocking the weakening of synaptic connections. Other studies seem to indicate that dopamine highlights novel or startling events that attract attention even though the events are not related to the reward. Not all research supports this attention-getting role for dopamine, however. Some studies seem to confirm the original dopamine-as-pleasure-seeker theory. No one is sure why the results of these experiments conflict; one reason may be that some researchers

record the electrical activity of dopamine-producing neurons when they fire and others measure dopamine levels in the brain as a whole.

Despite the lack of resolution, dopamine's effect on learning is worth further exploration. Looking at the connections between dopamine-producing cells in certain areas of the brain, some scientists theorize that dopamine helps frontal cortex neurons hang on to short-term memory longer. This may jibe with Cold Spring Harbor results showing dopamine's memory-strengthening effect in the hippocampus. Moreover, if dopamine signals are indeed attention-getting devices regardless of the specific event they're responding to, then dopamine's excess presence and overactivity in attention-deficit disorder (ADD) and in schizophrenia would explain some of the symptoms of those disorders.

For example, schizophrenics make bizarre associations and fixate on stimuli that normal people often find irrelevant—but such fixations may be due to sensory input entering a nervous system that is in a highly aroused state. People with ADD have trouble concentrating at work or at school because they are easily distracted by things most people can simply ignore, such as voices in the hall or a passing car.

When it comes to mental disorders, the varying roles of genes and environment are difficult to disentangle. A research team led by Robert Post of the National Institute of Mental Health discovered that environmental stress, as well as drugs of abuse, can turn on the genes linked to depression and other mental problems by activating a gene called C-fos. In his book *Inside the Brain*, science writer Ronald Kotulak describes that find as one of the most far-reaching discoveries in recent years. C-fos codes for a protein that, in turn, activates other genes that make abnormal receptors and connections to other cells. These abnormalities cause breakdowns in the brain's communication network, potentially leading to problems such as seizures, depression, and manic-depressive illness.

As Kotulak notes, nowhere does this link between environment and disease seem more evident than in current statistics that show a veritable epidemic of mental problems among the young people of the United States. In the last 25 years, according to some reports,

---

FOLLOWING PAGES: Backed by cross-sectional images of their brains, identical twins display differences that are more than skin-deep: The twin on the right suffers from schizophrenia.

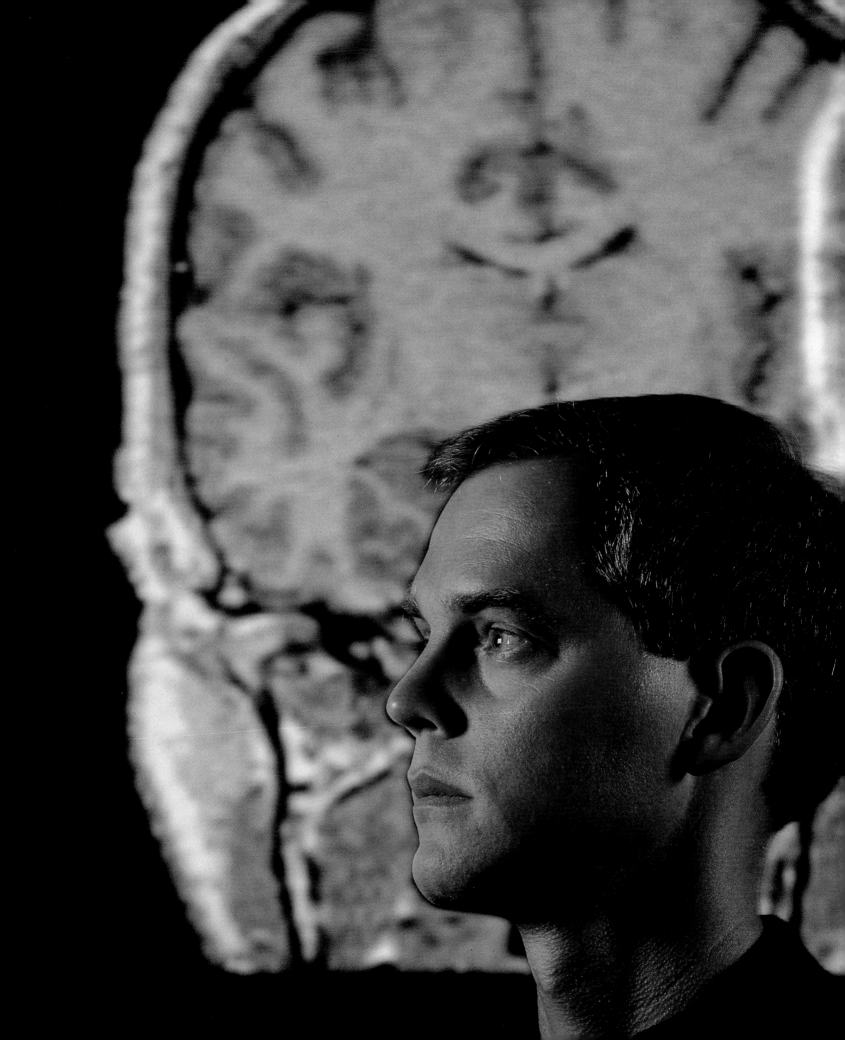

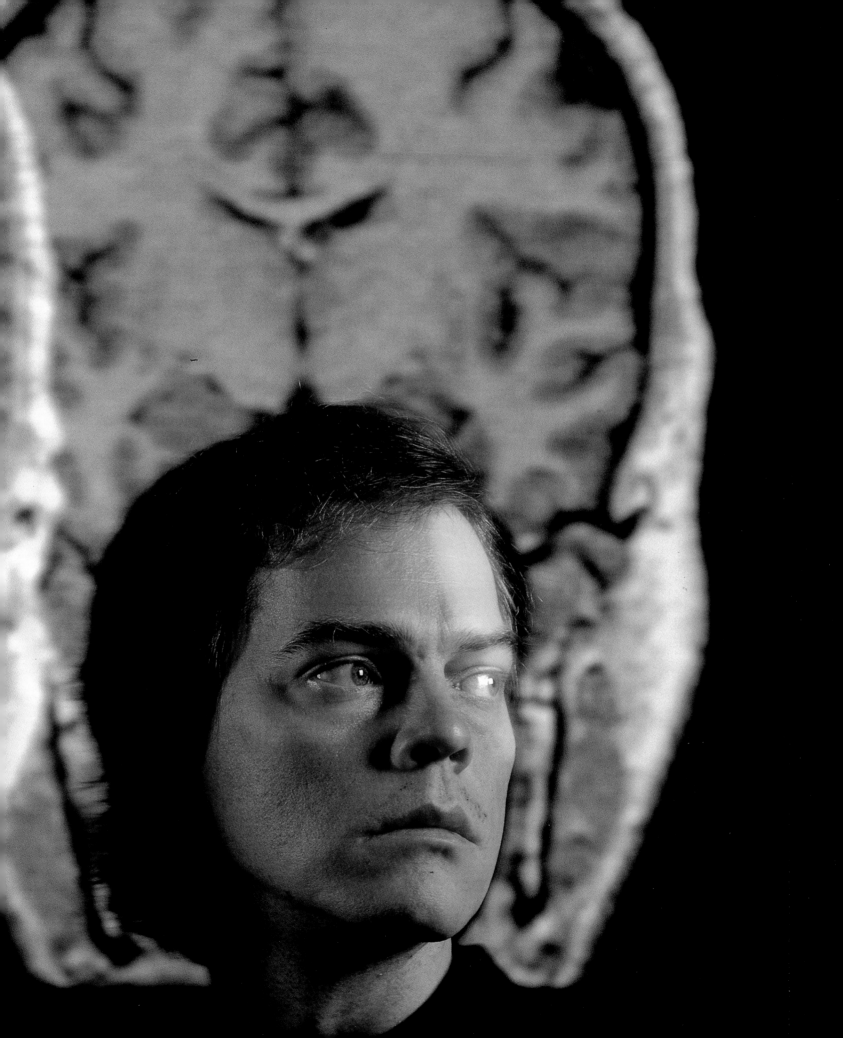

the rates of depression, drug and alcohol abuse, suicide, and crimes of violence have doubled, affecting both the inner city and the suburbs.

Researchers at the University of Chicago have fingered stress as the prime force that bends a young person's brain toward violence and aggression. Children raised in violent and abusive surroundings are at risk for developing hair-trigger responses virtually from the cradle onward. With stress hormones cranking up their systems, their hearts beat faster and they become extremely aggressive. In effect, their brains get wired differently: They are neurologically primed to interpret many nonverbal clues as threatening and to react angrily to innocuous situations. Given that stress has also been linked to the destruction of neurons in the hippocampus, these children are at high risk for suffering learning and memory problems as well.

Another source of stress is major depression itself, which has been found to raise the level of stress hormones such as cortisol, with their subsequent effects on the hippocampus and other parts of the brain. Depression is a problem not only for the person who suffers it, but also for infants of depressed mothers. The brains of such infants show markedly lower levels of electrical activity in areas of the brain involved with feelings of joy, curiosity, and other positive emotions. The longer their mothers stay depressed, the longer the connections between the infant brain cells that regulate positive emotions go unstimulated; as these brain cells wither away for lack of reinforcement, the likelihood the children will develop depression grows. The good news, however—and a reassuring confirmation of the adaptability of the young brain—is that if the mother's depression is treated in time, the neuronal damage in the infants can be reversed.

It's also true that there are effective treatments today, both pharmacological and psychological, not only for people with depression but also for those with many other disorders. Research is showing that psychotherapy and behavioral therapy can and do work—at least for some disorders. For example, people who are agoraphobic, who have come to fear going outside their homes, and who suffer panic attacks when they do, can be helped to overcome this problem by a combination of psychological therapy and medication. Medicine can usually counter the physiological symptoms of the panic attacks, such as sudden feelings of terror, dizziness, shortness of breath, and heart palpitations. Therapy can begin to override, if not erase, the emotional memories that have taught the victim, by associating the terrifying feelings with the places the feelings occurred, that the entire world is dangerous. To unlearn this lesson, the patient is taught how to calm himself with relaxation techniques and is systematically exposed to increasingly anxiety-producing situations. As medication and therapy forestall the panic attacks during each exposure, the patient's panic response is eventually extinguished. Most sufferers of panic disorder show marked improvement after a combination of medication and cognitive-behavior therapy. That therapy works the same way drugs such as lithium and Prozac do—and in the same way that learning does—it effects physical, neurological changes in the brain.

THE IMPORTANCE OF SLEEP

We've now come to understand some symptoms of mental disorders as imbalances of the various neurochemicals that the brain uses to carry out its normal functions. Intriguingly, though, significant shifts in several of these biomolecules occur for all of us every night—whenever we make the transition from waking, to sleeping, to dreaming. One of the most characteristic features of dreams is that they delude us into thinking we're awake, even though the most extraordinary and even impossible things are happening: We're flying without benefit of an airplane. We're standing in front of a classroom in our underwear—or naked. The hallucinatory quality and emotional intensity of dreams can sometimes wake us up in tears or drenched in sweat.

For millennia, human beings have interpreted dreams in many ways—as portents, hidden truths about ourselves and others. Now sleep and dream research is revealing what goes on in the brain that might account, at least in part, for the bizarre nature of our nighttime reveries. J. Allan Hobson, director of the Laboratory of Neurophysiology at Harvard, suggests that dreaming, though not pathological in itself, is a kind of psychotic condition that we enter into every night. Understanding how the neurological states that normally arise during sleep may sometimes arise during waking, he says, may help us gain insights into the tortured perceptions of people with schizophrenia, for instance.

Most dreams occur during periods of sleep characterized by rapid eye movement, or REM. Dreams occur in non-REM (NREM) sleep as well, but they tend to be briefer and less extraordinary in content. Hobson and others have compared the parts of the brain that are active during waking, during NREM sleep, and during REM sleep. When we're awake, for instance, the cerebral cortex—the part of the brain that handles language, memory, and planning, as well as motor output and sensory perception—is extremely active. During NREM sleep, the cortex is largely inactivated, but it revs up when we enter REM sleep. Not all of the cortex is reactivated, however. The frontal regions responsible for working memory and executive functions remain shut down.

Not surprisingly, research suggests that the on-off switch for the cortex is chemical. Whether we're asleep or awake, the brain's autonomic functions carry on, monitored by the brain stem. When we're awake, some cells in the pons secrete neurotransmitters such as norepinephrine and serotonin.

During REM sleep, these substances are drastically reduced. Instead, a different chemical system, generated by another cluster of cells in the pons, starts putting out quantities of the neurotransmitter acetylcholine in amounts equal to or greater than those found in the waking brain. Acetylcholine activates the nerves that move the eyes and send messages to the upper brain and inhibits muscle motion.

Individual sleep patterns can vary depending on age, level of fitness, and daily activity. In general, we go through regular cycles of REM sleep interspersed with four stages of NREM sleep in a repeating cycle that usually lasts 90 to 100 minutes. During NREM sleep we are deeply unconscious, especially early in the night, when we linger longest in the deepest stage of NREM, Stage IV. We're very hard to rouse from this stage and can be quite confused.

After we've been asleep for some 60 or 80 minutes, we cycle back up from Stage IV, through Stage III, to Stage II, and Stage I, during which periods we have some muscle tone but minimal cortical activity. Then the rapid

eye movements begin, continuing for periods of five minutes to an hour. After REM the brain starts to drift back down; in time it cycles back up. As night ends, the REM periods get longer and more intense while the NREM periods become shorter and shallower. Instead of descending all the way down to Stage IV NREM sleep, we tend to dip only to Stage III and then climb back up to Stage II and an REM period. Indeed, we spend most of the night in Stage II NREM sleep.

Studies done with cats show that during REM, even though the cat is curled up asleep, an electrical storm is taking place inside the pons: Electroencephalograms show enormous spike-and-wave complexes that look like the brain waves of an epileptic seizure. The pons cells are firing wildly, and these action potentials are being broadcast all through the brain, including the visual system, causing the eyes to move. Acetylcholine secreted by neurons in the pons activates nerves that cause a loss of muscle tone. The limbs go limp and virtual paralysis results, explaining the frequent sensation in dreams of trying to run and being frozen in place. Signals from the pons travel to the amygdala, that center of emotional memory, generating the intense anxiety typical of our fight-or-flight response. Signals also radiate upward to the cerebral cortex, where memories and input from the senses come together. At this point, the theory is that the cortex tries to make sense of it all—but without the benefit of the norepinephrine and serotonin it normally works with. A dream like this might be likened to a Rorschach test. Random neurons fire random signals; the cortex fills in the blanks.

If we live 70 years, the amount of time spent in a state of brain activation associated with hallucinatory dreaming and intense emotions adds up to something like 50,000 hours—or nearly 6 years. Evolution would thus seem to have placed a high priority on REM sleep. The question is: Why?

One answer is that REM sleep helps brain development. Infants, for example, spend about 12 hours a day

One aspect of a mirror-image world, this clock symbolizes the challenges facing those who suffer dyslexia. The condition, which causes letters and numbers to appear reversed, likely stems from disrupted development in the cerebral cortex.

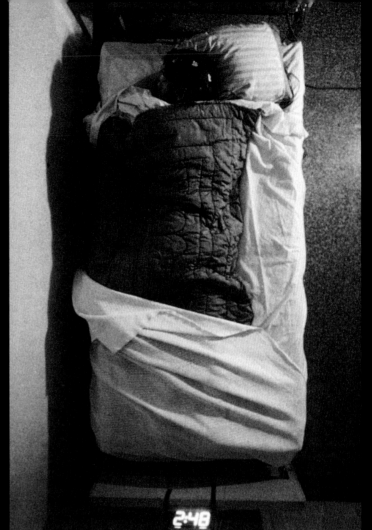

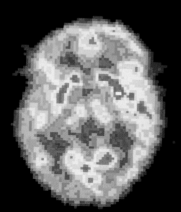

NON-REM SLEEP

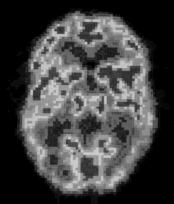

REM (RAPID EYE MOVEMENT) SLEEP

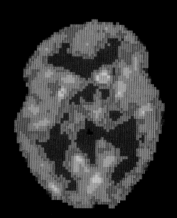

SLEEP DEPRIVATION

Lending his zzzzz's to science, a volunteer in a sleep study performs the subtle shifts in posture essential to a good night's sleep (top) while sensors monitor activity in his brain, eyes, and heart, as well as in the muscles of his throat and neck (opposite, top to bottom). PET scans of the brain (above) reveal different levels of neuronal activity in specific areas during various states of sleep. Colors range from reds (most active) through yellows and greens to blues and purples (least active)

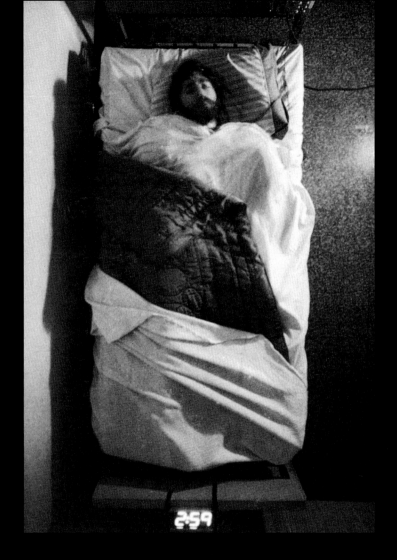

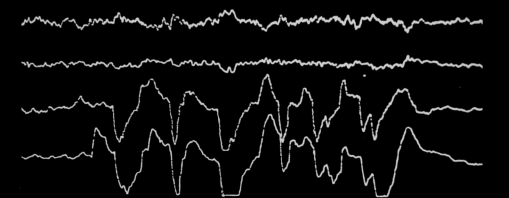

in REM sleep. The electrical activity in their fast-growing brains is somehow helping the brain lay down critical circuits. In addition, research with animals shows that the hippocampus, so important to memory, is dramatically active during REM—as active as it is when animals explore new environments during waking.

One popular theory, for which there is good experimental evidence, is that REM sleep fosters long-term memory consolidation. If subjects tested on a learning task requiring considerable attention are allowed to sleep as usual afterward, in the morning they perform as well or better on the same test than they did the night before. If they're deprived of REM during the night, however, they perform in the morning as if they were taking the test for the first time.

Scientists have known for a long time that depriving humans of either NREM or REM sleep can be extremely damaging. Many sleep-deprived individuals become psychotic and paranoid: They hear voices; they don't function well. Recent studies with REM-deprived rats show that REM sleep is essential to life. Deprive rats of REM sleep long enough and they die—their metabolism goes haywire, making it impossible for them to maintain their body weight or control their body temperature.

Animal experiments also have shown that amphetamines or other stimulants can shut down the REM system. REM deprivation creates a REM debt. It's known that many drugs can suppress REM sleep in addicts; when such drugs are taken away and the addict goes into withdrawal, his REM sessions can be very intense, even psychotic. This is what causes the tremors and hallucinations of an alcoholic's delirium tremens, and it may contribute to the psychosis of amphetamine and probably cocaine withdrawal.

Studies in 1980 focused on the chemistry of sleep itself, looking for the chemical on-off switch. The

compound adenosine, for example, has been found to affect alertness. Too much of it in the brain, and we become drowsy and fall asleep. In fact, the morning cup of coffee apparently works because caffeine prevents adenosine from binding to brain cells.

Not enough is known yet to expect insomniacs to be able to reach for adenosine sleeping pills anytime soon. For one thing, the compound has several other effects, including lowering body temperature, altering blood pressure, and potentially damaging the heart. But researchers in many fields of neuroscience are coming up with devices and technologies that hold out solutions, if not for insomnia then for a number of other disorders of the brain and central nervous system.

WHAT'S NEXT

As we near the end of the Decade of the Brain, as the 1990s were designated by Congress, progress in neuroscience continues at a remarkable pace. Various labs are pushing the boundaries of our understanding of the brain's miraculous workings. Geneticists are zeroing in on the genes that may put people at risk for Alzheimer's, Huntington's, and other neurodegenerative diseases. Brain implants and devices known as neuro-prostheses are helping people with a disorder called severe essential tremor and letting congenitally deaf

Thermograms of this sleeping man—clad only in briefs—reveal how skin temperatures vary during different types of sleep. Periods of dreaming and REM sleep coincide with warmer temperatures, evidenced by heightened reds and whites.

children hear. And specialists are honing techniques for the bloodless excision of cancerous brain tumors.

Essential tremor, a little-understood neurological disorder that usually runs in families, can cause severe shaking. A tiny brain implant known as the deep-brain stimulator now allows many patients with severe essential tremor to carry out normal living activities they have long been unable to perform by themselves. Implanted deep in the thalamus, the device consists of a tiny electrode with a wire that runs under the scalp and down to a pulse generator implanted in the chest.

The implant delivers mild electrical stimulations that block the brain-generated signals that cause tremors. It can also significantly reduce tremors in Parkinson's patients. However, because the implant does not treat other symptoms of this disease, such as rigidity, difficulty in moving, or mental deterioration, its effects on the patient's quality of life are limited. Because the long-term effects of constant electrical stimulation are unknown, scientists are moving cautiously. Parkinson's affects both sides of the body, but usually one side more than the other. The deep-brain stimulator has been approved for use only on one side of a patient's brain; studies are now under way to evaluate the safety of implanted electrodes on both sides.

Neuroprostheses, or cochlear implants, hold out hope for the hearing impaired. Unlike an ordinary hearing aid, a cochlear implant connects directly to the brain's hearing nerve, bypassing damaged or malfunctioning nerve cells in the inner ear and the middle ear. The devices vary in design, and the process is complex. Basically, sound enters a microphone worn over the outer ear and is sent via a wire to a processor carried on the body. The processor translates those sounds into electrical signals and relays them to electrodes placed in the inner ear that directly stimulate the auditory nerve. The auditory nerve relays the signals to the brain, as in normal hearing. One of the most important aspects of the implant is that it enables the wearer to hear his or her own voice. Self-awareness in speech is critical for learning how to control speech pattern and volume, especially in children who are born deaf. Such children have a more difficult time learning to interpret sounds than do adults who lose their hearing later in life. The sooner a child's hearing problem is diagnosed, the easier it is for him to learn to hear and to speak.

Technology is also coming to the aid of patients with brain cancer. Not only are brain-scanning devices such as MRI and PET making it easier to locate tumors deeply buried in brain tissue, but also high-tech surgical devices are promising to replace a spectrum of traditional surgical and radiation procedures. The gamma knife, for example, destroys tumors bloodlessly by beaming low-level radiation at the target from many different angles. Boron neutron capture therapy, a technique developed at the European Union's Joint Research Center in the Netherlands, reduces the risk of damaging healthy tissue. Patients are injected with a compound that contains boron, which accumulates only in tumor tissue, then are exposed to neutron beams from a reactor. As the neutrons are absorbed by the nuclei of the boron atoms, those atoms emit damaging short-range particles that kill the tumor while leaving nearby healthy cells essentially intact. The technique is still experimental, but this and similar efforts in Japan and at Brookhaven National Laboratory in New York show promise in fighting a disease that has been a death sentence.

A completely different approach is being taken by researchers homing in on the genetic causes of brain disorders and disease. The gene abnormality that causes Huntington's disease, for example, was found in 1993 by an international team, culminating ten years of intensive work. Rapid, jerky movements—called chorea—loss of control of bodily functions, deterioration of memory and thought that results in dementia, and eventually death characterize Huntington's. Scientists, long frustrated in their effort to pinpoint the gene's normal function, recently have found clues to how the defective version produces the disorder. The defect, identical to one occurring in other genes that cause half a dozen diseases similar to Huntington's, consists of a kind of stammer that inserts extra copies of the amino acid glutamine into key proteins. These proteins eventually migrate into the nucleus of brain cells, causing them to die. Researchers are hopeful that these new insights will lead to drugs for treatment of these various diseases, which affect an estimated 60,000 Americans.

As investigators identify more and more of the specific genes involved in brain disorders, gene therapy becomes possible—at least for some illnesses. For example, preliminary trials in humans are testing forms of gene therapy for brain cancer. Since the late 1980s researchers have been working to improve ways of delivering modified genes to specific cells, the idea being to replace a flawed gene with a healthy one, thereby enabling the cell to produce the correct protein. The delivery vehicles of choice have been viruses due to their natural ability to infect cells; the viruses are altered so they cannot reproduce. Scientists are testing methods to help injuries to the nervous system such as spinal cord damage and stroke. They are trying, in animals, gene therapies designed to cure neurodegenerative diseases ranging from Lou Gehrig's disease to Parkinson's.

Gene hunters continue to find evidence that manic-depressive illness, also known as bipolar affective disorder, stems from multiple genes. The illness is a major public health problem. Scientists have identified new sites on five chromosomes that may contain genes that predispose for manic-depressive illness. Sufferers experience recurrent mood and energy swings with depression so severe that they face a 20 percent risk of death by suicide if untreated. Indeed, some 31,000 people commit suicide every year in America—more than die from homicides or AIDS—and approximately 80 percent of those suicides are associated with depression or manic-depression.

The latest finding concerning the inheritability of this disorder comes from a study of many generations of several large Amish families in which manic-depressive illness and other mood disorders occur. Researchers found evidence for susceptibility to manic-depression genes on chromosomes 6, 13, and 15. This suggests a complex mode of inheritance, similar to that seen in diseases such as diabetes and hypertension, rather than a single dominant gene. All the studies also suggest, however, that individuals with a genetic predisposition do not invariably develop manic depression.

As with most heritable traits, environmental factors can play a role in determining whether a genetic vulnerability will result in onset of the actual illness. To complicate the picture, other major mood disorders also typically occur in families that have manic-depressive illness, so at best, one can get only a general idea of one's risk for developing a disorder.

Many studies suggest that the genes that make one susceptible to manic-depressive illness are also involved in creativity, a finding that argues against coming up with gene therapies that replace the genetic culprits outright. Indeed, current drug treatments for this disorder, such as lithium, have some of the same drawbacks that any potential gene therapy would have—that is, creative artists who are also manic-depressive worry that the drug will destroy the source of their creativity. As Kay Redfield Jamison, professor of psychiatry at the Johns Hopkins School of Medicine, explains, "if the medication is working it's not just taking away pathology, it's taking away certain highly pleasurable, and occasionally highly productive states as well. As a result, the single most difficult clinical problem in treating manic-depressive illness is keeping people on the drugs. The compliance rate is discouragingly low."

THE POWER OF THE BRAIN

Clearly, the complex nature of our many individual genetic inheritances—and the utterly unique wiring of every individual brain—is more than reason enough to approach any kind of tampering with supreme caution. Too much dopamine? Too little serotonin? The recipe is far from clear—and recipe for what, in any case? If the beauty—and strength—of the human species lies in its diversity, we may not want to come up with the perfect set of genes. Novelists and screenwriters have shown us what such a world could look like, and it's terrifying.

Comprehending how our brains work, and giving newborns and growing children every opportunity to develop all the appropriate neural circuitry are very different matters, however. "Understanding that the brain can organize itself to do undesirable things is a profound insight," Ronald Kotulak points out in *Inside the Brain*, "akin to the discovery that germs cause infection."

Kotulak cites the studies of Michael Merzenich, a pioneering neuroscientist at the University of California at San Francisco. One of the first scientists to show that general learning can produce massive physical changes in the brains of adult animals, Merzenich says the brain reorganizes itself during learning:. "It's something that people don't realize. They don't think about the power that they have within themselves to change their brains." When someone has difficulty with language or can't read, Merzenich emphasizes, the problem is not that the brain

FOLLOWING PAGES: Combining yoga with big yuks, members of a laughing club in India gather on the beach for a mirthful workout. Proponents claim that regular bouts of laughter can counteract some effects of stress.

is defective, but that it has taken a different learning pathway. "Learning disabilities are usually learned."

The same may be said about everything else that makes up our mental, emotional, and physical lives—from our handwriting to what we find funny and what makes us angry, from our food preferences to how we react when a driver cuts us off. Not only our physical appearance and health but also our emotional makeup and temperament are the products of what our brains have learned and the ways that our biology and our experiences have pruned and sculpted trillions of neural circuits and connections in the grapefruit-size organ inside our skulls. At some point in our development—whether in the womb or the cradle, in first grade or high school—something happened, and our brains responded. Early on we learned to trust a loving caregiver or fear an abusive one. This morning we learned that doing 62 miles an hour in a 35-mile-per-hour zone can result in a very expensive stop on the way to work; now, without thinking about it, we'll slow down at a certain bend in the road.

Whatever we have learned, painful or pleasurable, our responses to the world have been colored by events and lessons. The good news is that we can hang on to the lessons that serve us in good stead and, with guidance and concentration, we can unlearn some of the habits and erstwhile skills that may now be counterproductive. Just as working out at the gym helps us build muscle, so does working out the brain build neuronal connections and, if recent research proves true, build neurons. What's more, giving the brain a workout can mean anything from doing crossword puzzles to learning to play an instrument or discerning the difference between a painting by Monet and one by Matisse. Judging by the way our dopamine-producing cells behave, it looks like our brains enjoy the challenge of learning something new. We may echo Lewis Thomas in not wanting to take over control of our brains— the task really would be beyond us, and besides, we'd never get to sleep. But staying flexible and exercising the phenomenal powers of our brains can help keep them— and us—in good working order.

....

Mind over matter: Dedicated worshipers in Sri Lanka (above) seem able to blunt the trauma of iron hooks inserted into their backs, while a sword swallower (opposite) masters the normal gag reflex of muscles that lead from mouth to stomach.

....

Shower Power: Maintaining a congenial network of friends and acquaintances throughout life, as the members of this swimming club do, may be one key to successful aging.

Aging

STEVEN N. AUSTAD

None of us can run as fast, throw as far, or hear as well when we are 50 as we could when we were 20. The reason is that we age. Aging is a gradual deterioration of our bodies. It affects every cell, tissue, and organ. Although aging itself isn't a disease, it makes us more susceptible to serious diseases as we grow older. The chance of dying from cancer, for instance, is some 200 times greater at age 80 than at age 20.

That's the bad news. The good news is that of all the millions of animal species on earth, humans have one of the slowest aging rates. We deteriorate very gradually over a number of decades, which is remarkably slow considering the continuous biochemical attack we are under. By contrast, dogs deteriorate markedly in less than a single decade, mice fall apart in a matter of months, and flies grow old in a matter of days.

Aging is not programmed in the sense that a hand or eye is programmed to develop in a fetus; aging does not proceed by a stereotyped series of steps. It is more unpredictable. It happens somewhat differently to different people and to different parts of the body. Even identical twins do not have identical health problems as they age, and they do not necessarily die from the same cause. Aging can best be thought of as the gradual failure of a program—the general repair program that nature designed to maintain the proper functioning of our bodies.

One way to measure the deterioration of aging is to examine the chance of dying at different ages in a large number of people. In a large group even a subtle decline in the efficiency with which the body works will kill some who for other reasons are particularly vulnerable. Aging has a characteristic death-rate signature. From birth to about the age of ten, the risk of dying progressively decreases. Young children are not aging, they are doing the opposite—getting better as they get older—although there seems to be no special term for what might be thought of as antiaging. But by about age 11 the risk of dying begins to increase—we begin to age—and the risk of dying increases at a predictable and accelerating rate throughout the rest of life. In the United States and other industrialized countries, the risk of dying doubles about every eight years after the age of 11—a fact that has not escaped the attention of life insurance companies.

Women live longer than men in every country in the world where the two sexes have even approximately equal access to health care. Can we say, then, that women age more slowly than men? The answer is no. For most of adult life, women experience the same rate of increase in the probability of dying as men. Men just die at higher rates at every age. Men die more frequently of just about any disease one can name, from flu to stroke to cancer. Men even die at higher rates in the womb and in childhood. We do not as yet understand the way in which females are better designed for survival than men. Men, but not women, also experience an exceptionally rapid increase of death risk between puberty and about age thirty. This has been called the period of testosterone dementia and is caused by accidents and violence, not by aging.

Anything—a bet, a debt, a population, or an investment—that doubles at a constant rate is said to change exponentially. Exponential growth can occur very quickly. For instance, compared with a youth of 20 years, we are 4 times more likely to die at age 36, 8 times more death prone at 44, 32 times at 60, and 256 times at age 84.

This doubling of death risk in about every eight years of age is surprisingly consistent across cultures and environments. It is not very different even under comparatively harsh conditions such as those of a World War II prisoner of war camp. The eight-year mortality-doubling time seems as much a part of our biological heritage as a big brain or an opposable thumb. For comparison, mortality rate doubles every three months in mice, every ten days in fruit flies.

HOW LONG CAN HUMANS LIVE?

The toll that aging takes on us can be appreciated by knowing that if, as we grew older, we remained at that pristine peak of health exemplified by a 10- or 11-year-old, life expectancy in the United States would be about 1,200 years instead of the 76 years it currently is. What's

Really more than 100 years old? Remote parts of South America, the Caucasus, and several other regions of the world claim exceptional longevity for their people. Often, such areas offer similarly spartan lifestyles that include farming, occasional food scarcity, lack of modern medical care, and—most notably—absence of reliable birth records.

more, one of us in a thousand could expect to live 10,000 years. We obviously don't stay at that peak, however. If our chances of dying double every eight years, even in countries with the best modern health care, what is the oldest that humans can possibly live to be?

There is no fixed upper limit to human life, but the chance of living beyond certain ages grows vanishingly small as death rate continues to climb with age. In countries such as Sweden, with excellent health care and reliable birth records going back 200 years, about 1 person in 2,000 lives to be 100 years old. The chances of a 100-year-old dying are about 50 percent per year in Sweden, so even if that rate did not continue to increase as people got even older, only about 1 person in 64,000 would be expected to live to 105 years, and fewer than 1 in 2,000,000 would live to 110 years of age. It is not surprising, then, that as of the last population survey not one of the inhabitants of Sweden had been documented to live as long as 112 years.

Despite these slim odds of an exceptionally long life in all places with reliable birth records, the world press reports with great regularity that in some newly investigated region in a remote corner of the globe people commonly live a hundred years, or considerably beyond. The most extreme claims come from a remote area of Azerbaijan in the Caucasus Mountains of western Asia. There one man, Shirali Muslimov, supposedly lived for 168 years; reports of ages in the 130s and 140s have been routine. Similarly, in the small Ecuadorian village of Vilcabamba almost everyone seems to live an exceptionally long life. One senior citizen, Miguel Carpio, said he was born 132 years ago, and 23 people claimed to be at least 100. None of these claims has been verified, and in all cases when birth records were later uncovered, it turned out that the people were either mistaken or telling tales. In Vilcabamba, for instance, a thorough follow-up study revealed that none of the reputed centenarians was a hundred years old. Their average age was 86, which is not bad for an area with limited access to modern sanitation and medicine but is certainly within the age range familiar to demographers of industrialized countries.

It would be surprising if humans today did not live substantially longer than in the past. We have better

sanitation as well as the undeniable benefits of modern medicine. Also, there are now many more people alive than ever before. So even if the odds of living to 120 years of age are vanishingly small, there are so many living people that someone is bound to do it.

The longest-lived person whose birth date we can believe was the Frenchwoman Jeanne Louise Calment, who lived to be 122 years old. She was born in Arles, in southern France, in 1875, and she died there in 1997. Ulysses S. Grant was President of the United States when she was born, and Mark Twain's *The Adventures of Tom Sawyer* was published when she was a year old. As a teenager, she met Vincent van Gogh.

Jeanne's father lived to the age of 93, her mother to the age of 86. The Calment family obviously possesses genes that favor long life. Although she maintained her faculties, especially a mordant wit, to the end, Jeanne Calment spent the last 12 years of her life in a nursing home, having become blind and so deaf that it was difficult to communicate with her. After breaking her hip at the age of 115, she was confined to a wheelchair.

Despite modern improvements in health and medicine, our conception of the practical limit of life has changed surprisingly little over the millennia. There have been major changes in life expectancy in historical times. But life expectancy, which is the average age at death of everyone in a population including babies, can be particularly misleading with respect to the ages at which adults actually die. Before modern sanitation and health care as many as half of all babies died before five years of age. Infant mortality of that magnitude depresses life expectancy dramatically and can give a false idea of how old adults live to be. For instance, Alexander the Great was generally considered to have died young, although at 33 he had exceeded the life expectancy of his time. And even though the life expectancy in ancient Greece was less than 30 years, we know of famous people such as Plato and Sophocles, who lived to be 80 and 90 years old, respectively. More than 4,000 years ago, the ancient Egyptians considered the practical limit of human life to be 110 years, and at least one pharaoh, Pepi II, probably lived into his 90's. So although there are unquestionably more old people alive now than ever

Like stars, telomeres—the protective tips of chromosomes that shorten when a cell replicates—glow yellow with fluorescent dye. Recent research has shown that when human telomeres are prevented from shortening by administering the enzyme telomerase, the cells appear youthful longer and undergo many more cell divisions than they normally would.

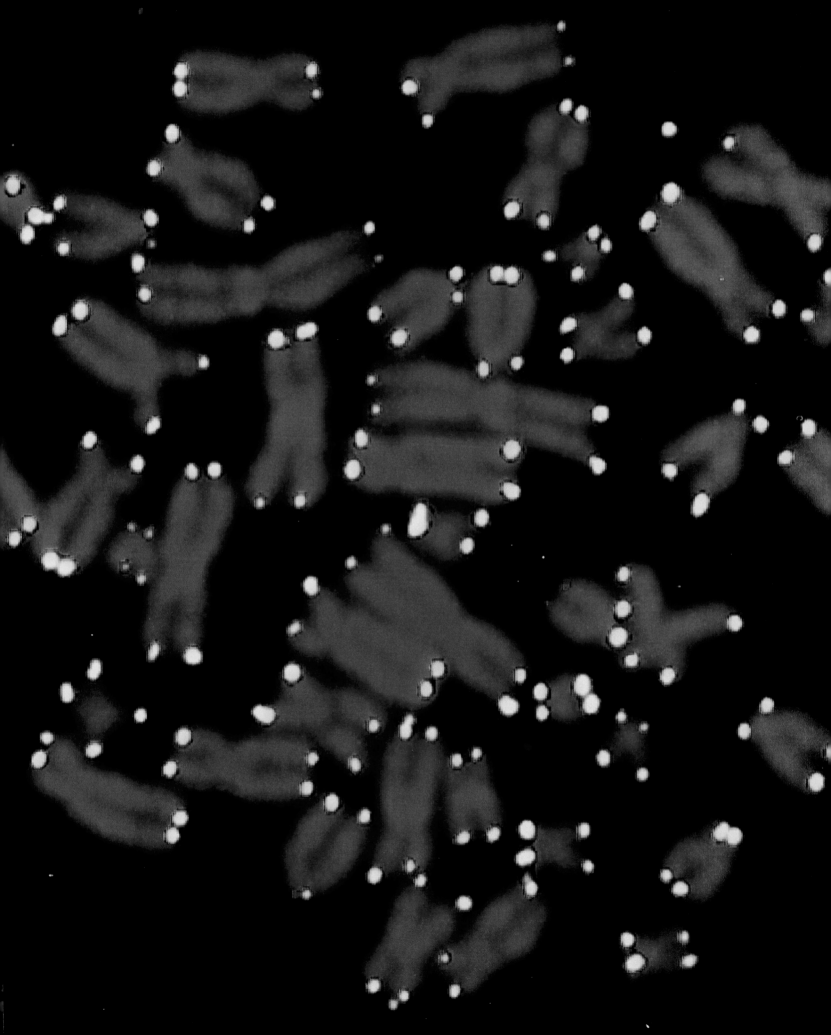

before, what is considered "elderly" has probably changed little over the course of human history.

....
WHY DO WE AGE?
....

We age because the fundamental processes of life are destructive. They put us under constant biochemical bombardment. Fortunately, nature has provided exquisitely effective mechanisms for protecting against, and repairing, the damage wrought by these destructive processes. These mechanisms aren't quite perfect though; eventually the destructive processes overwhelm us.

What are they? Let's start with the basics—eating and breathing. We need to eat and breathe to make the energy needed to sustain life. Our digestive system breaks down food into its component parts and disperses it through the body via the bloodstream. By far the most common of these breakdown products is the simple sugar glucose. Oxygen in the air we breathe is similarly distributed throughout the body by the bloodstream. It rides within a molecule, hemoglobin, that is specialized to carry oxygen. Once inside our tissues, oxygen helps dismantle—or oxidize—glucose, releasing the energy in the chemical bonds that previously held it together. In addition to energy, this process produces water, which we also need, and carbon dioxide, which returns to the blood and is exhaled through the lungs.

This process of energy production, called metabolism, is really like a controlled fire that burns at body temperature. Metabolism, like fire, cleaves apart fuel using oxygen as the cleaver. Deprive a fire of oxygen, and it dies, just as we do if deprived of oxygen. Fire releases energy in the form of heat as well as water vapor and carbon dioxide, which disperse into the atmosphere.

Both glucose and oxygen, the basic components of metabolism, have destructive sides that contribute to aging. We need oxygen to burn our fuel, but its use inevitably creates damaging by-products called free radicals that may be responsible for much of the damage of aging. Radiation also produces free radicals, which may be why exposure to excessive radiation shortens life and superficially mimics some symptoms of accelerated aging. Free radicals damage all the important molecules in our bodies.

Evidence that oxygen is damaging comes from studies in which animals raised in chambers containing more oxygen than the 20 percent in our normal atmosphere live shorter lives than animals raised in air. A rat kept in normal air will live two to three years, but the same rat can live in pure oxygen only about three days. In the 1940s, before the dangers of oxygen were known, the oxygen-enriched air in which they were incubated damaged the eyes of many prematurely born babies.

Free radicals are molecules with an unpaired electron. Electrons are most stable when they occur in pairs. When unpaired, they will rip electrons, or even whole atoms, loose from other molecules. In doing so, the electrons become stabilized, but only by producing another free radical, which will in turn rip an electron loose from another molecule, creating a sort of chain reaction. Free radicals are particularly damaging to molecules that make up cell membranes. They also damage DNA. Unrepaired DNA damage can affect any of the cells' normal roles, and can even lead to the uncontrolled replication of cells that we call cancer.

Fortunately for us, we have elaborate mechanisms for repairing damage to DNA. Biochemist Bruce Ames has calculated that free radicals damage the DNA in each of our cells some 10,000 times per day. Considering that our bodies consist of some 75 trillion cells, this adds up to as much as 10^{18} bits of DNA damage in our bodies each day, a number easily written but difficult to appreciate. Put another way, if each bit of damage in our DNA were represented by one inch, the total damage occurring daily within our bodies would be represented by the distance equivalent to about 75,000 round-trip journeys from Earth to the sun! Virtually all the damage is repaired, of course, which is why we can live as long as a century or more, rather than only a few days or weeks.

The more we study free radicals, the more problems of aging they seem to be involved in, from general loss of energy to cataracts and a host of diseases such as cancer and Parkinson's disease.

Like oxygen, glucose is both elixir and poison. It first became apparent that glucose might be involved in aging when scientists noticed that untreated diabetics, who have many times the normal amount of glucose in their blood, showed symptoms that looked like accelerated aging. For instance, some tissues in our bodies, particularly lungs, joints, and the walls of arteries, progressively stiffen as we age. In diabetics those tissues stiffen more quickly. Other common ailments of aging, such as cataracts, strokes, and heart attacks, also tend to occur earlier in life in untreated diabetics.

The reason for the acceleration of these symptoms is that we brown with age just as cooked meat browns—and diabetics brown even more quickly. Browning occurs when glucose in meat combines with protein during the cooking process. The same process occurs inside us. It occurs at a much slower pace, however, since our body temperature is much lower than cooking temperatures.

Chemically, browning occurs because glucose tends to stick to proteins at odd spots. Once having stuck, glucose acts like two-sided tape, sticking to other spots as well. If one of those spots is on a second protein, glucose binds the two together. If the second spot is on the same protein, it forms an aberrant loop or fold. Neither of these changes is conducive to the protein performing the chore that nature designed it to perform. What's worse, once these attachments are completed, they are not reversible. One technical term for these protein-glucose compounds is advanced glycosylation end products, but they are more often referred to by the clever acronym AGEs. AGEs accumulate with age.

It's easy to see browning in the long-lived proteins of our bodies such as tendons, which connect muscle to bone, or ligaments, which attach bones together. Tendons and ligaments are made from almost pure collagen, the most common protein in our bodies. Collagen forms nearly all the structure and support of our various organs,

....

Turtles, especially large species such as this green sea turtle, rank as the longest-lived vertebrates. Adults the size of dinner plates have lived to age 80; one tortoise reputedly served as mascot for a British fort on Mauritius for 152 years.
FOLLOWING PAGES: The body slows over time, but age is no barrier to reasonable athletic activity such as underwater ballet.

....

in fact. In a child, tendons and ligaments are glisteningly white, but as we grow older the white turns to yellow, and the yellow eventually becomes golden brown. Browned collagen is tougher but less flexible than young collagen, and this in itself can explain the stiffening of lungs and joints as we age. AGEs are most easily observed in collagen, but they are found elsewhere. AGEs may also play a role in trapping LDL cholesterol (the "bad" cholesterol) in artery walls, a step leading to atherosclerosis, or narrowing of the arteries. AGEs have also been found in the characteristic brain lesions of Alzheimer's disease and may play a role in its formation.

In the past few years it has also come to light that glucose not only attaches to proteins outside our cells but also can attach to DNA, the genetic material inside our cells. Glucose, then, can potentially disrupt the working of our genes. It may even cause mutations leading to diseases such as cancer.

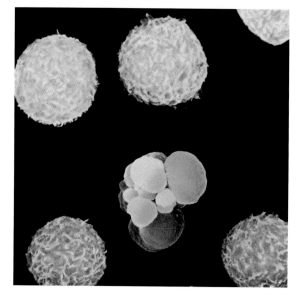

Perhaps the strongest evidence suggesting that browning is a serious problem in nature is that there are many types of simple sugars into which humans and animals could break down their food to use as fuel. Of all these simple sugars, glucose turns out to be the one that is least prone to forming browning products. Nature has designed our body to use the simple sugar least likely to cause us problems.

If the fundamental processes of life are inherently damaging, is there an absolute limit to how long animals can live? Some microscopic animals can go into a state of suspended animation called cryptobiosis and can probably survive for centuries, although this is not really the same as "living" all that time. The longest-lived among larger animals may be a type of clam, the ocean quahog, which is known to live as long as 220 years. Some deep sea rockfish live up to 140 years. Sturgeon are known to live longer than a century, and there are unconfirmed reports that pike and carp live that long, too. Turtles, particularly large tortoises, live very long lives. One famous giant tortoise called Marion's tortoise was reputedly kept as a mascot in a British fort on the island of Mauritius for 152 years until it died in 1918. Since it was an adult when it arrived, Marion's tortoise may have lived 200 years or more.

Considering that metabolism is involved in the damage of aging, it is not surprising that the longest-lived animals are cold-blooded. Metabolic rates of cold-blooded animals are many times lower than they are for warm-blooded ones such as birds and mammals. Humans may be the longest-lived mammals, although bats survive more metabolic expenditure per cell in their lifetimes. Birds generally live longer than mammals of the same size and expend many times the energy per lifetime as most mammals. Thus birds must have better defenses against metabolic damage than do mammals.

Both browning products and free radicals are the inevitable by-products of normal metabolism. In fact, these damaging processes can reinforce each other to produce more damage together than each would separately. That is, AGEs can combine with other chemicals to produce free radicals, and free radicals can speed up the formation of AGEs.

If the raw materials of metabolism are fundamentally involved in the damage of aging, it would make sense that aging rate and metabolic rate would be tightly related. People who exercise and thereby elevate their metabolic rate regularly might be expected to age more quickly than sedentary folks. In different animal species, cells might be expected to consume about the same amount of glucose and oxygen per animal lifetime before they suffer insupportable damage and cause the animal to die.

This is, in fact, a perfectly logical hypothesis. It has a century-long history in the annals of aging research, ever since German physiologist Max Rubner measured metabolism in five mammal species—guinea pigs, cows, dogs, horses, and cats—that ranged more than a thousand-fold in body weight and five-fold in longevity. Rubner discovered that each species indeed consumed about the same amount of energy per cell in a lifetime. Differences in longevity, he postulated, were due to some species, such as horses, using their allotted energy slowly, thus living relatively long lives, while other species, such as guinea pigs, used it faster and lived shorter lives.

This idea, for all its intuitive appeal, turns out to be wrong. We now know that some species routinely consume more than a hundred times as much oxygen per cell in a lifetime as other species. The flaw in the rate-of-living theory turned out to be that it ignored the fact that all oxygen-consuming organisms have elaborate mechanisms for resisting and repairing the damage of metabolism. In some species, such as humans, these resistance and repair mechanisms are highly efficient, so that our cells can withstand a great deal of metabolism in our lifetime. In other species, such as mice, the mechanisms are more rudimentary and inefficient, and the damage adds up much more quickly.

The mechanisms that resist or repair metabolic damage are of two general types. First, our bodies produce chemicals called antioxidants that defuse free radicals by transforming them into less harmful compounds. These antioxidants, which go by complex names such as superoxide dismutase, glutathione peroxidase, and catalase, are not the same as antioxidant vitamins, which must be obtained through the diet. Antioxidants are produced by our cells themselves. You can't increase your levels of these internally manufactured antioxidants by consuming them directly, even though some are sold in health food stores as dietary supplements. Superoxide dismutase, to pick the one that is most often found in stores, is a protein. Proteins are broken into tiny pieces in our digestive tract just as the protein in meat or eggs is.

Second, our bodies can resist metabolic damage by destroying and replacing cells or cell components that have been damaged. Components of our cells are being created and destroyed all the time. In fact, cells contain specialized structures called lysosomes and peroxisomes for the destruction of cellular materials that are damaged or are no longer needed. These structures might be thought of as the cells' chemical-waste disposal units. The importance of these units can be appreciated by noting that a mutation in the gene that codes for

these enzymes causes Tay-Sachs disease, an inherited malady characterized by seizures and blindness that usually kills its victims before the age of four.

Our waste disposal units get rid of damaged cell components, which are then replaced by turning on the genes that manufacture the components. If genes themselves are damaged, our cells have the capacity to repair the DNA. Remember that DNA consists of two complementary strands, so if one strand becomes damaged, the other can act as a template, or model, for its accurate repair. Unrepaired DNA damage typically leads to suicide of that cell and replacement by the division of neighboring cells, assuming the cell is in a tissue that is capable of continued cell division.

Cell suicide is known as apoptosis. As they die, apoptotic cells shrink, break into small fragments, and are absorbed by their neighbors. Cell suicide occurs constantly. It begins during development. As a hand forms from its paddle-shaped precursor, cells must die so that separate fingers can emerge. Apoptosis goes on throughout life. One layer of our skin continuously manufactures new cells. As these cells migrate to the skin surface, they commit suicide, then are sloughed off and replaced by new cells migrating up from below. If a cell's DNA is damaged sufficiently, it may become crippled or begin uncontrolled division, as in cancer. In either case, the DNA damage itself triggers an internal signal that begins the suicide process. Genes involved in governing cell suicide are also called tumor-suppressor genes, because they kill damaged cells that may be more likely to replicate uncontrollably and form tumors. When the tumor-suppressor cells themselves become damaged, cancer can result.

But even repair processes have their destructive sides. The capacity of cells to multiply, which allows them to replace cells that have been killed by external injury (called necrosis) or have committed suicide, can go awry, resulting in runaway cell division or cancer. Cancer is a major danger of aging in all animals that retain the power of cell division in adulthood. By contrast, the laboratory roundworm, which completely forgoes cell division after becoming adult, never gets cancer. Except

Surrounded by normal white blood cells, an individual cell resembles a cluster of colorful grapes as it undergoes apoptosis, or cellular suicide. Perhaps the suicidal cell was activated to fight a specific infection that later disappeared, negating the body's need for this defender. Virtually all parts of our bodies contain cells that are genetically programmed to die.
FOLLOWING PAGES: Does growing old together mean growing to resemble one another?

for a few reproductive cells—sperm and eggs—adult forms of this small transparent creature consist of exactly 959 cells, none of which ever divides. Scientists have looked long and hard for the occasional cancer in the worms, but have yet to find any. On the other hand, such animals have very limited powers to repair themselves once damaged. So their lives are short. Adults live only about three weeks.

Some cells in the human body, such as nerve cells and the muscle cells that allow us to move and our hearts to beat, forego their capacity to multiply about the time we are born. Cancer rarely develops among such cells. Once these cells are damaged, they cannot be replaced. Organs such as the brain, composed of nerve cells that can no longer divide, are particularly vulnerable to accumulated damage as we age. The opposite problem—uncontrolled division—is found in continuously dividing cells, such as those of the skin or the immune system. The most common cancers are those of cells that normally divide throughout life.

Skin cancers—melanomas and carcinomas—and immune-cell cancers, such as leukemias and lymphomas, are particular dangers associated with aging in humans. The action of hormones, which are chemical messengers with body-wide effects, often increase the risk of developing cancer by stimulating cell division in particular tissues in which cells would not otherwise divide, such as the breast, uterus, and prostate.

Thus aging is a consequence of normal biological processes that nature designed for our benefit but which also possess inescapable and damaging side effects or inherent dangers. Aging is a product of our genes, because our genes direct and carry out these processes.

AGING AT THE MOLECULAR LEVEL

A widespread folktale claims that because all our body cells are replaced every seven years, in some important sense we are not the same person now that we were seven years ago. Like most folktales, this one is part truth, part myth. The true part is that some cells, such as skin cells, are replaced regularly, as just mentioned. In fact, our skin surface is completely replaced about once a month. The mythical part of the folktale is that other cells, such as nerve and most muscle cells, are never replaced. They are as old as we are. Still other types of cells fall somewhere between these extremes. Liver cells, for instance, are replaced, but only when needed, as when existing cells die or are killed. Our bodies are actually made up of a combination of old, middle-aged, and young cells.

A second way that this folktale is misleading is that our bodies are made of more than just cells. We are also composed of various glues and support structures collectively called extracellular matrix, structures such as tendons, ligaments, cartilage, and bone. Although this matrix is secreted by our cells, once secreted it may stay around a very long time. Our tendons, for instance, are made from the protein collagen, and these protein molecules can be nearly as old as we are.

So even if we were entirely composed of new cells, we would still be held together by some very old proteins that might be subject to aging. In fact, our bodily proteins do age, due to a combination of wear and tear

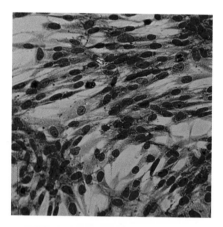

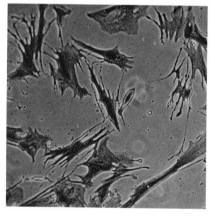

Artificially grown in a laboratory, skin cells divide a fixed number of times before stopping forever. At first they replicate vigorously, packing the surface of the dish (top). But as time goes on, they get larger and gradually lose their ability to divide, ultimately leaving sizable gaps in the culture (bottom).

from a lifetime of use as well as damage to their structure from AGEs and free radicals. Tendons and ligaments grow yellower, tougher, and less flexible with age, which helps account for the increasing rate of injuries among aging athletes.

Getting back to cells, can we distinguish individual cells that come from old people from those from young people? This is not the same as just distinguishing old versus young cells. Cells from the skin surface may be only a month old, yet we have no problem distinguishing an old person's skin from a young person's. If aging goes on at the cellular level, it would make sense that if we magnified any cell, old or young, to the size of a breadbasket and looked inside, we should be able to determine whether it came from an old or a young person.

Indeed, we can tell whether a cell came from an old or young individual, although the observations we would have to make are more sophisticated than you might expect. Recognizing old cells from old people is often easy. Many old cells accumulate deposits of a certain "aging" pigment called lipofuscin. Lipofuscin is the debris of cellular components destroyed within cells' chemical disposal units, specifically the units called lysosomes. The older the cell, the more lipofuscin it contains.

Old cells also have more damage to their DNA. Recall that DNA, our genetic material, is composed of four different chemical letters, the four nucleotides, arranged in a specific sequence more than three billion letters long. This DNA is located inside the cell nucleus and is arranged in 46 separate pieces called chromosomes. Any alteration in the inherited sequence is termed a mutation. Since free radicals damage DNA throughout our lives, and as most but not all of this damage is repaired, mutations gradually accumulate. So if we compared each of these three billion letters to the ones we were born with, we would see more and more differences in older and older cells.

But there is an even more obvious type of damage to the DNA in old cells, damage that occurs not in the cell nucleus but in the mitochondria of the cell. Mitochondria, the hundreds of sausage-shaped structures scattered throughout our cells, produce energy via the chemical breakdown of oxygen and various fuels. Since oxygen is a major source of free radicals, molecules inside the mitochondria should be in particular danger of free radical damage. Because mitochondria were once bacteria living independent lives before they happened to take up permanent residence inside other cells a

billion or so years ago, they still contain several small circular loops of their own DNA. In human cells each of these loops consists of 16,569 nucleotides, or 37 genes, involved in energy production. In recent years, it has been discovered that there is a gradual increase in the number of mutations found in mitochondria as cells age. Although some of these mitochondrial mutations are simply the substitution of one letter for another, other mutations involve the complete disappearance of as much as 20-40 percent of the DNA loop.

Young cells from old people are hard to distinguish from young cells from young people. It requires close examination of the chromosomes. Each chromosome is a bit shorter in young cells from old people because their telomeres have shrunk. Telomeres form "caps" of DNA on the ends of chromosomes. They keep the chromosome ends from sticking to one another—which would interfere with cell division—and are composed of a long sequence of units of the same six nucleotides repeated several thousand times. Whenever DNA replicates, as it must when cells divide, it cannot copy itself quite to the very end. Therefore, each time a cell divides, it loses about 15 of these repeated units. Over a lifetime of cell division, telomere shortening can be measured, but it requires sophisticated molecular techniques; even a long telomere composes less than one-ten-thousandth the length of an average chromosome.

For much of the past decade, telomere shortening was thought to be an important safeguard against cancer. Scientists believed that, once a cell continues to divide inappropriately—as it does in the early stages of cancer—its telomeres reach a critically short length, which cause the cell to halt further division. The development of full-blown cancer was theorized to occur only if a specific enzyme called telomerase was activated. Telomerase allows telomeres to maintain their length during cell replication and is generally active only in sperm and egg cells. Sperm and egg cells need to start off the next generation with full-size telomeres. Otherwise, over successive generations telomeres would get so short that even the cells of an embryo would be incapable of dividing. Reproduction—and life—would cease. However, recent work with mice genetically engineered to be incapable of producing telomerase has called this theory into question, as such mice still can get cancer.

So aging occurs in some fashion even in young cells from old people. The significance of these microscopic aging changes is not yet clear.

AGING AND THE SKIN

There is probably no more conspicuous difference between the young and the elderly than in the general appearance of their skin. Older skin is more wrinkled, more mottled, looser, more damaged in appearance, and more easily bruised.

The outermost layer of skin, the layer that peels off after a bad sunburn, is the epidermis. Its surface consists of a layer of dead, waterproof cells that are constantly being shed and replaced by new cells that migrate from the deeper epidermis. The epidermis also contains immune system cells, as well as cells specialized to produce the pigment melanin, which determines the color of our skin and our ability to tan. Beneath the epidermis, a thicker layer of skin called the dermis contains a welter of collagen and elastin fibers that give skin its strength and elasticity, as well as tiny blood vessels, hair follicles, and sweat and oil glands. Beneath the dermis is the so-called subcutaneous, or "beneath the skin," fat layer that helps insulate us from cold and attaches our skin to underlying tissues.

As we age, dermis and epidermis become less tightly bound, making our skin feel looser. The layers become more easily pulled apart, causing tears and blisters. Cells also divide less frequently and regularly with age. For the epidermis, this means that the dead cells of our skin surface will not be shed as quickly, that we will have more abnormal skin growths, and that wounds will take longer to heal. Also, we lose 10-20 percent of our pigment-producing cells for each ten years as we age. Even those cells that survive may not produce pigment as reliably as previously. Therefore, older skin becomes blotchier, does not tan as deeply or evenly as previously, and is more susceptible to sunburn.

In hair follicles this loss of pigment-producing cells leads to hair graying—or more correctly whitening—and the reduction of cell division rate leads to slower hair growth. We lose hair follicles as we age, and the remaining follicles narrow, so that even the elderly not subject to balding have progressively sparser, finer hair both on the head and the body.

The dermis of elderly people has lost about 20 percent of its previous thickness, due largely to the loss of fibers. Even the remaining fibers are not as effective at performing their function as they were previously. The stretching and twisting that accompany facial expressions eventually lead to the development of permanent facial lines in areas where we habitually wrinkle and unwrinkle our skin.

In other words, there is some truth to George Orwell's statement that by the time we are fifty, we have the faces we deserve.

Aging also leads to the progressive loss of tiny dermal blood vessels and sweat glands. Because humans control overheating by evaporating sweat from the skin and diverting blood from deep in the body to the skin surface—which is why we are flushed when overheated—the elderly gradually lose their capacity to cool themselves when exposed to hot temperatures.

The mortality statistics during any big city heat wave will attest to this fact. Some of this reduced cooling ability is also due to our brain's decreased ability to send out the proper signals to redirect blood flow. Older people also are more susceptible to colder temperatures because they lose insulating subcutaneous fat over much of their bodies, especially the extremities. This fat loss, combined with thinning dermis, makes elderly skin appear thin and almost translucent compared with the skin of young people.

Although skin ages by the same intrinsic processes as the rest of the body, it is also susceptible to accelerated aging from external factors. For instance, facial wrinkling is accelerated by cigarette smoking in direct proportion to the lifetime number of cigarettes smoked. Sunlight ages skin, too, and leads to skin cancers, which account for more than half the malignant cancers reported in the United States. Ninety percent of skin cancers, in fact, develop on that relatively small portion of the skin that is routinely exposed to the sun.

And yet, we require something that may ultimately be bad for us. Our skin needs sunlight in order to make vitamin D, a critical requirement for making and maintaining strong bones. The only major source of vitamin D besides sunlight on skin comes from dairy products, which are now supplemented, or fortified, with it.

Stretching a point, 75-year-old Foofie Harlan, a former dancer, shows that the best way to remain youthful may be to act that way. While all joints become less flexible with age, exercise and lifestyle can prolong agility far beyond one's teens.

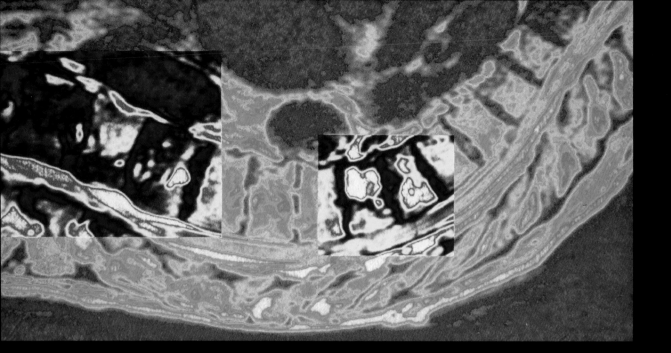

One of aging's dread diseases, osteoporosis can leave bones mis-shapen, the body all but paralyzed. A computer-enhanced x-ray (above) shows an osteoporitic spine; yellow blotches are vertebrae. One appears fractured; so little bone remains in some areas that individual vertebrae are difficult to recognize. Seriously degenerated hip joints now can be mended with metal replacements (below). In this x-ray, red indicates strong bone while weaker areas show as green-yellow. The femur on the left has little bone left.

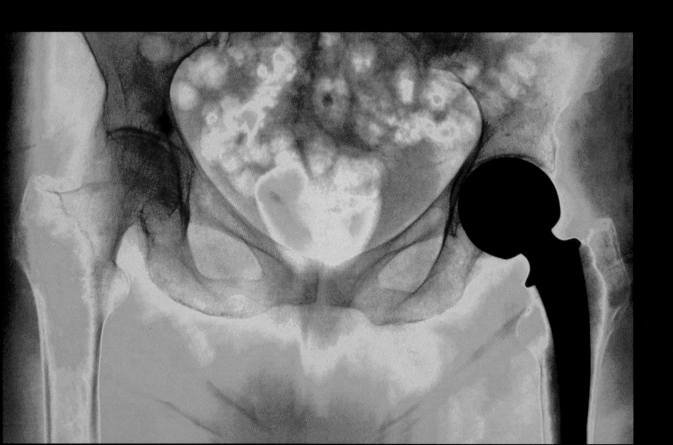

Aging, however, leads to a reduction in the skin's ability to manufacture vitamin D, even when exposed to the sun. Because the elderly are more frequently housebound than younger people and are less likely to have large areas of skin exposed to the sun even when outside, they are particularly reliant on dairy products for maintaining strong bones, a fact which needs consideration in dietary planning for the elderly.

The good news is that a lot of what previously was assumed to be inescapable skin aging isn't. Skin aging can be slowed by avoiding excessive exposure to the sun and by not smoking. Various skin ointments, usually derivatives of vitamin A, can make skin look younger, and cosmetic surgery is also an option.

behavior. For instance, it is now clear that weight lifting can increase a person's strength and endurance even through the 80s and 90s. Therefore, it is possible that a particularly sedentary young person might become a fanatical runner or weight lifter in later life, and could actually run farther or lift more weight at 60 than at 35. However, it is just as clear that someone who has always been a fanatical runner or weight lifter will not be able to run as far or lift as much at 60 as at 35. It is most accurate to think of ourselves as having certain physical potentials, which our behavior can allow us to approach to varying degrees. In this sense, strength and endurance potential inevitably decline after about age 30. But what we do with those potentials is entirely up to us.

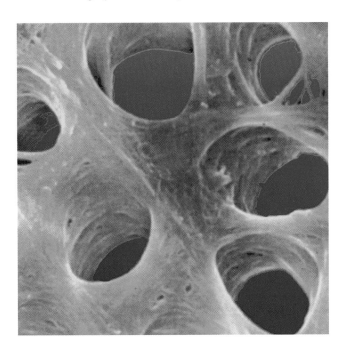

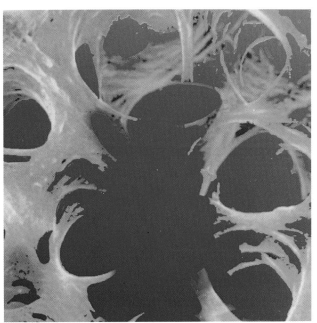

AGING OF THE SKELETON AND MUSCLES

Muscle strength and endurance decrease with age. Does this statement mean that everyone will be weaker and less fit at 60 than at 35? Not necessarily. The working of the human body can be modified enormously by

It is important to realize that fact because nearly all discussions of "normal" aging are really discussions about changes in physical potential rather than actual deterioration. Few of us ever reach our physical potential at any age; therefore, by hard work we might actually be able to improve with age. There are now probably many more runners who have completed their first marathon after age 30 than before. Because changes in physical potential are difficult to measure, aging research

Strong and lightweight, healthy, spongy bone found in the head of the femur and many other bones normally is riddled with holes like Swiss cheese (above, left). When osteoporosis develops, bone begins to vanish (above, right); even the remaining bone becomes less dense and more fragile.

summarizes changes in a large number of people under the assumption that important behavioral factors will cancel out. However, this may not always be true, and may give false indications of what is normal aging.

For instance, blood pressure increases after 30 in all industrialized societies studied, so rising blood pressure has been considered a normal part of aging. Yet a number of studies of so-called "primitive" peoples isolated from the technological world have shown that blood pressure does not always rise with age. What

The measurable decline in muscle strength in the aged is due to changes in both muscles and nerves, which must work together. As we age, some muscle fibers die. Some nerve cells die, too, causing the muscle fibers that they previously serviced to wither. Some muscle types, such as those used for running, are specialized primarily for rapid contraction. Other types, such as those used to maintain specific postures, are specialized for slow, sustained contractions. Aging seems to have a greater impact on fast muscles than slow muscles, which

has been considered normal aging actually may be a consequence of diet, exercise, or other lifestyle factors common to industrialized societies.

Muscles are composed of fibers formed from muscle cells that fused together. Most of these cells are non-dividing and cannot be replaced if lost. Like nerve cells then, muscle cells are as old as we are. The strength of any muscle is due to both the number and the might of its fibers.

may explain why sprint speed declines more rapidly with age than the ability to waltz for hours.

There is no more incontrovertible evidence of deterioration with aging than changes in athletic performance. However, it is important to note that there is less than a 20 percent change between the ages of 25 and 60 for relatively active people. In contrast, sedentary people show about a 30 percent decline in speed between their 20s and 60s. Continued activity and training can very

Still active in her golden years, weaver Eppie Archuleta has plied her craft in New Mexico for more than half a century. She carries on a seven-generation-long family tradition of weaving, which she is passing on to her granddaughter.

clearly have a substantial impact on athletic performance at any age. There are people in their 70s who are still capable of running a marathon in under four hours, a time superior to that achievable by most nonexercising people in their 20s. Bone strength, like muscle strength, decreases with age. Although it may seem inert, bone in a living creature is an active tissue that is continuously being molded and remodeled by the dual action of cells that create new bone—osteoblasts—and cells that dissolve old bone, called osteoclasts. During growth and early adulthood, bone creation outpaces bone resorption, so bone mass increases.

But as early as the mid-20s for some bones, resorption begins to outpace creation. Bone loss progressively increases with advancing age, beginning earlier and occurring faster in women than in men.

Women experience a particular acceleration of bone loss just after menopause, and this acceleration combined with their lower peak bone density means that they are considerably more likely than men to suffer the late-life collapses of vertebrae and fractures of wrist and hip characteristic of osteoporosis. The most graphic sign of osteoporosis is the "dowager's hump" caused by vertebrae that have gradually collapsed over years of normal activity such as bending and lifting.

Eight percent of American women currently 35 years old will suffer a hip fracture at some time in later life due to osteoporosis. After 50, the incidence of hip fracture doubles every five years, so that more than 85 percent of hip fractures occur after 65; three-quarters of these fractures occur in women. Although people do not typically die from hip fractures per se, up to 20 percent of elderly women die from complications within a year after the fracture.

In addition, because hip fractures often require lengthy hospitalization, and because many hip fracture survivors can never again lead independent lives, hip fractures among the elderly are estimated to cost Americans billions of dollars each year in medical and nursing home expenses. That cost is expected to increase substantially over the coming decades as the elderly population continues to grow.

Again, though, those people who develop high bone density early in life as a consequence of a good diet and weight-bearing exercise are much less likely to suffer osteoporosis later. Some current research suggests that diet and exercise late in life may also slow, and possibly even reverse, bone loss.

AGING AND THE SENSES

In *As You Like It*, William Shakespeare characterizes the "seven ages" of humans. The last age is "second childishness and mere oblivion,/Sans teeth, sans eyes, sans taste, sans everything." Is this a fair characterization? Shakespeare, of course, made use of poetic license. He also was in his 30s when he wrote these words, which probably were truer in the 16th century than today. But now that we understand and can treat periodontal disease, we no longer need to be "sans teeth." Current, very successful surgery for cataracts has helped us avoid being "sans eyes." Aging no longer needs to be as debilitating as it once was. Also, because of what physiologists call "reserve capacity," or the tendency of our bodies to do things considerably better than they absolutely have to, we can experience a considerable decline in something like visual acuity, mental processing speed, or muscle strength, and still have sufficient capacity to continue living normal lives even to the latest ages. To quote another English poet, Alfred Tennyson, on the subject of aging, "Though much is taken, much abides."

Our senses connect us to the external world. Like everything else, the senses deteriorate with age. But how much do they decline? And why?

Humans are particularly visual creatures, so we are fortunate that sight for the most part is well preserved into advanced age. Overall visual clarity, peripheral vision, and night vision all decline somewhat. The largest inevitable aging change is the gradual loss of the ability to focus on close objects, as is apparent by observing how far from their eyes young versus old people hold books. Close focusing deteriorates because our eye lens continues to grow throughout life. The lens, like an onion, is composed of concentric layers of cells. The normal cell contents are replaced with transparent proteins called crystallins, so that a good lens is as clear as fine glass. As long as we are alive, new cell layers are added to the outside of the lens. By age 70 the lens is about three times as large as it once was. To perform best, eye lenses must remain soft and flexible so that eye muscles can alter the lens shape to focus light from either near or far objects on the retina. The closer the object, the more these muscles must squeeze the lens into a spherical shape. As it grows, however, the len's additional cell layers make it stiffer and more difficult to squeeze. Finally it grows too stiff for close focusing, at which point we can get bifocals to do our close focusing for us.

The cells and proteins that compose the lens are not replaced as we age. They are as old as we are. Aged lenses grow somewhat less transparent, reducing night vision, and they develop a somewhat yellowish cast that affects color vision. Yellowish lenses do not transmit the so-called cool colors—blue, violet, and green—well, so these colors are more difficult for the elderly to distinguish. Warm colors such as red, yellow, and orange remain easily seen throughout life. Even when we are young, there are some colors of light—the near ultraviolet—that birds and insects see, but that we cannot. Our lenses filter them out. People who undergo surgical lens removal later in life because of cataracts can actually see a range of colors that even the youngest child can't perceive.

Our vision may not be so well preserved into old age if we develop specific eye diseases. When part or all of an eye lens becomes opaque instead of transparent, the affected area is called a cataract. By definition it reduces vision, although the degree of vision loss depends on the size and nature of the cataract. Until the mid-1970s, cataracts were a major cause of blindness in the elderly. Today surgery to remove the diseased lens and replace its function either with glasses or a clear plastic implant is successful more than 95 percent of the time.

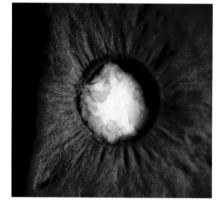

Glaucoma, increased pressure inside the eye due to a buildup of fluid between the front of the eye and the lens, affects 1 to 3 percent of the population over 40 years of age and is the second leading cause of blindness in the United States. Called the "sneak thief of vision" because it is painless and initially subtle, glaucoma can permanently damage the optic nerve and retinal nerve fibers. An early sign is the loss of peripheral vision. Most glaucoma can be treated with drugs, although some cases may require surgery to drain excess fluid. Diabetes mellitus and nearsightedness both increase one's risk of developing glaucoma. With increased vigilance through routine eye exams, and our modern array of pharmaceutical and surgical techniques, glaucoma also promises to be less daunting than it once was.

The leading cause of blindness among the elderly is macular degeneration. The macula is that part of the retina we use when looking directly at something. It allows us to see finer detail than any other part of our vision. To verify this, try reading something out of the corner of your eye. Macular degeneration usually results from our body's suddenly beginning to make new, unneeded blood vessels within the retina, and it destroys that part of our vision we use and need most. Peripheral vision is usually not affected. Macular degeneration is accelerated when these new blood vessels rupture. Although exposure to bright sunlight has been linked to macular degeneration, genes are involved, too. Treatments remain limited, the chief one a type of laser surgery that seeks to slow the disease's progress by sealing off or destroying the extra blood vessels. Hopefully, our understanding of the disease processes will improve now that a responsible gene has been identified.

There is probably no more common and unkind caricature of the aged than of someone cupping a hand to an ear and yelling "Huh?" Hearing loss is undeniably a problem of aging. While fewer than 1 percent of people under 17 have significant hearing loss, that number grows to 40 percent by 75 and continues to increase. One researcher has estimated that if life expectancy could somehow be increased to 140 years, virtually everyone would become deaf.

Hearing is generally measured by the ability to detect some combination of loudness, measured in decibels (named for Alexander Graham Bell, inventor of the telephone) and pitch, measured in hertz (Hz), or vibrations per second. Our hearing is most sensitive to sounds around 3,000 Hz (equivalent to the higher notes on a piano); 25 percent of us can detect sounds as soft as one decibel. A typical whisper is about 20 decibels. Sounds

A mature cataract (above) results when the eye's normally transparent lens becomes increasingly cloudy and, eventually, completely opaque. Such changes to the lens seem to be due at least in part to the same fundamental processes—oxidation and browning—that are involved in aging itself.

IMMATURE CATARACTS PRODUCE FUZZY VISION

GLAUCOMA ROBS PERIPHERAL VISION FIRST

NORMAL VISION

Hazy horizons: These photographic simulations approximate what people afflicted with various eye diseases actually see. Compared with normal vision (left), the spectrum of disorders includes cataracts, glaucoma, macular degeneration, and damage due to uncontrolled diabetes. Effective treatments now exist for all these conditions except macular degeneration.

MACULAR DEGENERATION DESTROYS THE RETINA'S CENTER

DIABETES CAN DAMAGE MANY RETINAL AREAS

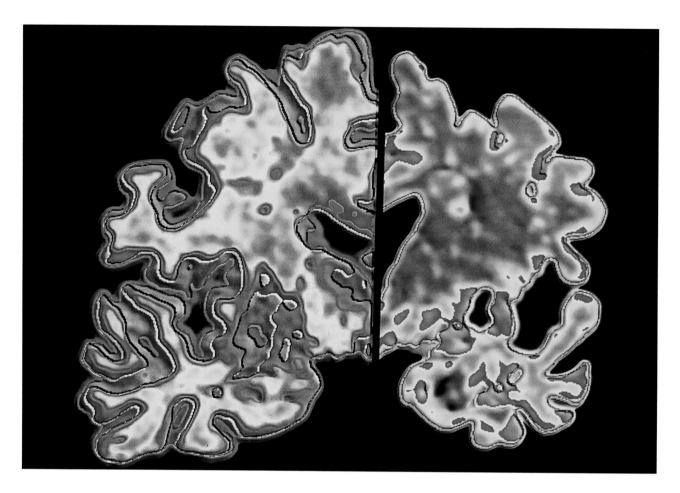

above 120 decibels are painful. At birth the human ear can detect sounds as high-pitched as 20,000 Hz, two and a half octaves above the highest note on a piano. However, we begin to lose our ability to hear the highest-pitched tones almost immediately, because nearly every part of the hearing apparatus is constructed of nonreplaceable parts, and the parts that perceive the highest pitches are the most delicate. A typical medical test of hearing involves frequencies up to only 8,000 Hz.

We sense sounds because they vibrate some of the 20,000 to 30,000 hair cells of our inner ear. Besides being irreplaceable, these sensors are capable of only limited self-repair. Damage to the hair cells is responsible for most hearing loss in the elderly. Hair cells occur in various lengths and stiffnesses. Just as different pitches are created by running a finger over comb teeth of different lengths and stiffnesses, specific pitches of sound in the air vibrate specific hair cells. Extreme vibrations caused by loud sounds are most damaging to hair cells.

Since higher frequencies are progressively more difficult to hear as we age, the most common hearing complaint is not being unable to detect sound, but being unable to hear speech clearly, particularly amid background noise. Spoken consonants are higher pitched than vowels; soft speech becomes an incomprehensible collection of vowels. Not all difficulty understanding speech is due to hearing loss itself, though. Among those suffering the same amount of hearing loss, older people have more difficulty distinguishing words than younger ones. The elderly also have more trouble understanding rapid speech and unfamiliar accents, which suggests that part of the problem is due to deterioration in the way our brains process the auditory information our ears receive.

Hearing loss can lead to depression and social isolation in the elderly, particularly since people with normal hearing seem to have less patience and empathy with hearing impairment than they do with visual impairment. Poor hearing is also associated with a host of other physical ailments; it is likely those are due to the psychological stress involved in difficulties communicating with other people. The fact that fewer than one person in three with hearing loss wears a hearing aid suggests that people generally, and the medical profession in particular, do not take normal age-related hearing loss with the seriousness it deserves.

Vision and hearing are not the only senses compromised by aging. All senses are affected to some degree. As we age, we have a less acute sense of taste and smell, more tolerance for pain (as our pain receptors deteriorate), and less ability to feel pressure applied to the skin. I like to think there is an upside to these changes as well. The elderly can eat chili that drops a child to its knees and can affect what appears to be an admirable stoicism when it comes to minor pains such as bee stings.

THE AGING NERVOUS SYSTEM

Though many aspects of mental function deteriorate during aging, none declines to the point of affecting daily life; some functions even appear to increase. As long as we don't develop specific afflictions such as Alzheimer's, we can continue to learn throughout life. No matter how old we are, nerve cells are still capable of sprouting new connections to other nerves, which neuroscientists believe are necessary for learning. Journalist I. F. Stone was in his 60s when he learned classical Greek, a task he compared to climbing Mount Everest on his hands and knees. Learning may not be as rapid as we age, but it still occurs. And our storehouse of knowledge, judgment, and presumably wisdom, can all continue to increase.

Part of wisdom is knowing how to compensate for changes that aging may bring. Successful aging to a large extent many be the willingness and capacity to change. One person who aged very successfully was the painter Ann Mary "Grandma" Moses. For many years she embroidered on canvas. When she was 78 she realized that her fingers had become too stiff to properly manipulate a needle, so she took up painting and became world famous. She illustrated an edition of *The Night Before Christmas* when she was a hundred. Someone else who, when handed lemons, proceeded to make lemonade was the painter Claude Monet. He completed what many authorities believe to be his best work, a series of paintings of water lilies, at 85, when his eyesight was failing. Monet's style, splotches of shimmering color that take apparent form only from a distance, is an appropriate one for someone with failing visual acuity.

Certain changes in brain structure occur during aging. The billions of nerve cells in the brain cannot be replaced if they die or become nonfunctional. Not surprisingly then, nerve cells (gray matter), as well as the nerves' supporting cells (white matter), gradually may be lost over time, although the loss is much less than previously thought. Some brain regions, such as the hippocampus, are very susceptible to cell loss, particularly in diseases such as Alzheimer's. But with our built-in reserve capacity, this cell loss needn't be critical. Even people who have major portions of their brains damaged by accidents or strokes often regain most functions as other regions compensate. So the numbers of connections that an individual nerve cell has with surrounding nerves may increase with aging, in partial compensation for overall cell loss.

There is also an uneven deterioration in the ability of the brain to perform certain functions. Language ability, for instance, seems little affected. In the absence of mental disease, recollection of word meanings and other acquired information as well as the ability to comprehend written text appear to be maintained until at least the 80s. On the other hand, mental performance requiring speed seems to decline slowly throughout life from a peak at about the age of 20. The degree of decline appears to depend upon the complexity of the task. Simple reaction time, such as how long it takes to push a button after a light goes on, slows perceptibly with age. It is composed of decision time—time until we begin to push the button—and motor time—time needed to finish pushing the button. The most noticeable age change is in decision time. Therefore, it isn't surprising that we show a more dramatic decline in our speed to perform more complex tasks (for example, press one button if a green light comes on, and another if a blue light appears),

Side-by-side computerized "slices" of brain tissue depict sections from a normal person (left) and an Alzheimer's patient (right). Obvious differences in size of various structures stem from nerve cell degeneration and death. Researchers have associated four gene variants with the risk of developing Alzheimer's, leading to hope that effective treatments may soon be within reach. FOLLOWING PAGES: Stroke victim slumbers in a nationally subsidized elderly care center in Japan.

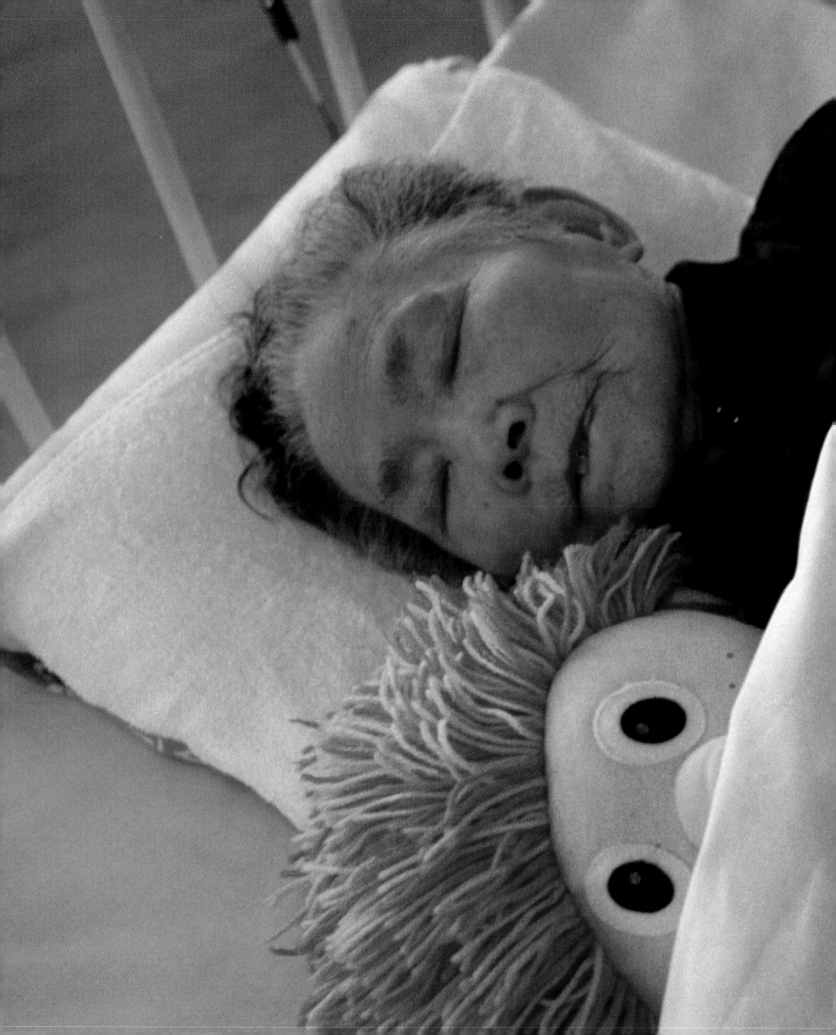

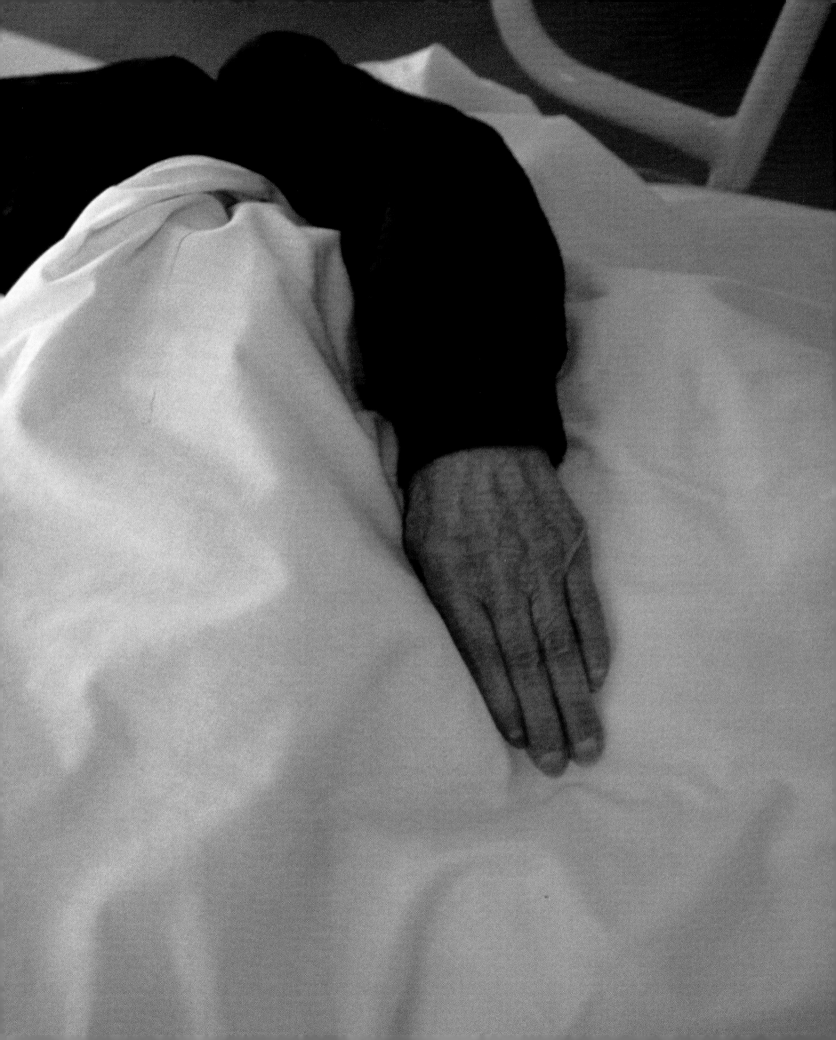

which by definition require more decisions. Since we know that the speed with which impulses are transmitted between nerve cells decreases with age, and that processing information requires nerve impulses to be transmitting across many nerves, there seems to be a straightforward explanation of declining decision speed.

A frequent complaint among the elderly is difficulty with short-term memory. It is not as easy to remember what they did yesterday compared to when they were younger. That complaint has some validity. Psychologists distinguish three types of memory—primary memory, which is extremely short-term such as remembering a telephone number long enough to dial it; secondary memory, which is what most people mean by short-term memory and deals with recollections such as what you saw on television last night; and tertiary memory, which is long-term memory that covers various facts you learned, and what you did, earlier in life. Primary and tertiary memory are much less affected by aging than secondary memory. My tertiary memory, for example, allows me to tell my children stories of my own youth that keep getting more and more interesting the older I get.

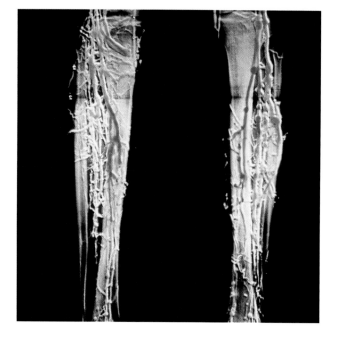

Secondary memory is usually tested by asking people to recall a long string of words or numbers within minutes to hours of having first seen or heard them. In this type of test there is a relatively sharp decline from the 20s and 30s throughout life. One experiment, asking people between 30 and 80 to recall two lengthy paragraphs 15 minutes later, showed that people in their 50s recalled about three-quarters as much, and those in their 70s about half as much, as people in their 30s. Of course, the older people probably remembered the important parts, having learned from a lifetime of experience what is important and what isn't.

Kidding aside, older people are better than younger people at some mental tasks. Older people typically show more "wisdom," defined as a recognition of the degree of complexity, difficulty, or uncertainty inherent in specific everyday problems, than younger people. Older people also often score better on what is called prospective memory, that is, remembering something to be done in the future. Of course, older people may use various tricks, such as writing down what needs to be done, in contrast to younger people, who are more likely to rely on memory alone. But knowing one's own capabilities is part of what wisdom is all about.

The changes we've described so far exclude diseases of the aging brain. However, probably no more than one-third of Americans over 85 are the "cognitively normal" individuals we have been discussing. Some 25-50 percent of people over 85 develop Alzheimer's disease or some other type of incapacitating brain malfunction.

In 1907 physician Alois Alzheimer applied some of the chemical dyes that were being developed by the new photographic industry to the autopsied brain of a patient who had died at 55 after five years of rapid mental deterioration. These dyes allowed Alzheimer to identify for the first time the characteristic physical signs of the dementing disease that has come to bear his name.

One sign is the presence of so-called neurofibrillary tangles, abnormal tangles of protein found inside nerve cells, particularly in the brain regions associated with memory. The second sign is a number of neuritic or senile

Of time and circulation: Varicose veins lace this leg scan (above) with purple ribbons. Such abnormally swollen blood vessels result when valves inside veins, which ordinarily help blood get back up the legs from the feet, fail. Recent research indicates that moderate consumption of alcohol (opposite) can lower the risk of heart attack.

plaques, a solid, complex, cluster of proteins lying outside nerve cells. Alzheimer's disease is not straightforward to diagnose, either in a living person or at autopsy. Diagnosis in a living person, in fact, is usually classified as either probable, possible, or definite Alzheimer's disease, based on the existence of progressive memory loss and hallucinations, plus the loss of reasoning ability, normal gait, ability to speak coherently, ability to recognize familiar people or places, or a number of other signs. The difficulty with diagnosing the disease even at autopsy is due to the fact that neurofibrillary tangles and neuritic plaques also occur in aged people without Alzheimer's disease or any sign of excessive brain aging, although the tangles and plaques usually occur in less profusion and in different brain regions in unaffected individuals.

Three genes on three different chromosomes have now been identified that cause familial, or inherited, Alzheimer's disease. A fourth, Apolipoprotein E, has three relatively common forms, called Epsilon2, 3, and 4 for short. Epsilon4 is associated with both a higher risk of developing the disease and a younger age at onset compared with the other gene forms, although many people carrying Epsilon4 never develop Alzheimer's disease. Researchers are making rapid progress in understanding the underlying disease processes, and drugs that slow or help prevent the disease are beginning to appear. Postmenopausal estrogen replacement and aspirin are commonly taken drugs that may help retard and prevent the disease.

In addition to those over 80 who develop Alzheimer's or other incapacitating brain malfunctions, another 30 percent have less marked, sometimes subtle, mental deterioration. Unless new medical advances can improve this situation, the continuing increase in life expectancy that is expected over the next decades will dramatically increase the number of elderly confined to institutions.

THE AGING CARDIOVASCULAR SYSTEM

The human heart is a masterpiece of nature's design. Beating nearly three billion times per lifetime, it continues to function perfectly decade after decade. It is difficult to imagine a human device of any sort that could operate so long and so reliably. Yet humans are rather unusual among mammals in that they often die of heart and artery diseases—so-called cardiovascular diseases such as heart attack and stroke.

Cardiovascular diseases, in fact, are responsible for about half of all deaths among the elderly in westernized societies, although the death rate from these diseases has been declining since about 1970 in the United States and is now some 30 percent lower than at its peak. Does this mean that, for all the amazing design of our heart, the cardiovascular system ages more rapidly than the rest of our body? Actually, there is little evidence to indicate that this system ages particularly quickly. Some early studies found almost no aging changes of any sort in the heart.

We now know, however, that because a resting heart functions at only a small fraction of its capacity, aging changes are difficult to detect. Aging becomes readily apparent when the heart is stressed, as by exercise or high blood pressure. Chronic high blood pressure changes the heart in ways very similar to accelerated aging. It shouldn't be surprising that heart function deteriorates as we grow older, because heart muscle is composed of cells (like brain cells) that do not divide, and thus cannot be replaced if they become damaged.

Still, our heart changes less than might be expected. One obvious change is size. The heart grows by about 0.3 percent each year between the ages of 30 and 90,

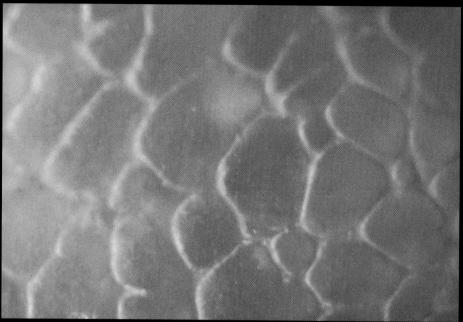

HEALTHY LUNG TISSUE OF A THREE-YEAR-OLD CHILD

How we live affects how we age: It's as obvious as callused hands and facial wrinkles. Tobacco smoke, for instance, contains free radicals—damaging molecules involved with aging. The lungs of a smoker—or even of a 50-year-old nonsmoker who has lived amid polluted city air—may be more damaged than those of an 80-year-old, country-dwelling nonsmoker. Similarly, the accumulation of fatty deposits on arterial walls (atherosclerosis) progresses at very different rates, depending not just on genes but also on lifestyle. Smoking, a high-fat diet, lack of exercise, and possibly even stress all can accelerate the atherosclerosis. As science identifies more environmental factors to aging, we may be able to slow its progress by making appropriate living choices even before new medical breakthroughs become available.

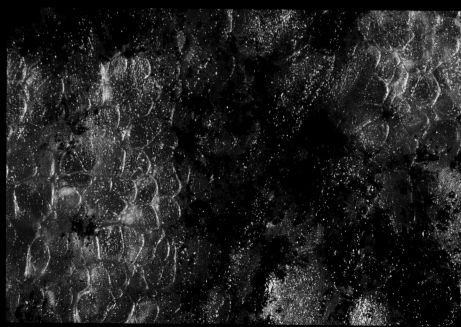

LUNG TISSUE OF A CITY DWELLER EXPOSED TO A LIFETIME OF POLLUTED AIR

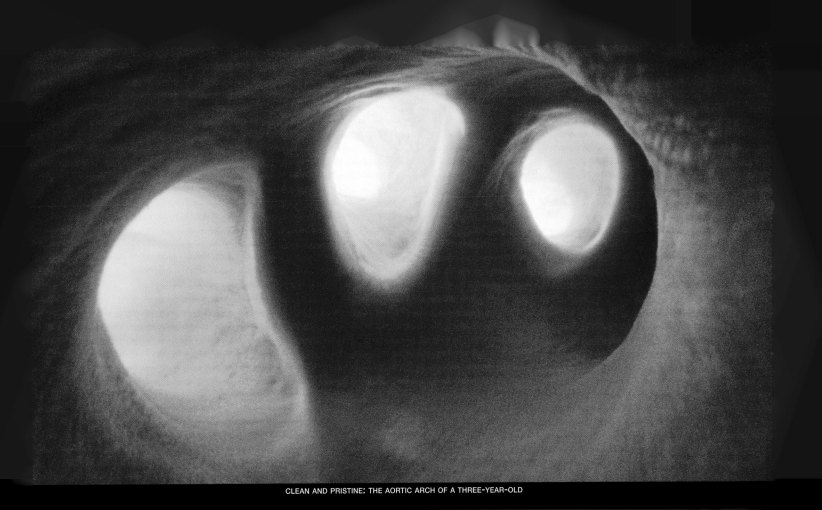

CLEAN AND PRISTINE: THE AORTIC ARCH OF A THREE-YEAR-OLD

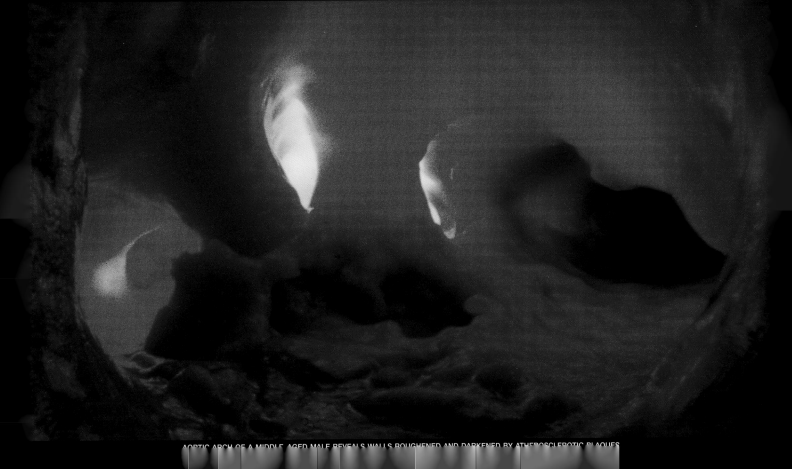

AORTIC ARCH OF A MIDDLE-AGED MALE REVEALS WALLS ROUGHENED AND DARKENED BY ATHEROSCLEROTIC PLAQUES

mainly due to increasing thickness of the left ventricle. The left ventricle pumps blood into the main artery, the aorta, and from there to the rest of the body. Thickening of the left ventricle may be due to a gradual stiffening and occlusion of the arteries, much as our cheek muscles would be enlarged and strengthened by repeatedly blowing air into a thick-walled balloon as opposed to a thin-walled one. Also, by age 60, the heart's pacemaker, which initiates cardiac contractions, begins to deteriorate. Specifically, fewer and fewer pacemaker cells remain, so that by age 75 only about 10 percent of the original cells still exist, though so long as these remaining cells work effectively the heart continues to function adequately. Finally, normal resting heart rate decreases with age.

Arteries change in more obvious ways than the heart. Arteries are like elastic, very thick-walled hoses, through which pressurized fluid—blood—continuously flows. The most obvious difference between young and old arteries is in their previously mentioned tendency to stiffen over time. Arterial stiffening, since it occurs more quickly in the arteries of uncontrolled diabetics, may be largely due to browning of the normal collagen fibers in the arterial walls. Associated with this stiffening is an increase in artery diameter. Further changes are associated with the development of the disease atherosclerosis, in which arterial walls are dramatically thickened and their inner channel becomes constricted. When the constriction begins to inhibit blood flow to a particular area, the disease becomes medically significant.

Atherosclerosis begins developing very early in life but progresses at different rates in different people, depending on heredity and lifestyle factors such as diet and exercise. The earliest signs of the disease are fatty streaks, deposits of cholesterol and certain white blood cells, under the lining of large and medium-size arteries. Since high levels of LDL cholesterol in the blood are consistently related to the speed with which atherosclerosis develops, it is thought to be that form of cholesterol that initially enters the artery wall. Fatty streaks have even been found in the arteries of infants and young children. After fatty streaks develop, muscle cells of the arterial wall may begin multiplying, and larger droplets of fats may accumulate. As atherosclerosis proceeds, muscle cells continue to divide, more fat accumulates, and collagen forms over the top of the lesion. In late stages of the disease, arterial plaques become complex aggregations of fats, collagen, muscle cells, and dead white blood cells. Blood clots may form on advanced lesions.

A heart attack is a sudden disruption of heart function, most often due to a lack of blood—that is, oxygen—supply to the heart muscle. Without oxygen, the heart cannot continue contracting, and cells begin to die if the blood supply is not restored. It might seem surprising that an organ continuously filled with blood can suffer from a lack of blood supply. However, blood inside the heart chambers themselves cannot penetrate into the heart's muscle cells where it is needed. Those cells are supplied by the coronary arteries. Heart attacks are colloquially known as coronaries because they usually result from a blockage of one of the coronary arteries, either because it has progressively narrowed due to atherosclerosis, or because it has partially constricted and a blood clot has lodged at the constriction. The decline in deaths from cardiovascular disease over the past three decades is due to many factors, among which are blood-thinning pharmaceuticals, coronary bypass surgery (in which new arteries are surgically implanted to circumvent the constricted artery), angioplasty (in which a device is inserted into a constricted artery to mechanically widen it), and changes in lifestyle.

Similarly, stroke, the other major cardiovascular disease, is a disruption of blood supply to a part of the brain, which also leads to cell death if the blood supply is not quickly restored. Like a coronary, stroke can occur from either the progressive constriction of arteries inside the brain, or from a blood clot that originated elsewhere lodging at a partial constriction of a cerebral artery.

Ironically, permanent damage to heart or brain tissue in survivors of heart attack or stroke may result as much from the resupplying of oxygen to the tissues as from the original oxygen deprivation. When oxygen is resupplied, the tissues may be bombarded with oxygen free radicals and become damaged by processes similar to those involved in the more gradual damage of normal aging.

The sum of changes in the heart and arteries leads to an overall decrease in maximum oxygen consumption during vigorous exercise. Part of this decreased capacity is due to a reduction in the ability of the heart to pump more blood when needed. That is, during extreme physical exertion the heart rate of older people increases less than that of younger people. Even during relatively mild exertion, such as squeezing a rubber dumbbell as hard as possible, the heart rate of men in their 20s typically increases by about 50 beats per minute as compared with only 12 beats per minute among healthy men in their mid-50s and mid-70s. A second part of decreasing oxygen use with age is that we typically have less muscle mass as we age, and the remaining muscle tissue cannot use oxygen at the same rate as previously. This decline may relate to cumulative damage to mitochondria inside each muscle cell. A third part of decreased maximum oxygen consumption relates to a declining efficiency of our lungs to extract oxygen from the air we breathe. Most treadmill tests are limited more by shortness of breath than by muscle exhaustion.

Despite these inevitable decreases in average heart, artery, muscle, and lung performance, it is now becoming clear that exercise can reduce the rate of all of these changes. Highly conditioned older athletes can consume more oxygen, and therefore perform greater muscle work until at least the age of 70 than lean but unconditioned young people. Also, even elderly individuals who begin conditioning programs can regain a significant degree of lost function.

AGING IN THE REPRODUCTIVE SYSTEM

Our ability to conceive children, like most everything else, diminishes as we grow older. But strangely, and very much unlike other aspects of aging, the sexes differ dramatically in the speed of the decline of their reproductive systems. Remember that although women live longer than men, it is because they survive better at every age rather than because their survival rate changes more slowly. Women, in other words, have a superior physiological design for survival compared to men, but the sexes age at the same speed. However, in terms of reproductive ability, healthy men can sire children throughout life. Men have become fathers as late as the age of 95. By contrast women, although they can perform sexually throughout life, are incapable of bearing children after about age 50 regardless of their state of health. In order to understand the details of how reproduction

deteriorates with aging, and in order to interpret the difference between the sexes, we need to understand something about the two types of sexual organs—the genitalia and the brain.

The role of the ovaries and testes in reproduction is well known; they produce gametes, sperm in men and eggs in women, which must unite to produce children. The genitalia also produce sex hormones—mainly testosterone in men, primarily estrogen and progesterone in women—which are required for puberty and the continuing production of functioning gametes after we become adults. However, it is less well appreciated that we cannot reproduce without appropriate input from the brain. One gland connected to the brain, the pituitary, produces hormones called gonadotropins, which are as essential to reproduction as the hormones made by the genitalia. Gonadotropins orchestrate gamete development and hormone production in the genitalia, while the sex hormones coordinate production of the gonadotropins. The process is a delicate and precise pas de deux, which if disrupted will also disrupt fertility.

Considering the details in men first, gonadotropin secretion stimulates certain cells of the testicles (or testes as they are medically termed) to produce testosterone, which enters the bloodstream and affects functions as diverse as beard growth, muscle development, and behavior in other parts of the body. Within the testes, however, testosterone's effect is to stimulate another cell type to produce sperm—about a hundred million sperm per day in an average adult.

As men age, they produce less testosterone, although the decline is subtle (about 1 percent per year) beginning at about 40. Men also produce fewer sperm per day as they age, a smaller proportion of their sperm are active, and children sired by older men are more likely to have birth defects, suggesting that sperm from older men contain more genetic mutations. Older men also report less frequent intercourse; their sexual response slows and their frequency of impotence (due to various diseases) increases. Only about one-third of men older than 80 are still sexually active. Whether these various behavioral changes are directly due to declining testosterone levels is unclear. It is clear though, that there may be no aspect

FOLLOWING PAGES: Eyes glued to a monitor, doctors watch the progress of a catheter as they insert it into a patient's femoral artery and then thread it through his circulatory system into the arteries of his beating heart. A tiny balloon at the catheter's end can be expanded to reduce blockages in the arteries caused by the age-related condition called atherosclerosis.

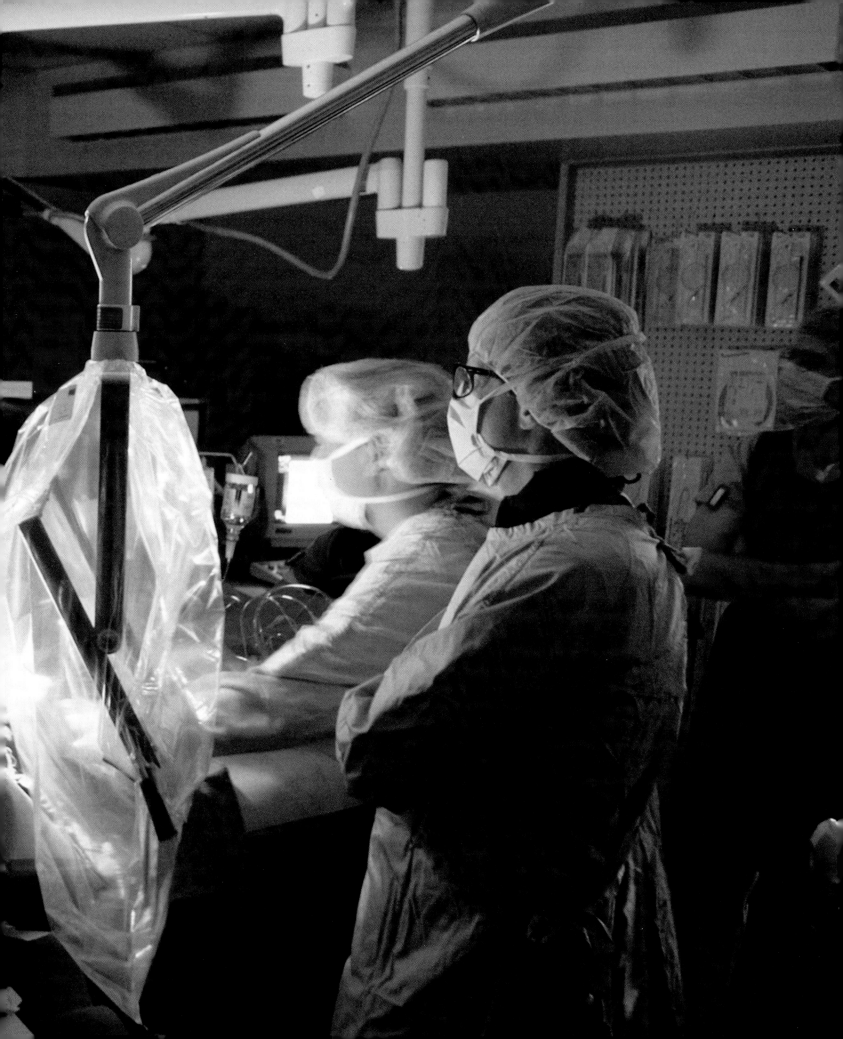

Angiograms—scans of blood vessels of a living patient—can reveal potential and actual blockages in the circulatory system. This angiogram (below) shows the coronary arteries spread like fingers over the top of the heart. Multiple constrictions in those arteries are due to atherosclerosis, a stage of arteriosclerosis characterized by fatty deposits on arterial walls. In many places, blood only trickles through. Normal arteries appear as thick pink highways; large orange streaks are ribs. Close-up of some arteries (right, top) shows reduced blood flow. A balloon catheter (middle) is inserted into the artery to widen it and restore blood flow (bottom). This procedure, called balloon angioplasty, is often used as an alternative to heart bypass

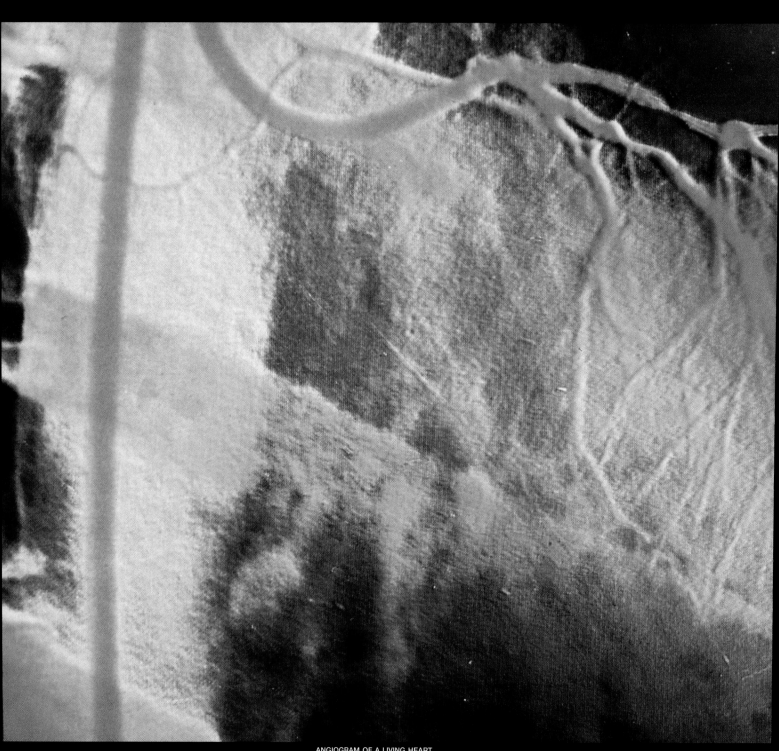

ANGIOGRAM OF A LIVING HEART

surgery. Diets high in fats and cholesterol, poor exercise habits, and increasingly sedentary and stressful lifestyles all contribute to the industrialized world's growing occurrence of atherosclerosis as its residents grow older.

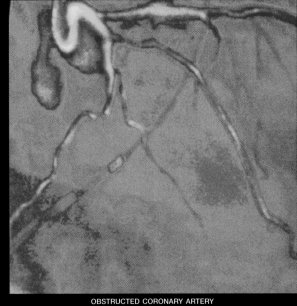

OBSTRUCTED CORONARY ARTERY

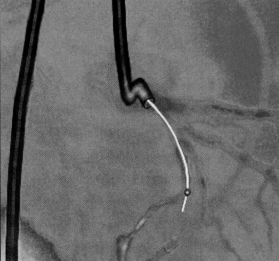

CATHETER REACHES THE OBSTRUCTION

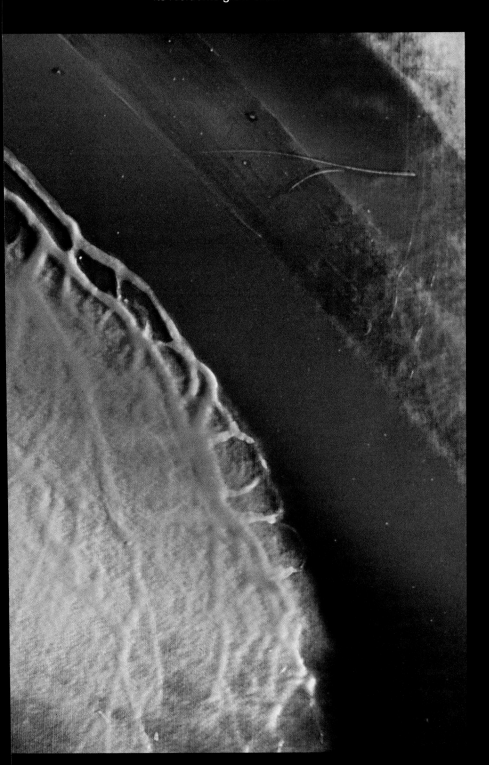

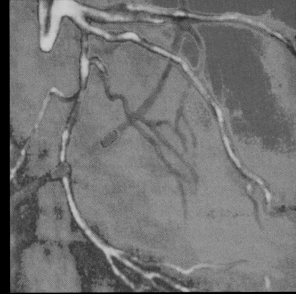

ENHANCED BLOOD FLOW FOLLOWING ANGIOPLASTY

of human physiology more variable than male sexual function. Some men in their 70s have higher testosterone levels, higher daily sperm production and activity, and more frequent sex than some healthy men in their 20s. Also, there is no decline, perhaps even an increase, in testes volume and gonadotropin levels as men grow older.

Men are nearly unique among species in commonly developing prostate problems as they age. Dogs are the only other mammal known to have similar problems. The human prostate is a gland that surrounds the urethra like a nut surrounds a bolt. The urethra is the tube connecting the bladder to the penis, through which urine and semen must pass. The prostate provides fluid and nutrients to the semen as it passes down the urethra.

The most common individual disease associated with human aging is benign prostatic hyperplasia (BPH), which occurs in up to 80 percent of men over age 70. BPH is noncancerous prostate enlargement caused by abnormal cell division. It frequently constricts the urethra, making urination difficult. Prostate cancer, a separate disease, is the most common cancer detected and the second leading cause of cancer death in American men. Although only about one man in three hundred over 50 is diagnosed with prostate cancer, autopsies detect cancerous cells in nearly one-third of men this age. Thus many have small prostate cancers, but many of these seem never to develop into significant medical problems.

The standard, digital rectal examination of the prostate does not reliably detect prostate cancer at an early stage or distinguish cancer from benign prostate swelling. A more expensive blood test detects a protein, PSA—prostate specific antigen—that is present in proportion to prostate size. Extremely high PSA levels in the blood generally indicate presence of cancer, but marginally high levels are seen with equal frequency in benign prostate swelling and local cancers. Because both digital exams and PSA tests are of limited use in detecting prostate cancer in its earliest and most treatable stages, a great deal of research is currently under way to develop better tests for this disease.

Reproduction in women also depends upon the interplay of hormones from the brain and genitalia. Gonadotropins from the pituitary stimulate development of eggs and their surrounding cells in a woman's ovaries. Cells surrounding developing eggs produce estrogen, which like testosterone has effects throughout the body, including development of breasts and bone growth. Estrogen is the chief hormone secreted by the ovary until ovulation, when the egg is shed. After ovulation the ovary secretes high levels of a related hormone, progesterone, which prepares the uterus to receive a fertilized egg. If no pregnancy ensues, these high levels of estrogen and progesterone inhibit further gonadotropin secretion. When gonadotropins decrease, the ovary stops secreting its hormones, the body stops preparing for pregnancy, menstruation occurs, and gonadotropin levels begin to increase again as the cycle repeats.

A woman contains all the eggs she will ever have by the time she is halfway through development in her mother's uterus, and they begin dying almost at once. At birth, a woman has only one-fourth of her original number, and by puberty she has only about one-thirtieth. Women continue losing eggs at a constant rate until about age 35, when for reasons that are not yet understood, the rate of egg loss doubles. Menopause, the ceasing of ovulation and menstrual cycling, occurs when there are too few eggs left to produce sufficient hormones to drive the menstrual cycle to completion. After that, no amount of gonadotropins can prod the ovary to develop more eggs or produce more estrogen or progesterone.

It isn't understood why women stop making new gametes before they are even born, whereas men produce them throughout life. It may have to do with the fact that each time a cell divides its chances of having genetic mutations increases. Eggs are the product of only about 33 cell divisions after fertilization, whereas sperm, even in a young male, are the product of several hundred cell divisions. As a consequence, sperm have many times more mutations that eggs. But men produce billions upon billions of sperm cells in a lifetime, and the several hundred million sperm in each ejaculation all compete to fertilize a single egg. Consequently, even if a fair fraction of sperm carry serious mutations, plenty of viable sperm remain to fertilize the egg. On the other

Picture of health, Jere Gottschalk lost body fat and gained muscle mass during a six-month experimental treatment in which he received human growth hormone; once off treatment, however, his body returned to its former state. Long-term health consequences of such hormone treatments are not known, but both humans and laboratory animals that produce too much growth hormone throughout life are known to have substantially shorter lives.

hand, women shed only one egg capable of being fertilized at a time, and no more than several hundred such eggs over a lifetime. Serious genetic mutations could easily have serious reproductive repercussions. The consequences of mutations during repeated cell division, in other words, are more serious for a woman.

Menopause itself, which occurs at an average age of 51 in industrialized countries, is not the sudden onset of infertility. It is only the final stage of a reproductive decline that has begun much earlier. Compared with a reproductive peak in their mid-20s, even 35-year-old women have more difficulty becoming, and remaining, pregnant. What's more, the chances that her baby will be genetically defective, or that a woman will die in childbirth, double by the age of 35. How much of this change is due to deterioration in the reproductive tract versus damage to the parts of the brain that respond to genital hormones, or produce gonadotropins, is not yet clear.

There are health consequences related to the loss of estrogen and progesterone associated with menopause. Estrogen, for instance, is known to stimulate the immune system, help regulate blood cholesterol and thereby slow the progression of atherosclerosis, maintain bone density, and have a variety of as yet unknown effects on the brain. Estrogen loss at menopause therefore increases the risk of cardiovascular disease, osteoporosis, and perhaps even Alzheimer's disease. There is a small acceleration in the risk of death after menopause, although women still survive better than men of equivalent ages. About one-third of women in western countries seek medical help for less serious problems due to estrogen loss, such as hot flashes, night sweats, insomnia, dry skin, incontinence, or painful intercourse. About another third of women report that they experience one or more of these symptoms but not severely enough to seek a doctor's help. The other third of women fail to experience any of these problems.

On the upside, loss of female reproductive hormones slows the exponential increase with age in the yearly risk of developing so-called gynecological cancers of the breast, ovary, or uterus lining. Women in the United States double the annual risk of breast cancer about every 3 years prior to menopause, whereas after menopause the risk takes 13 years to double again.

The various minor and major health effects of menopause can be alleviated by taking supplements of the hormones—estrogen and progesterone—previously produced by the ovaries. More than 90 percent of women who take these hormone supplements get relief from such non-life-threatening menopausal symptoms as hot flashes, insomnia, and depression. The risk of more serious effects such as cardiovascular disease also decreases with hormone replacement, as do the odds of developing osteoporosis or perhaps even Alzheimer's. The most thorough studies so far have revealed that postmenopausal hormone replacement therapy can decrease the yearly risk of death by more than a third.

In spite of the apparent benefits, fewer than one-third of postmenopausal women take hormone supplements, mainly avoiding them due to fears of increasing gynecological cancer risk. When hormone replacement was first begun, using only estrogen, uterine cancer rates increased as much as eight-fold. However, adding progestin—synthetic progesterone—to the estrogen has reduced the risk of uterine cancer to below that of women who do not use hormone replacement. Moreover, uterine cancer is relatively uncommon at any age. Women have been most worried by the possibility of an increased breast cancer risk. Breast cancer is common among elderly women, striking nearly one woman in ten at our current life expectancy. Monthly surges of estrogen and progesterone are clearly associated with the development of breast cancer, since men also have breasts yet account for less than 1 percent of all breast cancer.

The evidence on whether hormone replacement increases the risk of breast cancer is inconclusive. Of five recent major studies comparing users and nonusers of hormone replacement, two found no increase in breast cancer risk, three found a small increase. Since cardiovascular disease accounts for about seven times as many deaths as all gynecological cancers combined, hormone replacement would seem to be a wise health choice for most women. What's more, new hormone replacement methods such as skin patches, and even new hormones (synthetic mimics of estrogen and progesterone), are coming on the market constantly. The hope is that some of the newer synthetic hormones will be able to produce all of the beneficial effects without any of the disadvantageous side effects.

There are two theories concerning why women stop reproducing with nearly one-third of their life still ahead of them. One theory is that humans are physiologically designed to function for about the number of years they were likely to survive during the many Paleolithic millennia over which our physiology evolved. According to this idea, the profusion of women in their 60s, 70s,

80s, and beyond is a recent phenomenon—no more than a handful of centuries old—and is due to dramatic improvements in sanitation, hygiene, and the conquest of infectious disease by modern pharmaceuticals. Our physiology has not caught up with the changes we have wrought in our environment. Woman produce about as many eggs as they were likely to need over the bulk of human history. By stopping egg production when it was just sufficient to meet demand, a woman's chance of producing genetically mutated eggs was minimized.

An alternative, in many ways more intriguing theory explaining why women stop reproducing, is that menopause is nature's way of ensuring that women will be alive and well to help raise their last-born children, and their children's children. Called the good mother—or sometimes, the grandmother—hypothesis, this theory assumes that menopause is a consequence of the long period of child care that humans require. The logic of this hypothesis follows from the facts that childbirth becomes more and more dangerous as women age and that the odds of living long enough to rear previous children diminishes. Therefore, an age will eventually be reached at which a more successful tactic for passing on a woman's genes is to put all her effort into assisting her existing children to survive and reproduce, as opposed to continuing her own personal reproduction. Nature has therefore favored genes that turn off the reproductive spigot, and thereby free up maternal energy to devote to child care.

The jury is still out on which theory is correct. In support of the former, females of other mammal species, such as rats, also exhibit more rapid reproductive aging than males when they are maintained in the sheltered conditions of the laboratory. The difference cannot be explained by any long period of maternal care. Rats nurture their young for less than two months. In support of the latter theory, we know of other species, such as pilot whales and killer whales, in which several generations of genetically related females live together and in which there is a similar female postreproductive phase of life. Whale babies do require an extended period of care, and there is no reason to think that whale longevity has increased dramatically in the last few centuries due to improved sanitation or medical care. There seems to be no other reasonable explanation for the stopping of female reproduction than the good mother idea—for whales at least.

CAN WE CHANGE OUR RATE OF AGING?

For centuries humans have sought ways to slow the aging process. Our gullibility about supposed remedies to aging has been endless, and it continues unabated to the present, when so-called antiaging pills, potions, and total behavioral programs are a multibillion-dollar industry. We do, in fact, now know a great deal about how to reduce the risk of some diseases associated with aging, such as cancer and atherosclerosis. Reducing disease risk will increase life expectancy to a certain degree. But slowing aging is not the same task as reducing disease risk. Slowing aging means slowing the rate at which we lose running speed and hearing, the rate at which our ability to learn new languages or remember what we read declines, in addition to lowering the risk of disease. There is nothing currently available from the pharmacy, the health food store, or by credit card from a television advertisement that has been scientifically demonstrated to slow aging in humans.

Overwhelming evidence supports the conclusion that aging can be slowed by about 25 percent in laboratory rats and mice by simply reducing the number of calories eaten. To understand the significance—or lack of significance—of this finding for humans, it is important to know how these experiments are done. Laboratory mice and rats typically have excess food available to them 24 hours a day. The animals eat as often and as much as they wish, which is considerably more than a rat or mouse captured straight from nature and given unlimited food would eat. Humans, in their role as captive breeders of rats and mice, have selectively bred over dozens of generations for individuals that grow quickly and produce a lot of pups. Among rodents, individuals that eat a lot tend to grow faster and have more babies. Humans have therefore been

FOLLOWING PAGES: Eighty-five-year-old Carol Johnston holds the world pole-vaulting record—7' 6"—for his age class. We now know that—at any age—continued exercise can improve strength, endurance, bone mass, and one's sense of well-being.

selectively breeding for the rodent equivalent of gluttony for a long while. If you feed these rodents 60 percent or so of the calories they would like to eat, they clearly age more slowly. The same phenomenon is found in a number of cold-blooded animals such as fish, insects, and spiders although interpreting what the data mean is difficult. Insects probably live longer when starved because they slow their metabolic rate. The food-restricted rodents do not reduce their metabolism, at least if you calculate energy consumption per cell. However, they do produce fewer free radicals per amount of energy consumed.

Two important questions remain unanswered about food-restricted rodent research. First, what biochemical changes occur in food-restricted rodents that lead to slowing their aging rate? If we knew this, perhaps we could design pharmaceuticals that could perform the same key changes. Second, is this slower aging a general phenomenon, or is it an aberration of laboratory animals bred to be gluttonous? If it is a general phenomenon of mammals, it should also slow aging in people. There are currently two studies of food restriction under way using monkeys. The results so far are a mixed bag, with many, but not all, of the same changes being observed in the dieting monkeys that have been seen in dieting rodents. However, the studies have a long way to go, and we won't have any definitive answer for at least a decade or two because monkeys live as long as 40 years.

We don't have to wait that long, however, to be certain that even if a lifetime of dieting did slow aging in humans, it would have little impact on societal longevity. The existence of a multibillion-dollar diet industry is enough evidence to show that humans have too little control over their eating habits to ward off obesity, much less to reduce their food intake to 60 percent or so of a normal diet. Our only hope for benefit from the food-restriction experiments is if they point the way to pharmaceutical management of aging.

Of course, eating itself may be in for some pharmaceutical management soon. We are making considerable strides in understanding the physiology and genetics of our dietary habits. In 1994 a defective mouse gene that caused obesity was identified, and its human equivalent was located. Administering the normal gene product, called leptin, to these obese mice made them eat less, exercise more, and become svelte. Although human obesity is only very rarely due to this same defective gene, more obesity-causing mouse genes are being identified

all the time. Humans have equivalents of all such genes discovered so far. It is no longer inconceivable that we will be able to manipulate our eating habits using chemicals that our bodies naturally produce under the right conditions. Numerous diet products are in the research pipeline and will be available soon. If food restriction turns out to retard human aging, we may not have to rely on willpower alone very much longer.

Even if we can't reduce the calories we eat by 40 percent, what about the type of foods we eat, or the nutrients we take in pill form? If we are increasingly sure that free radicals, that is oxidants, are involved in the aging process itself, what about eating foods or pills that are rich in antioxidants? Could that retard aging?

THE ROLE OF VITAMINS

It has been well documented that eating a diet rich in fruits and vegetables, which are high in antioxidants and low in animal fat, is good for you. It reduces your risk of cancer and cardiovascular disease, the two major killers in western societies. So if everyone ate such a diet, life expectancy would increase, although because these diseases mainly kill the elderly, it wouldn't increase life expectancy as much as you might guess. Even completely eliminating all cancer and cardiovascular disease would add less than a decade to life expectancy.

There is no evidence, though, that eating this sort of healthy diet retards aging itself to any significant extent. What about taking antioxidant vitamins in pill form? The holy trinity of antioxidant vitamins are vitamins C, E, and A. But you can't learn about the possible benefits of antioxidant vitamins by asking people about their vitamin-taking habits and trying to correlate that with health and longevity. People who religiously swallow antioxidant vitamins swallow them in vastly different amounts. Also, vitamin enthusiasts likely differ from the general population in many ways, including their dietary, smoking, and exercise habits, as well as their tendency to maintain a vigilant watch over their health and seek medical advice at the first sign of a problem.

A number of clinical trials in recent years have been looking at the effect of antioxidant vitamins on specific diseases.

You can overdose on vitamin A, so for years it has been thought that the best way to supplement your intake was to consume beta-carotene, which your body converts to vitamin A. Because there are physiological controls over the rate of conversion, it was considered impossible to take too much beta carotene. Also, a number of correlational studies (the inadequate type of questionnaire studies described above) suggested that beta carotene supplements might prevent lung and stomach cancer. However, the three large clinical trials investigating beta-carotene supplements produced some worrisome results. In two of the trials, which focused on people at high risk for lung cancer (smokers and asbestos workers), those taking beta-carotene got lung cancer at a higher rate than the placebo-takers and also died at a higher rate. The third trial found the supplement had no effect. These results were disturbing enough to cause beta-carotene to be dropped from still other clinical trials.

Vitamin E in a test tube seems to have precisely those antioxidant properties that would combat cancer and heart disease. Moreover, the same clinical trial that found beta-carotene to be harmful to health found no similar harm with vitamin E supplements. But it found no beneficial effects on disease or death rate. There are at least four other clinical trials of vitamin E supplementation currently under way, so we should have more information in the near future.

Vitamin C has been popularized as a vitamin that, if taken in large amounts, will protect against colds. There is little evidence that it does so, although some modestly convincing evidence suggests that it may help recover more quickly. The evidence concerning vitamin C's effect on cancer is weak and conflicting. There is no information on its effect on aging. It has not been shown to be harmful, even when taken in large amounts, except possibly to two groups of people. People with a genetic predisposition to certain types of painful kidney stones may increase their risk of stones by taking a lot of vitamin C. People with a genetic tendency to store too much iron in their bodies might be harmed by vitamin C in unknown ways. In the presence of excess iron, vitamin C can change from an antioxidant to an oxidant—a producer of free radicals. No one knows what effect that fact might have over the long run.

The take-home message now is that eating lots of fruits and vegetables and reducing dietary fat are the best choices to help deter aging. Each fruit or vegetable will contain a lot of antioxidants in combinations that have been shown to help prevent common diseases of later life. Such a diet won't slow aging, but it won't have toxic side effects either.

Regular exercise has often been touted as a route to the fountain of youth. Virtually every medical guide to good health recommends it. There is no doubt exercise is beneficial. It is good for your heart, arteries, and lungs. It helps you feel better, gives you more energy, allows you to do more. Yet what about the effects of exercise on aging? The fact that people who exercise a lot consume more oxygen than those who don't suggests caution. If oxygen consumption inevitably creates free radicals, might exercise accelerate aging?

There is little evidence that exercise accelerates aging of any part of our body, although some people may die prematurely if they are in compromised health already and exercise too vigorously. It may be that exercise stimulates antioxidant production in key organs to compensate for increased oxygen consumption, even though there is no overall increase in antioxidant activity. There is also precious little evidence that exercise delays aging, except perhaps in muscles. The best study, one that has been following a group of 17,000 Harvard graduates through their lives, suggests that vigorous regular exercise may add a couple of years to life by helping to prevent premature diseases. Of course, someone had to point out that the amount of exercise required to achieve this two-year life extension would take about two years to perform.

There is no question, however, that people who exercise feel better and can do more in their later years, even if they begin exercise at a late age. Research on exercising rats, in which the amount of exercise can be rigorously controlled, shows the same result. Midlife diseases are prevented, but there is no increase in the maximum length of life.

A number of hormones that our bodies normally produce, but which inevitably decline with age, have been touted in recent years to have antiaging effects when taken as dietary supplements. The idea, which has a long and inglorious history, is that if something in your body declines with age, you have only to supplement it back to its youthful level to prevent aging. French physiologist Charles-Edouard Brown-Séquard may have begun this approach in the last century, when he injected himself with extracts of animal testicles to try to rejuvenate his flagging sex drive. There never has been evidence that any of these treatments slow aging in humans, and there are risks associated with the overproduction of virtually any hormone. Yet people keep trying and hoping to slow the aging process.

DHEA is produced in abundant amounts by the adrenal glands, which also produce stress hormones. No one knows the function of DHEA, although your body can turn it into a number of other hormones, such as testosterone and estrogen, and stress suppresses it. In rodent studies, DHEA has had some dramatic effects, although rodents produce vanishingly small amounts of it on their own.

Some human studies suggest that DHEA may improve immune response, muscle strength, and sleep patterns among the elderly, although other studies find no change or even a decline in immune response. There may be something to DHEA supplementation, but it is still too early to tell. Other steroids are well known to have cancer-promoting effects, so it would be prudent to wait until more evidence is available from research on humans.

Melatonin is a hormone produced primarily by a pea-size gland called the pineal in the middle of the brain. Five to ten times more melatonin is produced at night than in the daytime, and the chemical seems to have a role in setting body rhythms, affecting sleep patterns, and helping regulate seasonal reproduction in animals that breed only during parts of the year. In humans melatonin is produced in smaller and smaller amounts after about the age of five, and its daily secretion pattern deteriorates. Melatonin probably is effective as a sleeping aid and remedy for jet lag.

But despite the many wild antiaging claims made for melatonin over the past few years, there is absolutely no evidence that it slows aging or disease in humans. Ironically, the common laboratory strains of mice used in aging research cannot produce melatonin at all due to a genetic mutation. Yet they are relatively long-lived for mice, a surprising finding if the hormone melatonin plays the important role that has been claimed for it.

Human growth hormone has also received attention in the lay press recently, with claims that it can improve bone and muscle strength, cholesterol profile, and immune response in the elderly. Research is most convincing about the improvement in muscle strength and a slowing of the decline in bone mass. However, these have been short-term studies that could not have detected any long-term detrimental side effects. And there are long-known, long-term detrimental side effects to excess growth hormone. For instance, people who overproduce growth hormone as children grow up with a disease called acromegaly and typically die prematurely of cardiovascular diseases. Mice genetically engineered to produce excess growth hormone live only half as long as their normal relatives. Furthermore, dwarf mice that produce less than normal levels of growth hormone (as well as some other hormones) are particularly long-lived. None of these facts proves that long-term administration of growth hormone in moderate amounts will have the same detrimental effects in humans, but it does give reason for caution.

All the hormones discussed so far are actively being investigated. We should know much more about their effects in the elderly in the next few years. What about hope for retarding aging in the more distant future? The most promising research is probably in the area of gene therapy. In the not-too-distant future we will be able to replace defective genes and alter the activity of the genes we already have. This gives us an opportunity to fix things as they go wrong and to perhaps prevent their going wrong in the first place.

Now that we know some of the major processes—damage from free radicals and browning, the loss over control of cell division rate—that are major components of aging, we should be able to target the appropriate genes to slow the process. Such an experiment in fruit flies has already been carried out. The flies were genetically engineered to increase the activity of two of the normal antioxidants produced inside cells, and they lived about 25 percent longer than other fruit flies. These sorts of experiments need to be repeated, of course, on animals more closely related to us, such as mice, before we can take much hope that a similar treatment might work in humans.

Recently, however, the first successful trial in gene therapy was reported, so we are on the road toward genetic treatment for diseases. That means we are also on the road to being able to retard aging, even if we are still at the beginning of the trip.

Well into her sixth decade, entertainer Tina Turner still rocks and rolls like a teenager. Pursuing an active performance schedule that might daunt a far younger performer, she draws packed audiences and appeals to many ages.

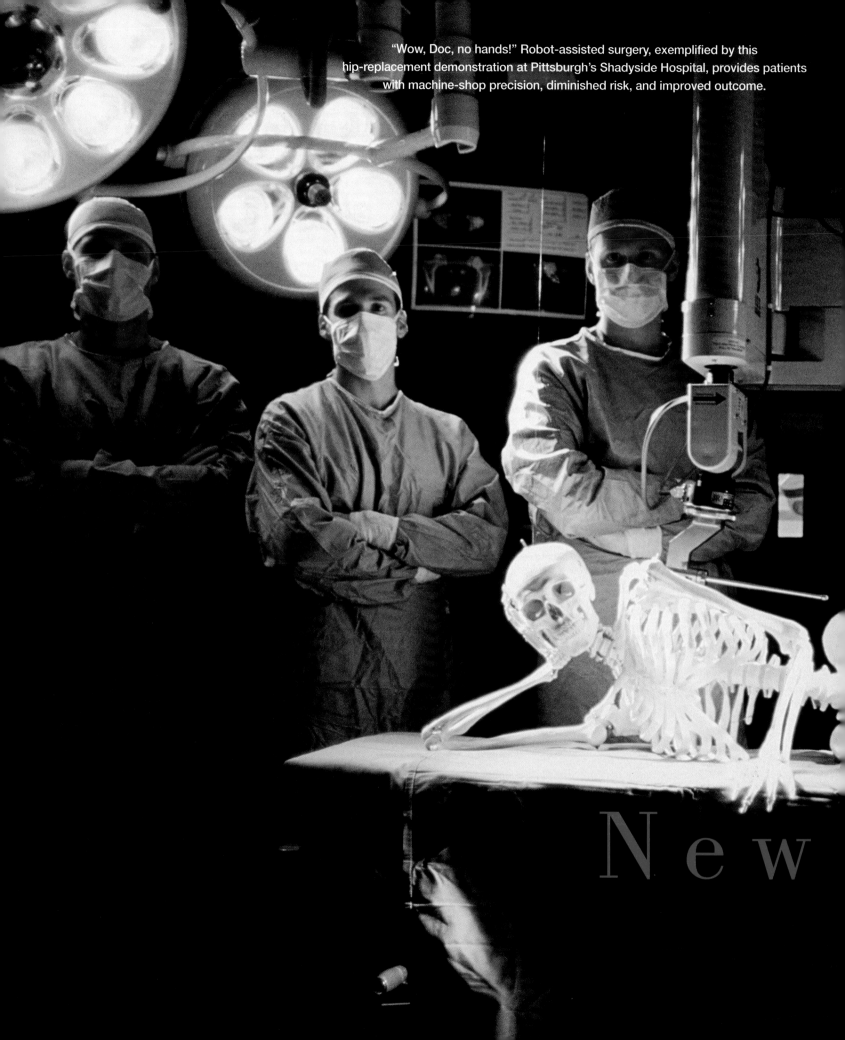

"Wow, Doc, no hands!" Robot-assisted surgery, exemplified by this hip-replacement demonstration at Pittsburgh's Shadyside Hospital, provides patients with machine-shop precision, diminished risk, and improved outcome.

New

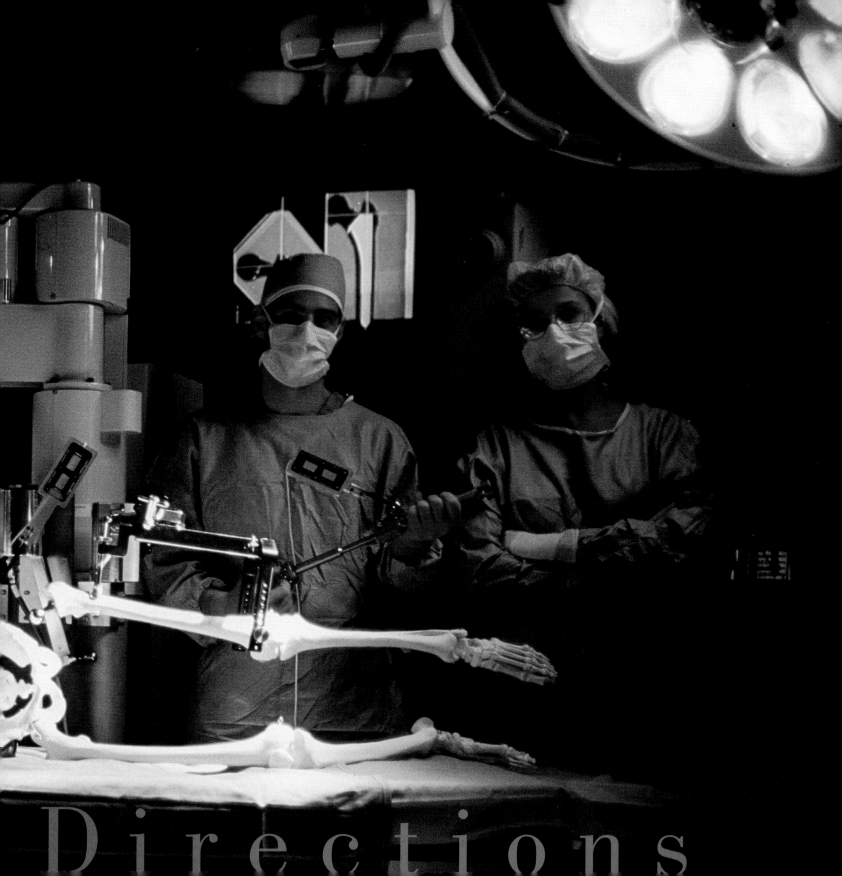

Directions

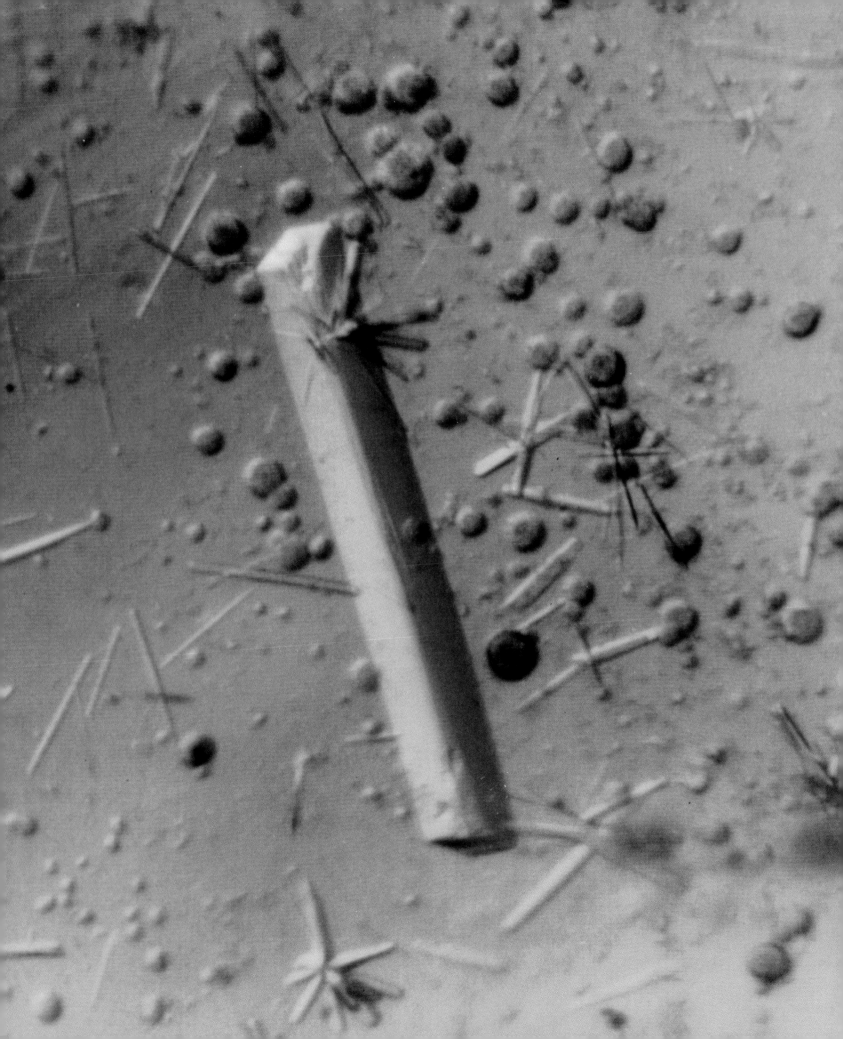

"What a piece of work is man!" rhapsodized Hamlet. "How noble in reason! How infinite in faculty! In form, in moving, how express and admirable! In action how like an angel! In apprehension how like a god!" We have seen scientific confirmation of that awestruck observation throughout the preceding pages. We are engines of phenomenal energy, masterpieces of architectural design and efficiency. A human being is, truly, a standing miracle—less a scientifically exact category than an intricate organic amalgam.

Maintaining the human body's integrity—healing it and keeping it healthy—is the responsibility, indeed the duty, of doctors and other health-care practitioners, scientists, and—given that we are what we eat and do or do not do—ourselves. The 21st century will not alter that. But there will be dramatic changes in the way our state of health is predicted, monitored, and maintained. Many of medicine's enormous achievements are already evident, thanks to the many research institutions that broadcast them to the world via journals and scientific symposia, and to newspaper science columns and televised hospital dramas. Organ transplants, artificial kidneys, cobalt femurs

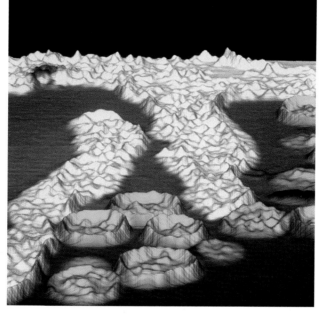

revolving in polyethylene hip sockets, coronary bypasses, and walk-in-walk-out surgery—mind-boggling to contemplate not so many years ago—are now routine. MRI, angioplasty, arthroscopy, and amniocentesis are household words; the laser, the light fantastic, cuts and repairs cleanly where scalpels cannot reach; genes responsible for diseases are isolated on a regular basis.

While these things are wondrous enough, they will yield to an even more exciting new order of treatments, techniques, and diagnostics. What has been accomplished is not destined for dismissal or disuse simply because we have found better replacements, no more than we ignore the invaluable lessons of history. Too much attention paid to what we may accomplish in the future does not help today's sufferer, and it can blunt the way medicine must be practiced with current tools. Still, medical science is a never ending quest that is continuously fueled by bursts of startling knowledge, refinements, and more efficient practical applications. What has gone before is sure to pale in comparison to what health care will be like, for better or for worse, in the new millennium.

Say hello, then, to the standard medical terminology of the future, a litany that defines an intricate technology

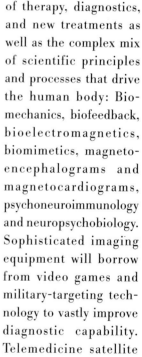

of therapy, diagnostics, and new treatments as well as the complex mix of scientific principles and processes that drive the human body: Biomechanics, biofeedback, bioelectromagnetics, biomimetics, magnetoencephalograms and magnetocardiograms, psychoneuroimmunology and neuropsychobiology. Sophisticated imaging equipment will borrow from video games and military-targeting technology to vastly improve diagnostic capability. Telemedicine satellite systems will routinely beam video images of organ abnormalities, obstructed arteries, and brain waves from underserved and remote rural areas to specialists in large medical centers. There will be robots that will, with more precision than a surgeon's hand, manipulate delicate instruments and automated cameras. When surgeons themselves grasp the tools of their trade, those tools will be designed to draw little or no blood, and the physicians will be guided by three-dimensional imaging devices that

Looking like space debris adrift in the weightlessness that spawned them, antibody crystals (opposite) specific to a deadly virus offer therapeutic promise. Grown aboard the space shuttle, their purity and size dwarf what can be achieved in earth-based labs. Hantaviruses, agents of hemorrhagic fever (above), sprawl into the semblance of a lunar landscape.

map their progress as they operate. And while it is unlikely that something as ironically normal as ill health and disease will ever disappear—on the contrary, lurking new viruses and other hidden, highly infectious agents are certain to emerge to plague humanity—a new generation of researchers is already focusing not only on the external symptoms of a disorder but also, increasingly, on the specific nature of each one, and on tracking the causative processes that lie deep within viruses, bacteria, and—in the place where trouble often begins for us—the genetic makeup of the human cell. In the future molecular medicine will have the definitive "book of life" to pore over. Now being written by the Human Genome Project, that book will transform the way we detect and treat diseases at the genetic level.

As scientists tinker with the very stuff of life, the courses of many familiar diseases are sure to be shortened long before they can do more damage; some will be transformed into paper tigers, still bearing their ominous names but rendered impotent. Some of these will have been detected and thwarted by gene therapies long before they wreak any havoc, or blocked by vaccines once deemed improbable; still others will be cured by super-drugs and novel systems

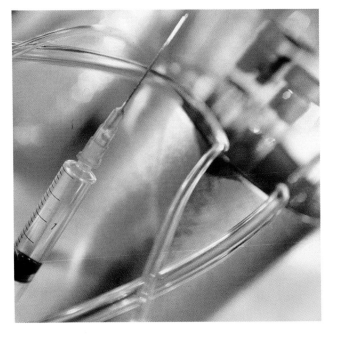

for delivering them. A few maladies will even be gone forever, their viral, bacterial, and fungal carriers bioengineered or chemically scoured off the face of the earth, with but a sampling of their fossilized remains stored as museum pieces in laboratory vaults. The words, 70 years ago, of William J. Mayo of the famed Minnesota clinic family, resonate today: "The glory of medicine is that it is constantly moving forward, that there is always more to learn."

ALTERNATIVE MEDICINE

Nowhere does that statement ring truer than in the current interest in what is popularly referred to as alternative medicine, or supplementary, adjunctive, or complementary medicine. Neither high-tech nor computer-driven, it will, without doubt, be a key consideration—albeit sometimes warily watched by the guardians of the orthodox—as medicine moves in new directions and as patients seek treatments perceived to be safer, cheaper, more effective, and more personal than the conventional. Alternative medicine is a wildly collective term, encompassing a bewildering array of

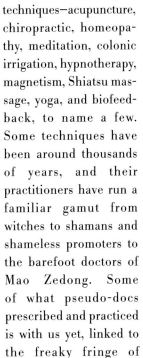

techniques—acupuncture, chiropractic, homeopathy, meditation, colonic irrigation, hypnotherapy, magnetism, Shiatsu massage, yoga, and biofeedback, to name a few. Some techniques have been around thousands of years, and their practitioners have run a familiar gamut from witches to shamans and shameless promoters to the barefoot doctors of Mao Zedong. Some of what pseudo-docs prescribed and practiced is with us yet, linked to the freaky fringe of medicine, and often still touted by amateur healers.

But science has wakened to a few techniques that seem promising and has begun to take a closer look. Some mainstream physicians show interest; such open-minded doctors ask only that their patients keep in touch during treatment. Medical schools not known for embracing the unorthodox are also curious. At Harvard Medical School—with 15 Nobel Prizes to its credit—students learn how alternative treatments might impact

On the way out? Test tubes and syringes, the traditional tools of drug development and delivery, are nearly passé as researchers turn to newer techniques: Computer-assisted drug design, innovative patches, pumps, and implants.

on clinical practice and research. At Harvard-affiliated Beth Israel Hospital, David Eisenberg, a Harvard professor who studied traditional Chinese medicine in Beijing, directs the Center for Alternative Medicine Research. Stanford offers a course on alternatives, and its Center for Research in Disease Prevention is one of ten centers nationwide that participate in a federally funded project that evaluates promising alternatives. Increasingly, too, health insurers are covering some unorthodox therapies if they are part of proven treatment programs or if patients are referred by physicians. Perhaps the biggest boost for unorthodox treatments, however, came in 1992 when, by congressional mandate, the Office of Alternative Medicine was set up as part of the National Institutes of Health. By 1996 NIH had reviewed proposals for some 40 studies on alternative medicine and had determined that they were worthy of funding.

There is demand for such services, no doubt. Patients swear by many of the alternatives, legitimate or not, or at least are willing to try one or two on the theory that, like mother's chicken soup, they won't do any harm and may do some good. One out of three adult Americans uses some form of alternative treatment. We spend an estimated $13.7 billion a year out of our own pockets—and most of us still consult our medical doctors. In one study conducted by Harvard's Eisenberg, 61 million Americans used diet, massage, relaxation, chiropractic, acupuncture, spiritual healing, herbal medicine, or homeopathy. Those who visited therapists returned to them an average of 19 times a year—some 425 million visits—more than the number of visits patients make to specialists and general practitioners combined. Eisenberg also found that patients most often chose alternative therapy for chronic conditions that resisted standard treatment, such as back pain, insomnia, headaches, anxiety, and depression. Worldwide, the number of people who rely on nonconventional methods is staggering: Perhaps only 10 to 30 percent of human health care is delivered by biomedically oriented practitioners. The remaining 70 to 90 percent ranges from self-care according to folk principles to care given in an organized health-care system based on an alternative tradition or practice.

But use and interest are one thing, efficacy is another. Serious assessment of any medical intervention must rely on well-designed and controlled laboratory experiments, controlled randomized clinical trials, and peer review. While some reliable scientific studies have been carried out, many of the experiments on alternative therapies have been loosely designed and influenced by anecdotal testimonials from satisfied patients. It is understandable why critics such as the American Medical Association and many scientists feel that some of the alternative practices more closely resemble witchcraft than medicine, and that any improvement or cure afterward is a placebo effect, or merely chance. As Stanford professor Paul Berg, a Nobelist in chemistry, has put it, "Quackery will always prey on the gullible and uninformed, but we should not provide it with cover from the NIH."

Be that as it may, not all alternatives are worthless or wishful thinking. A few, such as acupuncture, herbal medicine, and meditation, have been practiced for centuries and have demonstrated enough success to warrant serious investigation. Some of the modern techniques, such as biofeedback, also merit research. A mind-set that accepts the demonstrable regardless of the source is necessary. "I am concerned when the terms 'quackery' and 'con artist' are used to describe complementary medicine," Kirsti A. Dyer, a California medical doctor, wrote in the *Journal of the American Medical Association*. "These emotionally charged labels create a negative image and block investigation of any positive aspects in this area. Physicians need to help expose those practices and practitioners who may be detrimental to our patients. At the same time, we can incorporate those interventions that are beneficial."

The origins of acupuncture go back thousands of years in Chinese history. One of the most thoroughly researched of the so-called alternative medical practices, acupuncture may have begun as a way to drive demons out of a sick person's body by puncturing the skin with needles, but it soon evolved into a complex therapeutic system that drew on the interplay of mystical philosophy and speculative anatomical theory. Acupuncture was based on the ancient belief that a nebulous "vital spirit," a life force called *qi*, circulated in all living things, and that when this flow was disturbed, disease resulted. Inserting fine needles into the skin at various bodily points associated with specific organs presumably corrected and regulated the vital flow, restoring health. Today, more than 3,000 conventionally trained U.S. physicians incorporate acupuncture into their practices; some 62,000 medical doctors are among the more than 88,000 practicing acupuncturists in Europe. While many still twirl needles by hand, others have succumbed to modern technology, using heat or impulses of electromagnetic energy to stimulate the acupuncture points.

There is no sound evidence that the "vital spirit" exists, but modern science suggests some possible parallels: Researchers at several reputable centers see similarities in the bioelectrical-magnetic fields that flow through the human body and may play a role in disease. According to some researchers, the acupuncture points have certain electrical properties, and stimulating them may alter the body's chemical neurotransmitters, or in some way prod into action the body's defenses against disease. It is also suggested that acupuncture stimulates brain cells to release morphinelike endorphins, and this may be why—along with inhibiting the transmission of pain impulses through nerves—the technique provides at least short-term relief from migraines, lower back pain, and menstrual cramps. In several controlled studies, acupuncture also showed evidence of effectiveness— although the mechanisms are unclear—in treating bronchial asthma, bronchitis, stroke-induced paralysis, osteoarthritis, chemotherapy-induced nausea, bladder instability, and even drug addiction. In 1997 an NIH consensus panel addressed acupuncture's role in health care and concluded that there "is clear evidence" that it is effective for postoperative and chemotherapy nausea and vomiting, nausea of pregnancy, and postoperative dental pain. The panel added that a number of other pain-related conditions may be responsive to acupuncture as adjunct therapy, as an acceptable alternative, or as part of a comprehensive treatment plan.

As ancient as acupuncture, meditation is also Eastern born. It is still practiced virtually unchanged today, or with modern refinements. Though the primary purpose of meditation in China, Japan, and India was religious, its positive influence on health has long been appreciated. There are numerous variations on the theme—yoga, prayer, faith healing, biofeedback, and imagery—but each depends on the well-established interconnectedness of state of mind and the physiological processes of the body, as well as the enormous power of each to affect the other. In times of stress, we now know, the sympathetic nervous system works like a car's accelerator, flooding the bloodstream with the hormone epinephrine. According to meditation, psyche, the mind, joins its emotional state to chemistry—and can cause

soma, the body, to create psychosomatic illness. Blood vessels are constricted, and blood pressure rises; the heart beats faster, the pupils of the eyes dilate, the muscles tense, and gastric juices and the involuntary muscle movement that propels food along, peristalsis, are inhibited, impairing digestion. Stress and emotions can trigger a heart attack, cause hives to erupt, ulcers to flare, asthma to worsen, and perhaps even put one at risk of cancer and arthritis. Indeed, it has been suggested that three out of four illnesses begin in the mind and that half the problems of convalescing begin there. Conversely, organic causes are often at the root of abnormal behavior. Schizophrenia, humankind's greatest mental crippler, may be one example of how the soma might affect the psyche. It's been suggested that an inherited chemical defect, or perhaps an abnormal amount of a normal substance that weakens the membranes of neurons and other cells, produces poisonous substances that affect the brain much as overdoses of drugs and alcohol do; the result may well be marked disturbances in perception and drastic changes in thought, personality, and behavior.

That meditation and relaxation techniques can alleviate many disorders, physical and emotional, is well documented, and the familiar advice, "take deep breaths," is no quack cure. Most of us are aware by now—whether because we are devotees of the health guru of the moment or the health columns—that controlled breathing is one of humanity's cheapest and most effective medicines, capable of working more wonders than a church full of saints. A mainstay of yoga—the ancient Hindu system of meditation, postures, concentration, and diet for achieving bodily and mental control and well-being—controlled breathing is a fundamental element of all relaxation techniques and enables the practitioner to break the chain of interfering, everyday thought, to achieve what the Dalai Lama has called "the clear light of mind."

For many years, however, the therapeutic potential of meditation was known only through travelers' fantastic tales of yogis' ability to influence their own heartbeats, to go for long periods without nourishment, and to withstand pain. Eventually, influenced by neurology's keen interest in the link between mind, will, and bodily functions, researchers conducted thousands of studies

Sliced by the digital knife of an MRI, the skull reveals glowing details. Such electronic dissection has revolutionized the study of anatomy—and scrapped its traditional caveat: Once a cadaver is dissected, it cannot be reassembled and taken apart again. FOLLOWING PAGES: Virtual hand emerges eerily from the three-dimensional CT scanner that recreated it.

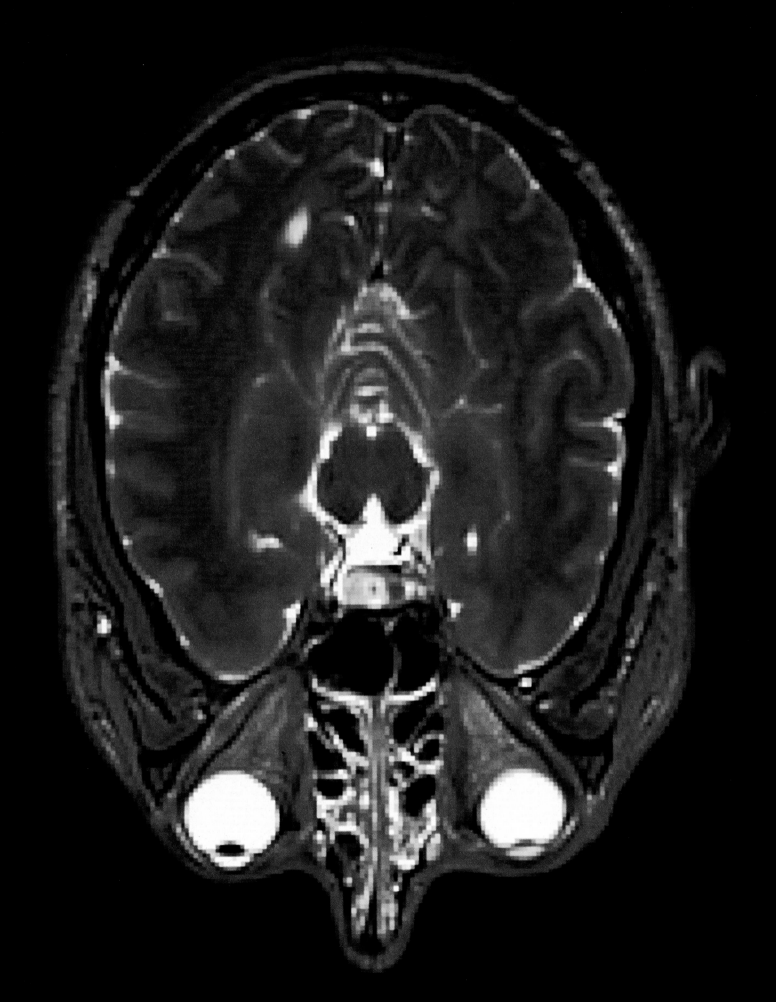

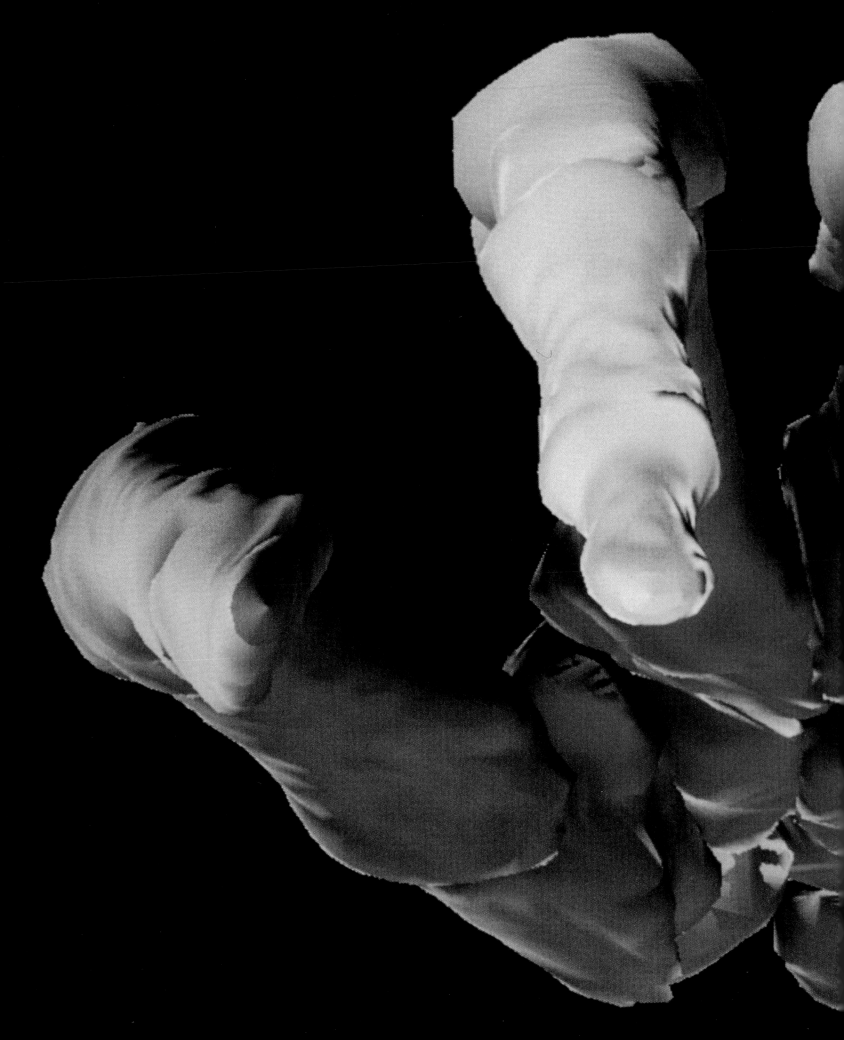

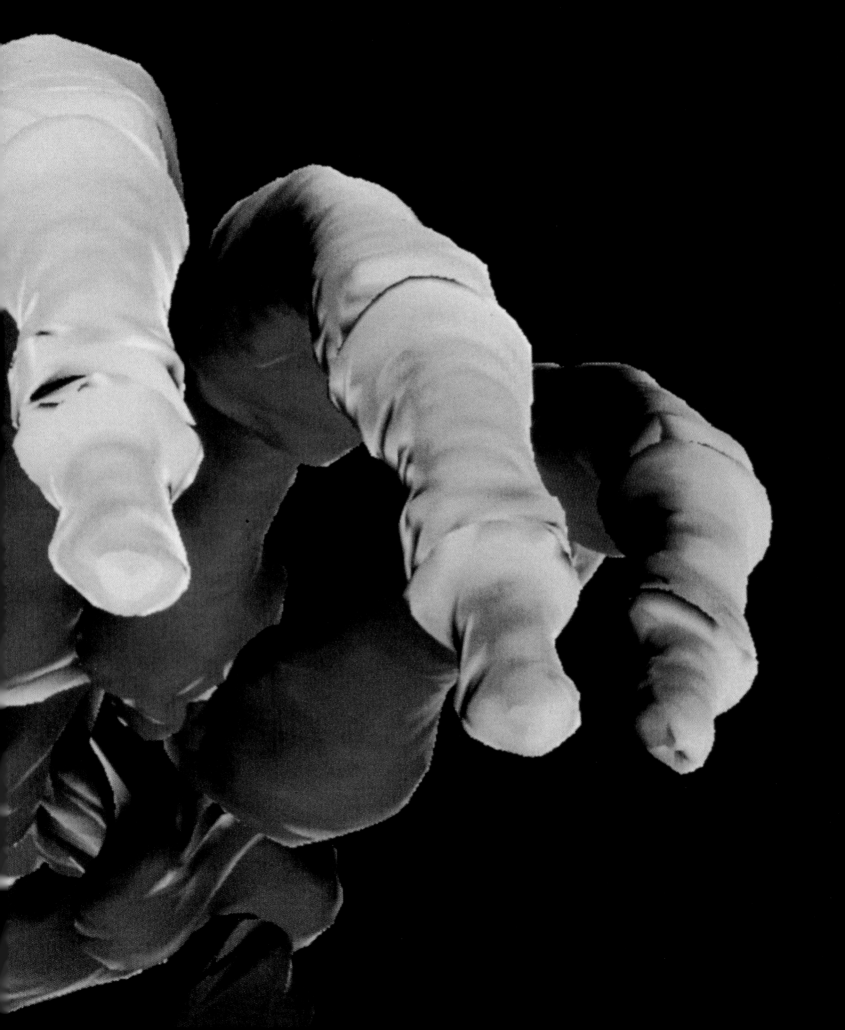

that confirmed the anecdotal. Today, scientists agree that those who practice relaxation techniques can, indeed, control their blood pressure, heart rate, anxieties, skin resistance, body temperature, and brain waves.

Biofeedback is a high-tech way to do that. Electronic monitoring instruments feed a patient physiological data he is ordinarily unaware of—temperature, blood pressure, gastrointestinal functioning, brain wave activity—enabling him to "see" a heartbeat on a screen or graph, or "hear" brain waves in the form of a series of high-pitched beeps. By "observing" such usually unconscious occurrences, the patient is essentially able to control involuntary processes, thereby influencing physical and mental well-being. In the case of brain waves, each is associated with some mental state. During periods of deep meditation and relaxation, for instance, Zen and yoga practitioners generate alpha waves almost exclusively. Beta waves appear to be emitted during anxious moments, delta during sleep, and theta in periods of creativity. By listening to these waves, amplified electronically, a patient is often able to regulate mental state and, some scientists believe, relax without using medication and improve memory and concentration. Some researchers suggest that the power of prayer and faith healing, like some forms of meditation, might be physiological in that they may protect the body against the negative effects of the stress hormone, epinephrine. "Medicine is a three-legged stool," says Herbert Benson, associate professor of medicine at Harvard Medical School and director of the Mind/Body Medical Institute. "One leg is pharmaceuticals, the other is surgery, and the third is what people can do for themselves. Mind-body work is an essential part of that, along with nutrition, exercise, and how to change your way of thinking."

Herbal medicine is another traditional alternative with a growing modern following. Mythology, ancient lore, and religion are rich in accounts of the health-bestowing properties of plants and plant products, and so, too, is the 20th-century pharmacopoeia. In fact many of the drugs and other medications commonly used today have herbal origins. According to the NIH Office of Alternative Medicine, about a quarter of the prescription drugs dispensed in the U.S. contain at least one active ingredient extracted from plant material. Morphine, for example, was initially processed from opium poppies; curare, which has been used as a muscle relaxant, comes from a vine; and atropine, an antispasmodic, comes from belladonna. The widely used nonprescription medication

aspirin was first extracted from willow bark. If animal, vegetable, and other natural sources are included, the list of additives becomes longer. Insulin comes from animal pancreases, penicillin is made from bread mold, and carnitine, used in treating hyperthyroidism, comes from the human liver. Hundreds of phytochemicals, found in fruits and vegetables that have disease-preventing and health-promoting properties, are under intense scrutiny by food scientists and manufacturers. Several are being investigated for their anticancer potential, among them broccoli's sulforaphane, tomato's lycopene, and resveratrol—present in red grapes, grape seeds, and skins, and in some 70 other plants. Folic acid, a B vitamin in the phytochemical folate, is plentiful in citrus fruits, green leafy vegetables, and peanuts.

Some 4 billion people, 80 percent of the world's population, rely on herbal medicine, and in China herbals—from astragalus root to stimulate the immune system to zea corn silk for urinary tract infections—are the backbone of medical care. To tap the enormous diversity of natural biological materials and to identify plants with possible medicinal value, scientists literally ransack—"bioprospecting," they call it—rain forests, jungles, and fields. One ambitious project funded by the National Institutes of Health, the National Science Foundation, and the U.S. Agency for International Development has researchers and pharmaceutical company representatives working hand in hand with shamans, traditional village doctors, and healers in Suriname, South America. Extracts from plants are shipped to the U.S. drug manufacturer Bristol-Myers Squibb for testing. David Kingston, the project's principal investigator, is a natural-products chemist at Virginia Polytechnic Institute and State University who has worked with the natural cancer-fighting agent taxol for several years. Searching for evidence of anticancer activity in 1996, his team obtained "hits"—evidence of positive therapeutic action—from some 80 extracts out of nearly 2,000 tested. That works out to 30 to 40 promising plants.

Some of the herbs that show promise as anticancer agents are familiar to all of us. Rosemary and turmeric, widely used as spices, are among these. Their chemopreventive potential was evaluated recently, leading researchers to conclude that extracts inhibit tumor development in several animal organ sites.

Other active botanicals are known by name only to scientists. One comes from the ginkgo tree. In 1997 researchers at the New York Institute for Medical

Research reported in the *Journal of the American Medical Association* that an extract from the leaves of young ginkgo trees slowed the progress of Alzheimer's disease in some patients. The extract has been used as an herbal medicine in China for thousands of years. Another medicine is camptothecin, from the *Camptotheca acuminata* tree found in China. Several forms of the drug are in use or under investigation. One, CPT 11, is the first new drug for colon cancer to have received federal approval in 20 years. Another, topotecan, is considered one of the best drugs available for ovarian cancer.

Perhaps the herbal preparation that most seized the public imagination in recent years has been St. John's wort. Its showy yellow flowers bloom over a long period, but there seems to be much more to the plant than meets the eye. Used for centuries to heal wounds and burns and to alleviate the pain of sciatica, St. John's wort is currently being studied for its apparent power to treat mild to moderately severe depression. If that power is confirmed, it would mean a drug that is far cheaper, and has fewer side effects, than currently prescribed antidepressants. A review published in 1997 in the *British Medical Journal* was optimistic: Given the variety of plants that have already contributed to modern medicines, it would be odd if no others turn up.

HOMEOPATHY AND CHIROPRACTIC

While acupuncture, meditation, and some herbal remedies are generally embraced by mainstream medicine, chiropractic and homeopathy, popular though they may be, travel a sometimes rockier road in their search for total professional acceptance. Compared to the other alternatives, both are relatively new schools of therapy.

Homeopathy is the older of the two. Founded in the 18th century by German physician Samuel Hahnemann, it is based on the notion that like cures like—that is, that microdoses of substances known, in large amounts, to cause illness can actually treat that illness by prompting the body's natural defenses. In that respect homeopathy is not unlike immunization or allergy treatments. There are differences, however. Homeopathic prescriptions, for example, are based on a patient's personality as well as what ails him. Usually, only a single remedy is prescribed to cover all symptoms, mental and physical. Also, patients receiving homeopathic care often feel worse before they feel better because the medicines often stimulate, rather than suppress, symptoms.

For these and other reasons, homeopathy has come in for its share of scorn, none the least of which was leveled by the formidable man of letters and once dean of the Harvard Medical School, Oliver Wendell Holmes. Homeopathy, he wrote, "is a mingled mass of perverse ingenuity, of tinsel erudition, of imbecile credulity, and of artful misrepresentation, too often mingled in practice...with heartless and shameless imposition." Current critics argue that doses of homeopathic medicines—nontoxic compounds derived from plants, animals, and minerals—are too diluted to do any good, that they are no better than water at worst and are placebos at best.

Nonetheless, homeopathy has endured. It is practiced worldwide, notably in Europe, Latin America, and Asia. The U.S. has some 3,000 practitioners, including a substantial number of physicians, nurses, veterinarians, chiropractors, and naturopaths. The homeopathic drug market in the United States, which is regulated by the U.S. Food and Drug Administration, is a multimillion-dollar industry and is growing at the rate of 20 percent a year. So far as the dilution issue is concerned, homeopathic researchers argue that electromagnetic energy in the medicines may interact with the body on some level. Others suggest that even after dilution a medicine retains its structure and leaves an imprint.

Chiropractic is a wholly different method, requiring no medicines, no surgery, and—except for diagnostic x-ray—no radiation. A descendant of the age-old technique of bonesetting, chiropractic was introduced in 1895 by Daniel David Palmer, an unschooled healer and self-taught anatomist. It is based on the premise that misaligned vertebrae in the spine put pressure on spinal nerves and contribute to a variety of disorders. The primary aim of treatment is to restore the structural integrity of the body, especially that of the spine, by manipulating the vertebrae. While many conventional physicians question chiropractic's ideas—notably those of fundamentalist practitioners who blame spinal misalignment for disorders that stray far from the backbone—chiropractic has proved immensely useful in treating acute low-back pain and other muscular and neurological problems. Athletes, construction workers, and deskbound secretaries and students swear by it.

More than 20,000 chiropractors are licensed in all 50 states; the AMA, once chiropractic's chief antagonist, sanctions referrals, and chiropractic services are covered by health insurers, workmen's compensation programs, Medicare, and Medicaid.

In the last analysis, the alternative therapies—unfortunately, "alternative" is a misleading term that implies equality with standard treatments—are by no means cure-alls. While some work, others fail miserably, and doctors advise caution before a patient abandons conventional treatment for them. Still, open-mindedness is imperative. Perhaps the comment of Harvard's Eisenberg is apropos to all of the unconventional therapies that are certain to be put to sterner tests in the new millennium. "It is conceivable," he has observed, "that physical manipulation or acupuncture needles or active ingredients in particular remedies are in fact physiologically powerful. But if for certain patients, in certain instances, faith or belief can cause changes in physiological function in a reproducible way, then we have to bring together the best scientists and ask them to figure out why."

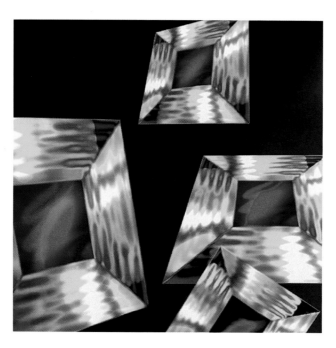

18th-century wit observed, by dismissing the doctor. Diseases and their symptoms are complex, though they may appear as deceptively simple as the common cold. Diseases and symptoms are as abundant as the hairs on our heads, and as varied as the biochemicals and elements that make up the human body. This array of diseases calls for a variety of treatments—sometimes nominal, presumptive, experimental, or heroic. Even Hippocrates, the Father of Medicine, knew the limitations of one treatment over another. "Those diseases that medicines do not cure," he noted, "are cured by the knife. Those that the knife does not cure are cured by fire. Those that fire does not cure must be considered incurable."

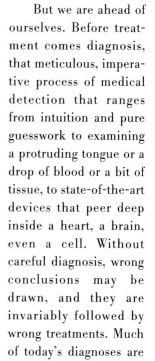

But we are ahead of ourselves. Before treatment comes diagnosis, that meticulous, imperative process of medical detection that ranges from intuition and pure guesswork to examining a protruding tongue or a drop of blood or a bit of tissue, to state-of-the-art devices that peer deep inside a heart, a brain, even a cell. Without careful diagnosis, wrong conclusions may be drawn, and they are invariably followed by wrong treatments. Much of today's diagnoses are the result of technological advances that, in many instances, far outdistance what a mere physician can do with his hands, eyes, and ears. This is not to disparage the sometimes uncanny accuracy of a skilled diagnostician who knows exactly where and how to probe, palpate, and auscultate. The new diagnostics are a quantum leap beyond all that, and they have transformed the art of practicing medicine so dramatically that the grandfather of diagnostic machinery, the simple, clunky x-ray machine, is a box camera in comparison. The x-ray still manages to do its job, but in a sense it is akin to the paper

DIAGNOSTIC TECHNIQUES

When we seek to preserve our health, it is not enough, of course, to merely meditate, dose ourselves with herbal medicines, or attempt to fix what ails us with the prick of an acupuncture needle. Nor will we be healed, as one

From energizing supplements like the molecules of vitamin B-5 (above) to acupuncture that transforms the body into a pincushion, alternative therapies rely sometimes on science, sometimes on coincidence, and often on wishful thinking.

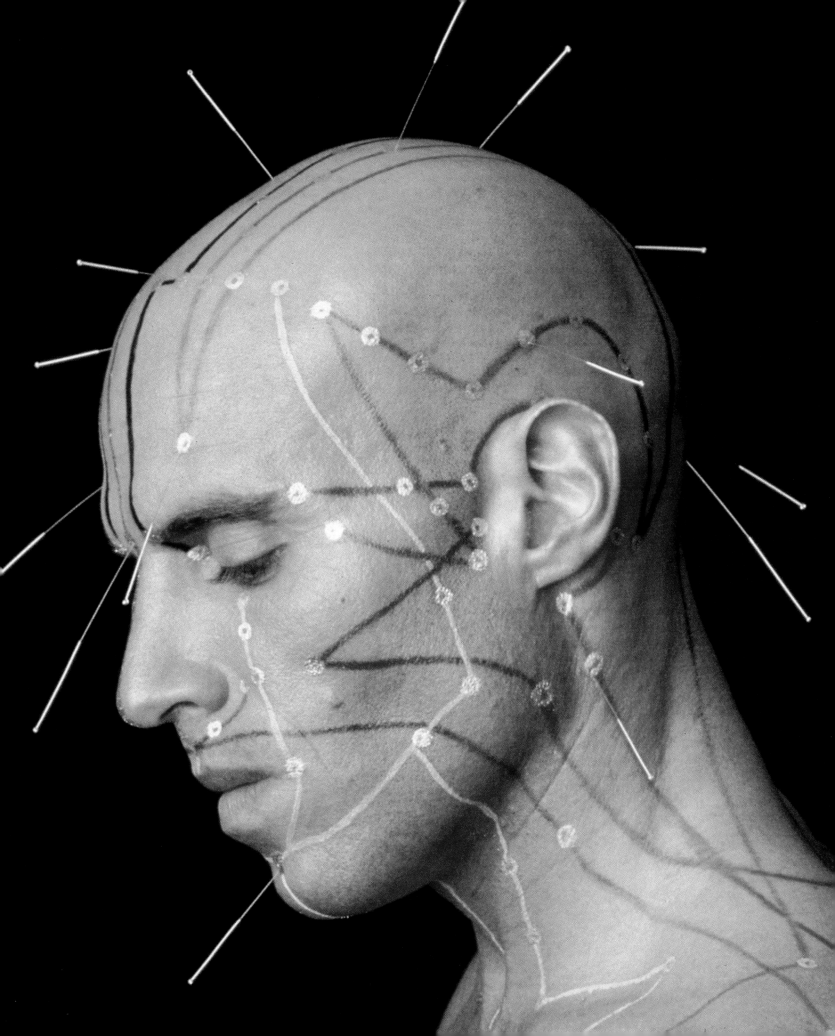

cylinder that early physicians pressed to their ears to hear a heartbeat. The days are also numbered for some of our other means of diagnoses, particularly invasive ones such as endoscopic procedures, which necessitate inserting long instruments through the body's natural openings to view internal organs. These procedures, along with blood tests, bone-marrow extractions, and needle biopsies—all of which require sticking something into a patient—will one day be replaced by the realm of noninvasive digital medicine, computer algorithms, and virtual diagnostics.

CASE REPORT. January 2, 2050

A patient arrives at the SuperCon Medical Center, takes off his shoes, and steps into one of the pairs of magnetized, superconducting clogs lined up by the door. A nurse tells him to follow a blue track to the MagLab, a metallic room that will shield him from outside magnetic fields. The patient floats in effortlessly, and soon is lying on a low examining table. The patient's neurologist enters the room and presses a button. The examining table rises off the floor and glides to a bank of sensors that resemble salon hair dryers. The table stops under one and gently settles to the floor. The sensor, lined with ultra-sensitive superconducting detectors, now descends slowly to fit neatly around the patient's head. The doctor throws a switch, and in a few seconds the faint magnetic fields that emanate from the patient's brain cells are precisely mapped and displayed on a video screen.

"I can give you a clean neurological bill of health just by eyeballing this reading," says the doctor. "The fields are uninterrupted. That means no neurotrauma, epilepsy, tumors, emotional depression, multiple sclerosis, Alzheimer's or Parkinson's diseases, and no proneness to stroke, schizophrenia, or migraine."

The patient's table rises again, and he is transported out the door and into an adjoining room, where a magnetocardiologist and a magnetointernist guide the table under another superconducting sensor, this one shaped to fit the entire body from the neck down. The table is linked to an expert diagnostic and prescription system and a digital telemetry system. "You saw that old sci-fi movie *Fantastic Voyage?*" the internist asks. "Well, we're going to take a little trip inside your body without the submarine, and without one of those long tubes they used back in the dark ages of the 20th century. We'll be doing a virtual computed endoscopic-celluloscopic fly-through. We'll see as much as if we were actually in there with floodlights." The machinery hums, and again,

the magnetic fields that flow out of the patient's body are monitored. Organ systems appear on screens in 3-D, along with detailed analyses of enzyme and mineral levels, blood, urine, electrolytes, and microstructures.

"Electrical action of the heart, normal," says the heart specialist, working a joystick and making a virtual pass through a coronary artery. "No danger of a heart attack for you." "Kidney, bladder, and liver function, check," says the internist, perusing a screened graph. He turns to another monitor. The image is slowly traveling the length of the patient's intestines, pausing to enable the physician to examine the shiny, mucous lining and scrutinize adjacent structures. "No lesions, no polyps," he says. "There's a slight break in the field from your duodenum that's indicative of the beginning of a small ulcer, but our built-in prescription writer here says all you'll need is a little over-the-counter acid-reducer." He smiles at the patient. "See you next year."

SEEING INSIDE THE BODY

The new diagnostic machines are gleaming museum-quality pieces of designer metal and plastic, sculpted monitors on which vividly colored images of internal organs pulsate and elaborate graphics unfold, monitored by beeping, whirring, humming computer clusters. The souls of these machines are a dazzling conglomeration of electronics, low-temperature superconductive magnets, silicon chips, array coils, sensors, actuators, the protons in the nuclei of hydrogen atoms, radiowaves, infrared sources, and ultrasonic wave-generating piezoelectricity. Wilhelm Conrad Röntgen, the German physicist who discovered x-rays a century ago, would have been overwhelmed—but keenly interested since the invisible beams of electromagnetic radiation are at the core of much of the new generation of scanning equipment.

While x-ray, or roentgenography, gave doctors their first noninvasive look inside a living body—it was first used for that purpose in 1895 by Canadians John Cox and Robert Kirkpatrick to find a bullet in a patient's leg—its photographs seem best suited for bones. The soft tissues of our organs, the cartilage in our noses, and the gelatinous disks that separate our vertebrae appear as misty shapes. The same haziness applies to nerves, veins, plaque in hearts and brains, and many tumors. X-rays from older machines can be dangerous, especially when a patient is overexposed to the cell-altering beams. Scientists had to find a better way.

That better way began after World War II with diagnostic ultrasound, a spin-off of the sonar that determines water depth and detects submarines, shoals of fish, and other underwater objects. As used in medical diagnosis, ultrasound can be a noninvasive, nonradiative, real-time technique that enables physicians to "see" inside the body and get a "feel" for an organ's structure, texture, and contour. A probe, pressed against the skin, sends high-frequency sound waves into the body, where they bounce off targeted organs. As they return, the echoes are converted into two-dimensional images. Except for the machine-generated imagery, the process works on somewhat the same principle that night-flying bats use to identify obstacles in their path.

One of the most reliable and cost-effective imaging devices known to medicine, ultrasound seems to have few bounds beyond its inability to travel through bone. It can assess the status of virtually everything from a kidney cyst to a racehorse's damaged tendon to a fetus's development, the latter use familiar to all future parents who have watched on screen their unborn child's tiny, beating heart. Ultrasound is a standby in emergency rooms, where it tells doctors whether an abdominal cavity must be drained of fluid; it can detect the plaque that, when accumulated in the carotid arteries, causes 80 percent of strokes; it also can warn well in advance of an impending attack in a person who has no symptoms. As technology advances, ultrasound's uses are expanding. One variation, echo-cardiography, can determine the size, shape, and motion of the heart, and can identify abnormalities. Doppler ultrasonography, which can measure the speed and turbulence of blood flowing through an artery, is used to detect vascular disease, its pulses focusing on the major veins and arteries of the head, neck, arms, and legs. When color-flow imaging is added, the differences between benign and malignant conditions become evident. With 3-D functional imaging attached, ultrasound takes a view of an organ that existed only in the surgeon's mind or in the anatomist's atlas, and screens it for all to see. Intuition and imagination are thus made real, therapy is tailored, and the most important person in the operating room—the patient—benefits. One day, handheld ultrasound units will be as common as stethoscopes, invaluable in rural and remote regions and in ambulances. Their pulsating, sound-generated images will be sent ahead to a medical center via telemetry for analysis—20 telemedicine projects in 13 states and the District of Columbia are now being evaluated—and a treatment plan will be set in motion long before the patient arrives at the hospital.

Another better way of seeing things was born in 1972. Computerized axial tomography, CT for short, is a marriage of x-ray and digital technology that scans organ systems and bodily cavities. To scan the brain, for example, instead of a static x-ray camera an x-ray tube revolves about the head, emitting and detecting rays and converting the pictures into a digital computer code to make high-resolution images accurate to tenths of a millimeter. Differences in density between normal and abnormal tissues are revealed; fine details of bone structure appear clearly, and the location of tumors, cysts, and other signs of disease can be pinpointed.

Positron emission tomography, or PET, was next. Similar to CT scanning, PET uses trace amounts of radioisotopes to measure blood flow through tissue and to determine if a patient's biochemical processes are functioning properly.

MRI, magnetic resonance imaging, is the new techie on the hospital block, a linchpin of today's diagnostics that technicians are continually improving or joining with other systems. Like the CT scan it is expensive—millions of dollars to install, and costly for patients and their insurers—but it saves valuable time and, often, the need for further diagnostic scrutiny. An MRI machine, into which a patient is rolled on a table, is usually a tunnel-like affair—actually a "doughnut" filled with liquid helium that cools superconductive magnet systems, which maintain high currents and intense fields while expending relatively little energy. Essentially, MRI reads radio signals returning from the hydrogen atoms in water molecules within a given tissue; the signals are fed into a computer that converts them into detailed images of soft tissues, as in the brain and spinal cord. Using the color-enhanced pictures—actually cross-sectional "slices"—a doctor can differentiate between normal and

FOLLOWING PAGES: "Why has not Man a microscopic eye?" the poet Alexander Pope asked. He does now, thanks to new tools of microsurgery that enable ophthalmic surgeons to cut and sew nearly invisible blood vessels and nerves.

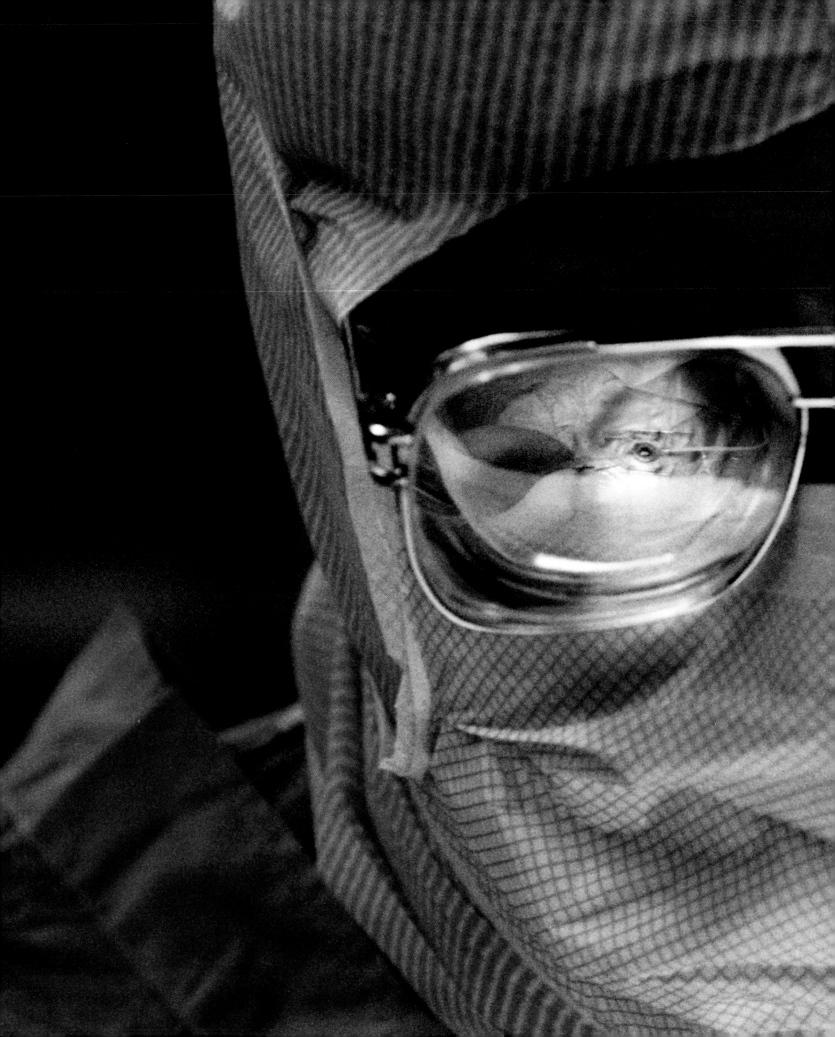

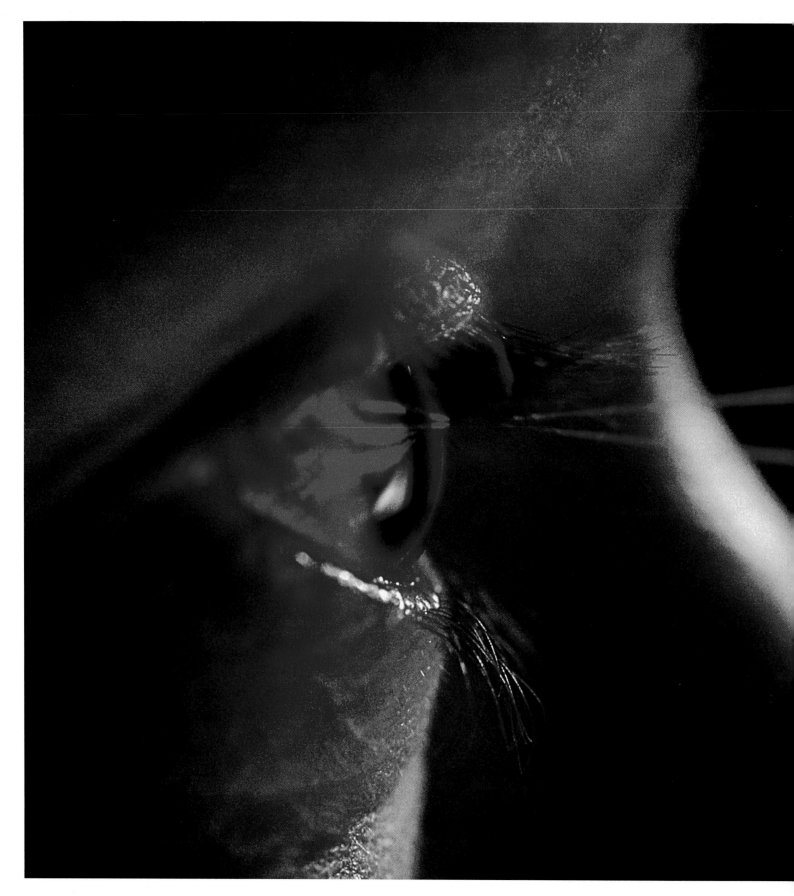

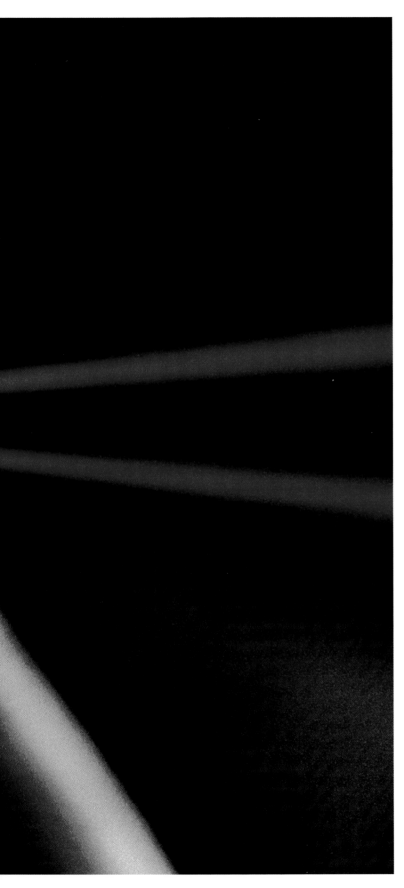

abnormal regions. For instance, multiple sclerosis plaques in the brain, not easily visible by any other means, are defined by the MRI, as are brain lesions caused by vitamin B_1 deficiency, tiny tumors hidden in the heart and pituitary gland, and torn knee cartilages.

Even structures that resisted early MRIs are now visible. Nerves produce weaker signals than organs and soft tissues, which makes their MRI images resemble those of tendons, blood vessels, and ordinary fat. Diagnosis and treatment must rely on images of structures near the nerves, such as bone, muscles, and ligaments, as well as on electrical tests and physical examination. A technique developed at the University of Washington School of Medicine promises to change that, and, ultimately, the way pain and nerve-related disorders are diagnosed. Called MR neurography, this new technology links the MRI to modified computer programs and signal-enhancing devices that magnify weaker nerve impulses that normally are overwhelmed by adjacent tissues. The image signals from other tissues are filtered out until only signals from nerve tissue remain, making the nerve fibers the brightest structures in the image. This technique can also block out tissue that sheathes the nerves, enabling individual nerve bundles to leap into sharp focus. Moreover, it displays images both of normal nerves and, through bright regions called hyperintensities, nerves that are compressed or injured. Detailed images have been produced of a spate of nerve problems, among them, herniated disks, sciatica, and mass lesions. "As the surrounding tissue disappears," explains Aaron Filler, who helped develop the technique, "we are left with a three-dimensional picture of the nerve, like the smile of the Cheshire cat in *Alice in Wonderland*.

One of the major barriers the MRI has had to overcome is any patient's panic at the thought of lying still inside what is really a giant thermos bottle for an hour or so. To deal with this drawback, manufacturers now use permanent magnets or electromagnets that generate lower fields and don't require helium. Such machines are easier to shield than the standard behemoths: New-generation MRIs are user-friendly—more spacious and less claustrophobic.

....

Seeing the light: Intense, concentrated beam of a surgical laser, focused to a width smaller than a pinpoint, can sculpt a damaged cornea, vaporize unwanted tissue and bloated blood vessels, or reconstruct a detached retina.

....

Another MRI twist that suggests a new treatment approach in the decades to come is an advanced brain-imaging technique known as fMRI, the "f" standing for functional. Researchers at Harvard Medical School and Massachusetts General Hospital switched it on recently, not to examine tumors or track a nerve, but to identify brain circuits that might be activated during the distinct experiences following a cocaine addict's rush and craving. Previous studies with other techniques provided insights into how and where cocaine acts in the brain by recording nerve-cell activity in the brains of laboratory animals treated with drugs. Other studies of the effects of drug abuse on the human brain have been consistent with animal-research findings, but the techniques used could not provide detail and specificity.

Enter the fMRI. By fine-tuning the instrument and training it on subjects as they experience a rush or craving, researchers are able to come up with a brain map of cocaine-induced euphoria and dependence. The fMRI produces a highly detailed map of brain activity while individuals have these different experiences and exhibit distinct patterns of activity in different parts of the brain associated with each experience.

Besides laying out in exquisite detail the brain circuitry involved in drug use, the study offers the potential for enormous practical application in treating a major public health problem. As Alan I. Leshner, director of the National Institute on Drug Abuse, which funded the study, explains, "It suggests specific brain areas that might be targeted in developing new medications to either block individual aspects of cocaine's effects, like the rush versus the craving experiences, or as a broader treatment for cocaine abuse and addiction."

The heart, too, has been opened to view by MRI technology. The technique widely used to scrutinize the insides of arteries—cardiac catheterization—is a valuable diagnostic tool, but it is invasive. Patients are given a local anesthetic, and a long, thin tube is inserted through a tiny incision in the arm or groin. To guide the tube to its destination, the cardiologist uses a fluoroscope, an x-ray machine that projects images on a screen. In a right-side catheterization, the catheter is sent through a vein into the heart, through the right atrium and ventricle, and out through the pulmonary artery. For the left side, the catheter must be passed through an artery into the aorta and then the coronary arteries, the left ventricle, or both. Dye is then injected into the heart and arteries, and, as it filters through, abnormalities or obstructions become

visible. An MRI enables doctors to peer into an artery without catheterization, and with astonishing efficiency.

VIRTUAL ENDOSCOPY

Even more astonishing is the advent of the virtual endoscopy experienced by our hypothetical patient in the 21st-century Supercon Clinic. No figment of a science fiction writer's imagination, virtual endoscopy is among the newest and most exciting examples of diagnostic imaging, one that relies on CT and MRI images and high performance computing to provide simulated visualizations of specific organs. It is modern medicine's version of a video game. Endoscopy is a general term for tests that allow doctors to peer directly at internal organs or other structures through a flexible fiberoptic tube. The endoscopic tube, with its lighted scope, may be passed through natural openings in the nose, mouth, anus, bladder, and vagina. Or a narrower lighted instrument, a laparoscope, may be used to examine pelvic or other abdominal organs through a small incision in the abdominal wall. Endoscopy that examines the large intestine is called colonoscopy, a highly accurate way to detect inflammatory bowel disease, gastrointestinal bleeding, and the cancers of the colon and rectum that kill nearly 55,000 Americans every year. Special instruments attached to the tube enable surgeons to simultaneously remove or destroy polyps or lesions and collect biopsy samples. There are risks, although slight, of perforation, infection, and hemorrhage. Since air is injected through the tube in order to inflate the folds of the colon for better viewing, the procedure may be uncomfortable, and the patient is often sedated and tranquilized.

Virtual endoscopy eliminates the need for invasive conventional endoscopic procedures. Instead, taking a cue from nonmedical practices such as flight simulation, virtual endoscopy converts CT and MRI scans into three-dimensional representations of anatomical structures that are computed and transformed into an animated video sequence. Sitting at a computer workstation, the virtual endoscopist can simultaneously view inner anatomy and, with a computer mouse, manipulate the "flight" path—in effect perform a fly-through of the body—in a realistic way. The technique provides viewing control and options that are not possible with real endoscopy, such as direction and angle of view, scale, and the capability to shift immediately to new views. A wide range of anatomic structures are accessible to virtual endoscopy, including the stomach, esophagus, heart, spinal canal, inner ear, biliary and pancreatic ducts, and large blood vessels—indeed everything that surgeons can now see with conventional techniques, and more. Technical problems must be overcome, but the technique shows significant promise.

Simply opening the living human body to scrutiny, as CT, MRI, and ultrasonography have done, is not always enough. Indisputably, disease diagnosis is a far more exact science than it was even a decade ago; treatments as well as cure rates have improved because of it. Essential, though, is not just diagnosis but early diagnosis. In the case of cancer, for example, more than half of all new cases occur in nine screening-accessible sites—breast, tongue, mouth, colon, rectum, cervix, prostate, testis, and skin—where early detection translates into treatment more likely to be successful. The relative survival rate, which is the percentage of people who are alive five years from the date of diagnosis, is around 80 percent for these cancers. According to the American Cancer Society, if all Americans participated in regular screenings, this rate could increase to more than 95 percent.

Invariably, the root of cancer and many other diseases lies in the cells, a world far tinier than that of the structures that MRI, CT, and ultrasound visualize.

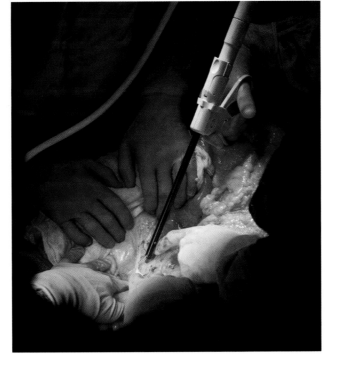

Mindful of this, scientists have approached closer to the cells, not only to obtain a better understanding of the molecular basis of genetic diseases, as the Genome Project promises, but also to lay the groundwork for earlier and more precise detection and, ultimately, more effective therapies. In a sense the research of biologists is akin to that of physicists working toward a better understanding of the world of subatomic particles. Just as revealing the innermost levels of the atom is vital to an understanding of matter, unraveling cellular mechanisms is essential to fathoming life itself.

Signaling the trend toward cell diagnosis are new methods of measuring heart-cell damage caused when blood flow stops and restarts, as occurs in a heart attack. Working with rabbit and dog hearts, scientists at Johns Hopkins and Northwestern Universities have demonstrated that MRI can quickly assess such damage. The method, first shown to work on animal heart cells and later on beating animal hearts, produces sodium maps of the heart. Cardiologists need to know which parts of the organ are alive, damaged, or dead after an attack. In a heart attack sodium accumulates in dead heart cells, and the maps show dead cells as brighter and living cells as darker. When the technique is applied to humans, it will enable doctors to appraise damage far more quickly and directly than echocardiography and stress tests, which are indirect measurements of something gone awry.

Other methods of checking cells for signs of disease must go deeper still, into the chromosome. Aberrations

Minimizing blood loss during surgery means fewer transfusions, quicker recovery, and shortened hospital stays. A so-called harmonic scalpel—actually a vibrating laser—simultaneously cuts tissue and clots blood (above). FOLLOWING PAGES: Improving on conventional scattergun radiotherapy, a hulking linear accelerator adjoins an operating room, delivering precise arcs of radiation to zap a brain tumor.

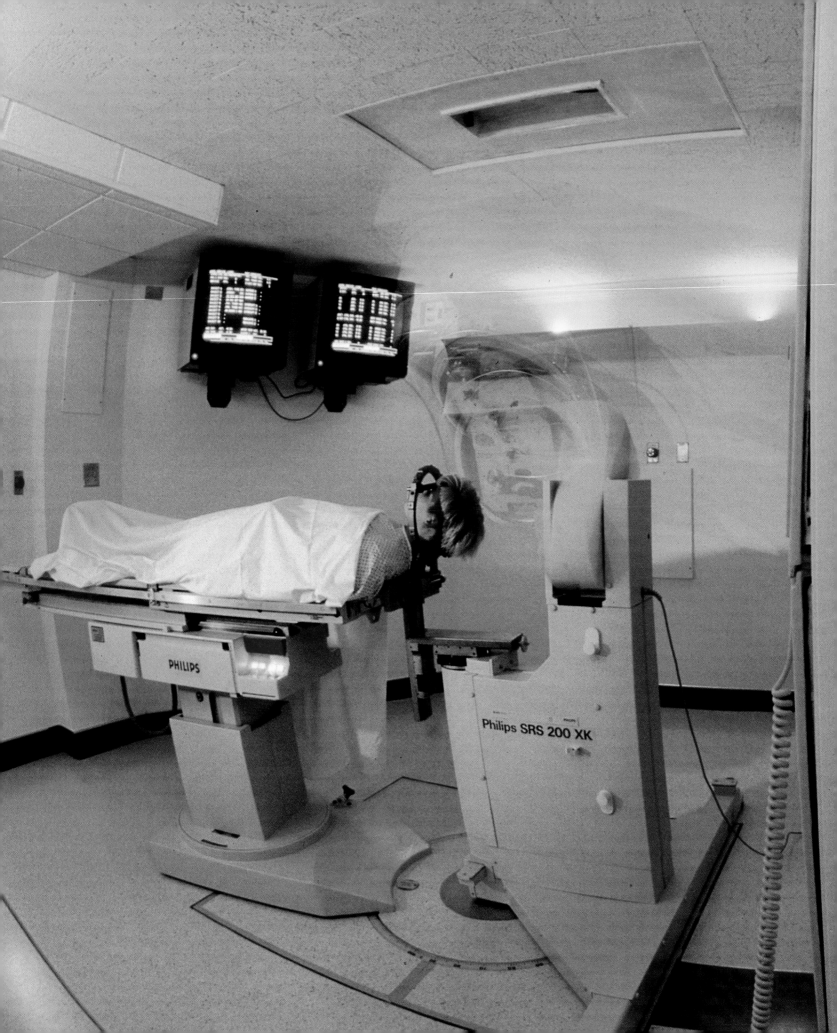

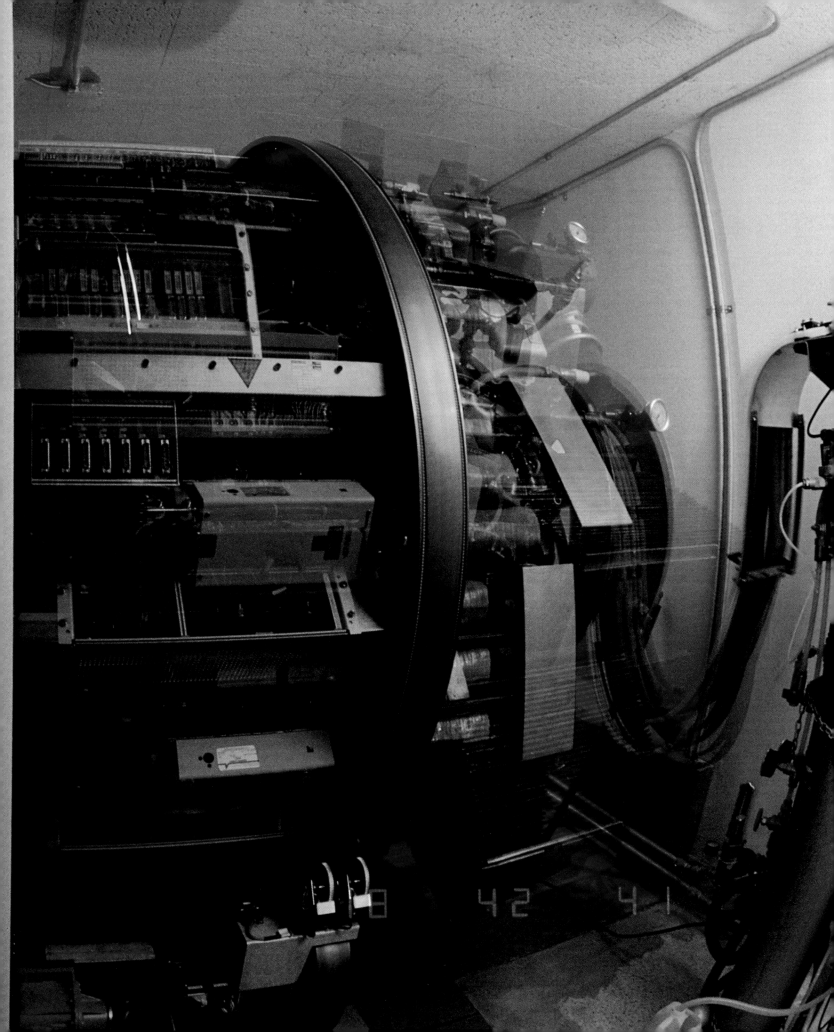

of that gene-bearing body are linked to leukemia and other cancers as well as to birth defects such as Down's syndrome. Today's most widely used test for detecting chromosome defects is the Giemsa, or G-banding, method. By staining chromosomes with a dye called Giemsa, lab technicians produce a karyotype, or chromosomal arrangement, that shows a distinctive banding pattern for each chromosome. If there is an aberration, a karyotype usually reveals it, but not always. Parts of chromosomes may be translocated, or switched, from one site to another on a chromosome or between one chromosome and another; other chromosomes may be deleted or duplicated in whole or in part. G-banding may miss a subtle translocation.

One way around that is with a new technology named SKY, for spectral karyotyping. The name is appropriate because the technique incorporates spectral analysis, a tool astronomers use to separate the rainbow-like components of light from distant stars. Developed by scientists at the National Human Genome Research Institute, SKY is one more example of how the Human Genome Project is spinning off technologies with almost immediate benefit to diagnosis and treatment. SKY uses molecular probes that attach themselves to parts of chromosomes and glow when exposed to light. The tagged portion of each chromosome appears in a specific color, creating a multicolor pattern that vividly distinguishes one chromosome from another, revealing aberrations that G-banding routinely misses. In a study of 15 patients with various forms of leukemia, teams led by Thomas Reid at the institute located chromosome defects in leukemia cells that went undetected with G-banding—in every case. Future tests should determine whether SKY can be used to detect chromosome aberrations in other diseases, such as birth defects. If the new technology proves out, it will augment and perhaps even replace the current G-banding technique, which is now performed half a million times a year in the U.S. and Canada.

SKY is a high-tech diagnostic, but often in medicine, a low-tech one cannot be gruffly dismissed. Nor should it be. All that is sometimes required to create a diagnostic technique are a few reliable old tools, some knowledge and ingenuity, and perhaps a bent for tinkering. Consider laser capture microdissection. A powerful new technique with a state-of-the-art name, it is remarkably down-to-earth. A way of extracting cells from tissue so they can be examined for signs of cancer, laser capture microdissection is the product of a group of NIH researchers who

constructed it out of some fairly humble components: A standard laboratory microscope, a low-energy laser, and a piece of "transparent ethylene vinyl acetate polymer thermoplastic," a rather pretentious way of referring to the same everyday plastic that goes into the seal on a container of yogurt.

Usually, doctors must painstakingly examine tissue biopsies of cells for suspicious, early signs of cancer. Cells must be tediously extracted by either yanking them free manually or following a complicated and inefficient process that requires isolating and culturing them. Laser capture microdissection is a one-step technique that essentially works on the basic principle of a point-and-shoot camera. A tissue biopsy, which typically contains hundreds of types of cells, is examined under the microscope. When a group of tumor cells is spotted, a button on the side of the microscope activates a laser, which flashes a beam of light. As the beam passes through plastic film that has been placed above the tissue to focus on the abnormal cells, it heats the plastic, giving it the thermal quality of a piece of Scotch tape; the cells stick to the plastic directly above them and are easily extracted for analysis.

The device has successfully extracted cells and even enzymes from all tissues in which it has been tested, including breast, prostate, and lymph nodes. One of the tasks of NCI's Genome Anatomy Project is to define patterns of gene expression in normal, precancerous, and malignant cells, and this new technique may one day help record these very patterns of expression. Lance Liotta, an NCI scientist and one of the researchers, assesses the situation: "Having this technique is the difference between being able to investigate a crime in progress and going back two weeks later to the scene of the crime when much of the evidence has vanished, as we typically do now. Laser capture microdissection gives us access to the disease, in a sense, while the crime is still in the planning stages...."

Ultimately, diagnostic techniques both high- and low-tech require a human being to operate, read, and analyze the pictures and the data that have been gathered. Nowhere is this more evident than in the reading of a simple electrocardiogram (ECG), the standard test used to diagnose heart attacks in patients seen for chest pain in hospital emergency rooms. An estimated 25 percent of ECG readings are misjudged or overlooked by the physician, and patients may be sent home without a correct diagnosis. To the rescue now come so-called artificial neural networks, a new computer-based method of reading an ECG. Neural networks, under intense study in Sweden and elsewhere, are designed to "think" like human doctors and technicians who must draw on experience and knowledge to make decisions. To teach a neural network how to recognize heart attacks, scientists at the University Hospital in Lund, Sweden, exposed a computer memory to thousands of ECG readings—from people who had suffered heart attacks and from those who had not—far more than any cardiologist could possibly read in a lifetime. The neural networks proved to be 10 percent better at identifying abnormal ECGs than the most experienced cardiologists on staff.

FIXING THE BROKEN BODY

Treatment generally follows diagnosis, a fact of life known to patient and doctor. The purpose is to forestall, ameliorate, or cure, and there are myriad possibilities, some as simple as waiting out the course of a cold. For the most part, though, treatment today involves a mind-boggling assortment of surgeries, microsurgeries, transplants, implants, radiation, drugs, and, on the near horizon, gene therapies. Driven by medical technology, the science of health restoration is a far cry from what it was in the days of the 13 Colonies. Then there were no stethoscopes, thermometers, or hypodermic syringes; germs and viruses were yet to be identified. Some 3,500 "doctors" ministered to patients, but less than 400 had legitimate medical degrees.

There was but one modus operandi, a three-pronged approach that applied no matter the physician or, in most cases, the affliction: Open a wrist vein and draw off a pint or two of blood, provoke vomiting with herbs and foul-tasting medicines, and induce a bowel movement with strong purgatives. George Washington, lying on his deathbed at 67, was bled three times, given massive doses of calomel, popular as a fungicide as well as a purgative, and dosed with tartar emetic, a poisonous salt also used in dyeing. An application of blistering poultices finally finished the old general off. "Difficult as it may be to cure, it is always easy to poison and to kill," Elisha Bartlett, a prominent medical writer of the time, observed drolly.

CASE REPORT. February 1, 2050

A patient, a 42-year-old male with rectal bleeding, is preadmitted to the Digital Medical and Surgical Center via telemetry communication from the referring physician. Virtual endoscopy rules out hemorrhoids, cancer of the colon and rectum, proctitis, and ulcerative colitis, but on the return fly-through the scanner finds several ulcerated diverticula, small saclike protrusions in the wall of the ascending colon. Transfusion and surgery are recommended. Because the patient has religious reasons for refusing a blood transfusion, a team swiftly injects an all-purpose drug—contained in a skin patch on the patient's arm and sped into his body by pulses of ultrasound—that reduces his body's demand for oxygen, provides him with iron and vitamin supplements, and goads his bone marrow into producing red blood cells. A polymer-based artificial blood that does not compromise the patient's religious beliefs is on standby. A hematologist sitting at a digital telemetry system connected by remote to the patient watches the level of hemoglobin—the vital, oxygen-carrying compound in the patient's red cells—slowly rise to close to normal.

The patient is positioned under an intraoperative image guidance system that projects MRI views of the affected area of the colon on screens suspended over the surgical team's heads. Another screen, using three-dimensional images derived from the patient's scans and ultrasound data, re-creates the exact operation the surgeons will perform—simulating anatomical structures, organ movements, correct position of surgical instruments, and the safest surgical approach. A surgical navigation system hums, enabling surgeons to project the simulation directly onto the patient's abdomen; crosshairs zero in on the problem area, and the operation—viewed remotely by the referring physician in a nearby city and by two specialists a thousand miles away through an interactive teleintervention hookup—begins.

While a digital device that resembles an airliner's flight-data recorder tapes every move—information to be used later for education or to assess what, if anything, goes wrong—the head surgeon makes a two-inch incision in the abdomen with a harmonic scalpel, a vibrating laser that simultaneously cuts tissue and clots blood. Another surgeon, eyes on a screen image and pumping a foot pedal, guides a robot-held laser-auger that makes four pencil-diameter holes near the incision. On voice command, another robotic "hand" inserts an ultrasound probe into one of the holes, and 60-second bursts of sound are sent into the colon to relax the muscle walls and ease the surgery. A tiny fiberoptic camera attached to a laparoscope is guided robotically through the second hole, an electrocoagulation-tissue-vaporizing device and a laser tissue-welder through the third, and a miniature cutting tool into the fourth.

Fortunately, the site of the bleeding is a small section of colon. "Coagulate, minor resect coordinate seven, anastomosis, seal, implant," commands the chief surgeon, speaking into a microphone. Inside the colon the instruments pick up the order and perform their programmed tasks. The bleeding diverticula are sealed by the laser, and what blood still flows is suctioned out, cleansed, and returned to the patient. A small segment of colon is clipped off with the cutting tool, vaporized, and intestinal continuity is restored by laser-sealing.

"You'll be fine," the surgeon tells the patient, who has been intently watching the procedure on a monitor. "We'll just stick a Band-Aid over those holes, and you'll be home in a couple of hours, back to work in two days. A prescription med will help relieve some of that abdominal discomfort—you'll sprinkle it over your cereal—and one for stress, which will be a sticky patch you can apply anywhere. We've already implanted a tiny, time-release, colon-specific coagulant that'll ward off any future bleeding in possible weakened areas. Should be good for eight, ten years."

SURGERY: CUTTING AND PASTING

Surgery is, and undoubtedly will remain in the foreseeable future, a crucial element in a goodly proportion of treatments. It cures diseases that cannot be cured by any other means; it repairs where drugs cannot. What surgery accomplishes is often quick and visible, whereas what drugs and radiation achieve is often slow and hidden. Surgery and the arena in which it is practiced are the stuff of high drama, be the procedure a simple appendectomy or a complicated heart bypass, and when an operation is a success, we proudly display our scar, praise the surgeon's skill, and marvel over a quick recovery.

What will change, and already has, is how surgery is performed. While surgery remains, as second-century Greek physician Galen saw it, "the ready motion of steady and experienced hands," those hands are becoming robotic extensions of human physicians, no longer the best wielders of surgical instruments. As science increases its influence over the healing arts, the astonishing successes achieved by modern medicine are sure to be replicated in the new millennium, an evolution that will necessitate even more extraordinary efforts. The surgeon's hands will need extra guidance to keep pace with newer, more sophisticated systems and techniques, the products of close collaboration between physicians, engineers, and a host of other scientists. None of these innovations is meant to replace the doctor, only to give him or her more precision, and by so doing to improve the patient's chances, reduce complications, speed recovery, and lessen the need for repeat surgery.

The dream, then, of the 21st century, already partly realized, is for "smart" tools of the same sophistication as those used in our imagined resection: Voice-controlled surgical instruments, surgeons' helmets inside which real-time, see-through images of a surgical site are projected, surgery simulators, image-guided implants, robot-held laparoscopic cameras, teleintervention that allows a surgeon to operate by remote control many miles away from an underserved operating room, and automated precision milling to achieve a perfect fit of bone and implant during hip replacement surgery. Even now, at Boston's Brigham and Women's Hospital, surgeons can step inside a giant MRI and operate on a patient, their moves tracked and guided by computerized systems as they watch live, three-dimensional images of brain and body—even images of the surgeons' own hands and scalpels. At the Jet Propulsion Laboratory in Pasadena, California—a facility known more for rocket science than medicine—a ten-inch-long robotic "arm" with a three-axis "wrist" moves with six degrees of freedom to position and

Pushing surgery's envelope: A "littlest patient," a 30-week-old fetus partially removed from her mother's uterus, undergoes a diaphragmatic hernia repair. She will be returned to the womb to complete gestation. New minimally invasive fetoscopic techniques also permit operations on fetuses in the womb. FOLLOWING PAGES: Uneasy comfort comes to a Kazakhstan child wired for diagnostic tests to assess fallout damage from a former Soviet nuclear test site.

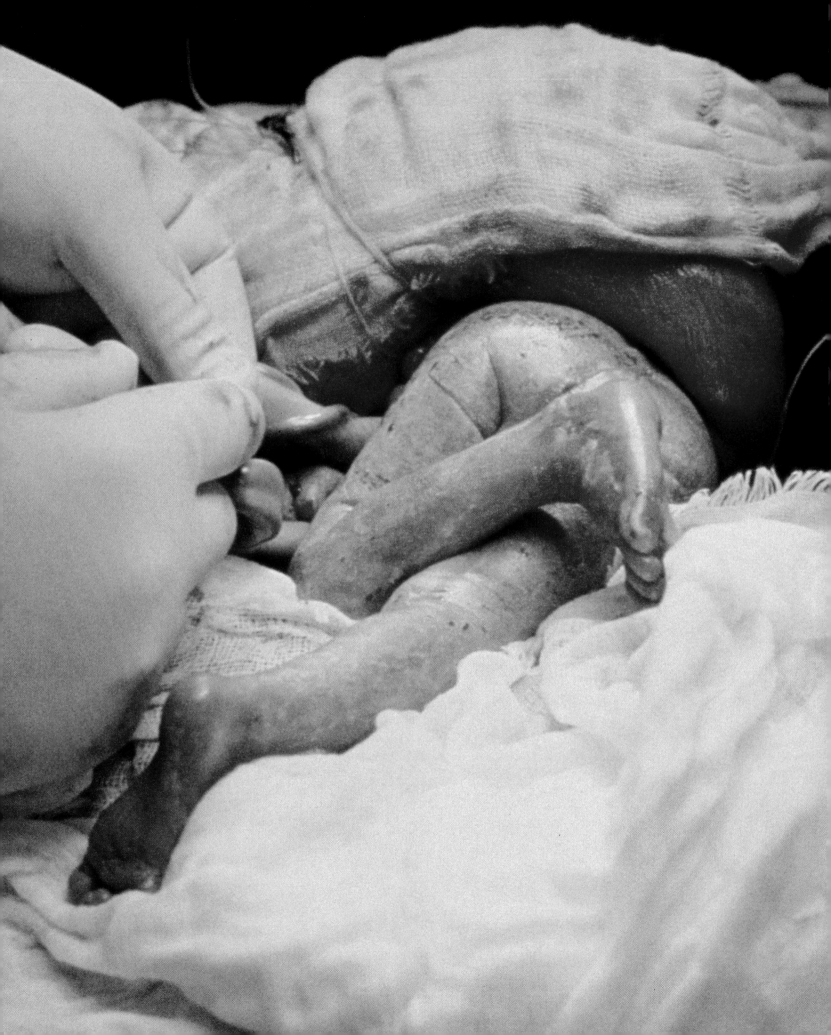

steady surgical tools in the most delicate regions of the human body; so finely tuned is the device that it can move instruments as precisely as 20 millionths of a meter.

Perhaps one distinguishing characteristic of 21st-century surgery will be increasing reliance on minimally invasive and noninvasive procedures as well as on microsurgery, which uses a powerful operating-room microscope, tiny instruments, and smaller needles and sutures to do everything from reattaching a severed hand or a single blood vessel to transplanting corneas, to connecting and sewing nerve fibers.

Lasers once used mainly to sculpt computer chips now smooth wrinkled skin and reshape corneal tissue to help those with astigmatism, nearsightedness, or farsightedness see well enough to pass a driver's test without corrective lenses. The laser's brawnier relatives, offspring of beams that can punch a hole in a diamond in a thousandth of a second and slice through a steel beam with an edge that never needs honing, now routinely blast scar tissue from the site of an operation, vaporize a tumor deep in the brain, spot-weld a tissue graft to a perforated eardrum, and cauterize a bleeding ulcer.

Minimal-access techniques, also known as "keyhole surgery," are already commonplace. They grew out of the now familiar arthroscopy, in which a surgeon peers inside, say, a damaged knee through a small hollow wand, and, using microsurgical tools, removes diseased tissue. Gynecologists have long used a similar technique to remove cysts from ovaries and fallopian tubes. Other early uses relied on x-ray views to guide the surgeon, as in percutaneous automated diskectomy, a technique used to treat some cases of herniated vertebral disks. The cartilaginous disks that cushion your vertebrae are susceptible to injury. When one "slips" or bulges, it can cause severe pain by impinging on various nerves. The standard operation, a laminectomy, requires general anesthesia, the dissection of muscle, and the removal of bone. Percutaneous automated diskectomy changes all that. With the patient under local anesthesia, a stainless steel tube, guided by x-ray, is slid into a two-millimeter incision in the back until the tip of the instrument rests against the troublesome disk. The surgeon threads a combination cutting-suction device about the diameter of a pencil lead down the tube and pushes it into the protruding area. At the probe's tip, a tiny pneumatically driven blade—a miniature guillotine—neatly slices off unwanted disk tissue, which is then flushed from the probe and sucked out through a porthole by an external vacuum. The procedure takes less than an hour, requires no stitches, and sends the patient out of the hospital with a Band-Aid over the tiny incision.

With the arrival of videolaparoscopy, minimally invasive surgery ensured its place among the most revolutionary medical developments of this century. Tiny instruments mounted at the ends of fine tubes are inserted into the body, along with a miniature videocamera that projects a magnified image of organs and tissues onto a TV screen. Surgeons, given a clear view of the operating field on the screen and relieved of the strain of working while staring through an eyepiece, now can devote their attention to manipulating the instruments. General surgeons and gynecologists alike have embraced the technique, and a wide range of operations—gallbladder removal, hernia repair, and hysterectomies among them—are now performed through the keyhole. By the turn of the century it is expected that the majority of some operations will be laparoscopic, perhaps even procedures as complicated as those involving the heart.

Already, "lap" surgeons have successfully repaired aortic aneurysms—bulges in the main arterial trunk that can burst and cause fatal hemorrhaging—using only five or so punctures in the side of the chest, each no larger than three-quarters of an inch. Conventional operations, done through a foot-long incision in the abdomen, often call for several pints of transfused blood and require that the patient breathe on a respirator for a few days. With laparoscopy, blood loss is minimal and patients are off the respirator right after the procedure. Coronary artery bypass surgery has also proved amenable to laparoscopy. In a given year more than 300,000 patients undergo the operation, which is designed to improve blood supply to the heart muscle. During open-heart surgery the patient's heart is stopped, and a heart-lung machine takes over. A long incision is made down the chest from neck to abdomen, the breastbone is sawed open and pried apart to allow the surgeon to reach the heart, and veins that will be used to bypass blocked arteries must be harvested, usually from the legs.

"God...rested on the seventh day," and so do countless individuals aware of the health benefits of relaxation techniques, from meditation to controlled breathing to a snooze in a technologically advanced "napping chair."

A few surgeons now employ several alternatives, which, though not yet routine, are almost certain to be the wave of the future. One is known as port-access surgery. The patient's heart is stopped during the operation, but instead of opening the chest, the surgeon makes small incisions—ports—in the wall between the ribs. Through these, with the help of specially designed instruments and catheters, heart valves can be replaced and coronary bypass operations performed. Another method avoids stopping the heart, and instead allows surgeons to work on a beating heart through small incisions under the breast and collarbone. Patients recuperate much faster after such procedures—two to three weeks instead of six to eight—and those who are elderly or who suffer from lung and kidney complications are better able to withstand the operation.

While it requires enormous skill to sever, say, a gallbladder from the liver and other structures, cut the cystic duct, and remove the gallbladder through a tiny incision, microsurgery is an even more daunting task. Surgeons must work with blood vessels and nerves so small that powerful binoculars and specialized lighting are required. Reattaching a single blood vessel, for example, may call for 20 to 30 sutures with thread about a tenth the diameter of a human hair. To close a wound or reshape a badly damaged limb, a surgeon must remove tissue from another part of the body and meticulously connect and sew nerve fibers and blood vessels. Moreover, some operations take up to 24 hours to complete, a day's work suitable only for highly skilled surgeons who are in better physical condition than many athletes. As physicians at the Penn State Geisinger Health System have said of microsurgery, "The margin of error is as thin as the thread itself." On the horizon are robotic systems that will make surgeon exhaustion a thing of the past

Microsurgery operates, of course, in a minute world, while conventional surgery deals with organs and tissues easily seen with the naked eye. But by and large, both are relatively straightforward, albeit often complex, affairs that do their jobs in a patient's inner spaces. Tissue is punctured and sliced open; organs and other structures are moved, removed, transplanted, or repaired; bones are sawed; skulls are perforated or opened outright; wounds are cleaned; and skin incisions are closed. Sir John Erichsen, a 19th-century English surgeon, acknowledged improvements and modifications in such practices, but also called for a cutoff point. "There must be a final limit to the development of manipulative surgery," he wrote in the British journal *Lancet* in 1873. "The knife cannot always have fresh fields for conquest."

True enough. While minimally invasive surgical techniques will have an essential role in future surgery, noninvasive, bloodless ones may well nudge them aside. Consider magnetic healing, once the darling of 18th-century dilettantes who purveyed all manner of contrivances that allegedly cured by stroking them. A few practitioners, like Swiss ophthalmologist Otto Haab, put magnetism to legitimate use: Haab invented a strong magnet to extract metallic particles lodged in the eye. Today, bioelectromagnetics, the study of the interaction of living organisms and electromagnetic fields—and of the electrical phenomena found in all living organisms—is a more complicated and serious pursuit. We know that electromagnetic fields at certain frequencies can have beneficial effects, such as treating osteoarthritis or promoting nerve and immune system stimulation, bone repair, wound healing, and tissue regeneration.

Ultrasound's medical reputation was made in diagnostics, but it shows great promise as a noninvasive alternative to some surgery. The use of shock waves to pulverize kidney stones has become standard, but now surgeons are looking at ultrasound as a way to blast tumors. One major drawback—how to focus sound waves so that they will destroy a cancerous target without damaging normal tissues—has already been overcome. Ferenc Jolesz at Boston's Brigham and Women's Hospital solved the problem with a dish-shaped transducer that focuses a number of weaker beams of ultrasound onto one tiny point to heat tissue to 132°F; at that level the concentrated beam can diminish or destroy the target, but it is low enough to avoid damage to adjacent tissue. Another experimental technique uses an invasive form of ultrasound to break up clots and restore blood flow in once clogged vessels. A catheter with a probe on its tip is threaded through an artery in the thigh, much as is done during cardiac catheterization, and snaked through the circulatory system to the blockage. Ultrasonic bursts are fired at the clot, liquefying it by breaking up its protein component, fibrin. In one highly successful demonstration of the technique, Israeli scientists opened blocked arteries in 13 of 15 heart attack patients. Cardiologists see the promise of coronary ultrasound thrombolysis, as the technique is called, as an adjunct to angioplasty, in which a balloon catheter flattens artery-clogging substances against the blood vessel wall. Coronary ultrasonic thrombolysis may also be used to treat heart arteries that are repeatedly obstructed, or to reestablish blood flow in brain or leg arteries occluded by fatty deposits.

Brain tumors may be especially good targets for non-invasive surgery because the brain, that extraordinary repository of thought, self-perception, and emotions, is extremely vulnerable to cutting and probing. The dramatic behavioral changes that a lobotomy can bring about are adequate proof of that. Even without such a drastic measure, brain surgery can leave its mark on the patient. While there is really no good place to suffer a brain tumor, among the worst is the lower portion of the brain, at the base of the skull. There, tumors often mesh with critical arteries and cranial nerves, making the growths notoriously deadly, the surgery challenging and risky. Injuries to a nerve can cause facial paralysis or loss of eye movement, and artery damage can trigger a life-threatening stroke. Armed with the newest tools and techniques of their trade, such as fMRI imaging, and approaching the tumors from unconventional directions—through the nose, for example, instead of slicing through the entire living brain—skilled surgeons can remove growths that are difficult to reach. No matter how successful, however, conventional brain surgery can still be devastating to a patient, and noninvasive methods are on the minds of virtually everyone who operates in this delicate arena.

One standard method of treatment that doesn't involve cutting into the patient is radiation, and surgeons have found various ways to deliver the cancer-destroying rays. Conventional fractionated radiotherapy has relied on relatively small doses of radiation to bathe a wide region, with the doses spaced to allow healthy tissue near to a tumor to recover between treatments. A far more accurate and safer method—called stereotactic radio-surgery—now focuses carefully targeted blasts of radiation entirely on the tumor cells. The gamma knife is one way to do this. Not a knife in the usual sense, it resembles a cross between a salon hair dryer and a space alien's helmet. Placed over the head of a patient and aimed by CT or MRI scans and specialized computers, the gamma knife delivers high, focused doses of radioactive cobalt through tiny holes in the helmet. Neurosurgeons at the Stanford University School of Medicine have successfully used another technique called LINAC radiosurgery in which radiation generated by a linear accelerator rotates around a patient's head to deliver radiation in a series of precisely computed arcs that converge on the target. In

one series 55 patients with skull-base meningiomas—tumors that arise from membranes covering the brain and put pressure on adjacent structures—were treated with the technique. Only two suffered permanent damage to cranial nerves; complete surgical removal of such meningiomas can result in damaged cranial nerves in 30 percent of the cases. Scientists are hunting for even faster and safer ways to kill tumors. On the horizon is the use of radiowaves to vaporize them. One device to be tested by the turn of the century is an MRI with microwave antennas that would cook a cancer into oblivion.

TRANSPLANTS: IMPROVING ON PERFECTION

Such emphasis on operations that are non- or only mildly invasive does not mean that all other forms of surgery will become dinosaurian when the 21st century dawns. One, which for a long time was unconventional, and which may serve us well for years to come, is transplantation. For years after the first kidney transplants between humans in the 1950s and the first heart and liver transplants in 1967, the public was mesmerized by operations involving organs and other structures. As the list of what was capable of being switched from one person's innards to another's grew—corneas, lungs, livers, pancreases, cartilages, bones, bone marrow, and, in an attempt to revive the battered immune system of AIDS patients, T-cell-producing thymus glands—transplant operations became as routine as space shuttle flights and lost their perennial position in the headlines. Refined surgical and tissue-matching techniques and more effective drugs to ward off rejection of transplanted organs by the recipient's immune system greatly increased the chances that the recipient would live for many more years, even decades. According to the U.S. Department of Health and Human Services, more than 90 percent of kidney and pancreas recipients are alive a year after surgery; for heart the figure is 82 percent; for liver it is 76.7 percent; and for lung, 68.4 percent.

There is a downside to the statistics, however. As the chances of a successful surgery and longer life expectancy increase, so too will the demand for a transplant, even though not everyone, because of age or condition, is a candidate. In a given week nearly 50,000 Americans await organs, some 33,000 for a kidney; of that total, 3,000 will die because suitable organs are not available. Indeed, according to the United Network for Organ Sharing, there are generally enough organs available to permit only some 20,000 transplants. Besides trying to convince people to sign organ-donor cards, medicine is exploring several alternative solutions, among them cell transplants, artificial organs, implants, and xenografts, the transplanting of animal organs to humans.

The most unconventional, indeed exotic, of these alternatives is the xenograft. Probably the first recorded animal-to-human transplant was attempted in 1905 by French physician Mathieu Jaboulay, who implanted a pig's kidney into one woman and a goat's liver into another. Neither patient survived.

Beginning in 1964 American surgeons started transplanting kidneys from baboons, which are genetically close to humans, into patients. The most noteworthy of these attempts involved a California newborn named Baby Fae, who lived 20 days after receiving a baboon heart. All the patients died within weeks of the operations, a not-uncommon occurrence since immunosuppressive drugs leave the recipient seriously vulnerable to infections.

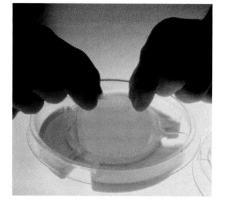

While primates bear a close genetic resemblance to us, there are drawbacks to the use of their organs in humans. Chimpanzees are now an endangered species. Baboons reproduce slowly, producing one offspring at a time, and their organs often are too small for human use. Turning primates into virtual organ factories presents an ethical quandary because they are so similar to humans in physical appearance, intelligence, and in their highly developed social structure. More important, nonhuman primates carry a variety of infectious agents, and these pose a grave threat to humans whose immune systems must be suppressed by antirejection drugs.

For these reasons, xenotransplant researchers are turning to pigs, which are easy to raise, produce large litters, and have organs comparable in size to those of humans. Already, pig heart valves have been used to replace damaged human ones, pig skin grafts have treated burn victims, and fetal pig tissue has aided Parkinson's patients. Although pigs do carry several types of flu viruses, they are generally healthier than many primates, and can be bred relatively disease-free. Moreover, since pigs are slaughtered commercially as an important food source for humans, distress over harvesting their organs is not as great as it would be for primates.

As with all transplants the problem of rejection remains. As difficult as it is to overcome in human-to-human grafts, rejection becomes a monumental obstacle in xenotransplantation. We all form antibodies in response to an intrusion into our bodies of anything perceived as "foreign." When an organ from a species as different as a pig is introduced, an even stronger response, known as hyperacute rejection, is unleashed. Bolstering the attack is part of the immune system called complement, proteins that prowl the bloodstream like picketboats looking for trouble. Complement is usually held at bay by regulatory "shield" proteins that coat and guard an organ system. But if the guardian proteins are disparate enough—as occurs when a pig organ is transplanted into a human—the recipient's complement joins forces with its antibodies to set off the violent rejection process that can kill the transplanted animal organ in minutes by choking off its blood supply.

To get around this formidable defense system, scientists are perfecting ways to coax pigs into expressing human complement-inhibiting shield proteins that render their organs less foreign and more amenable to transplant. Accomplishing this requires reprogramming a pig's genetic blueprint by inserting human genes that will produce proteins recognizable as human—and then

---

The superficiality implied by "only skin deep" belies the complexity of the body's largest organ, skin. Yet new versions, lab-cultured from living human skin cells (above), look and feel real. A boon for the severely burned (opposite), this and other human skin equivalents interact with wounds to speed the healing process. FOLLOWING PAGES: While transplantation continues to be an option for patients with severe cardiac disease, research proceeds on developing mechanical hearts.

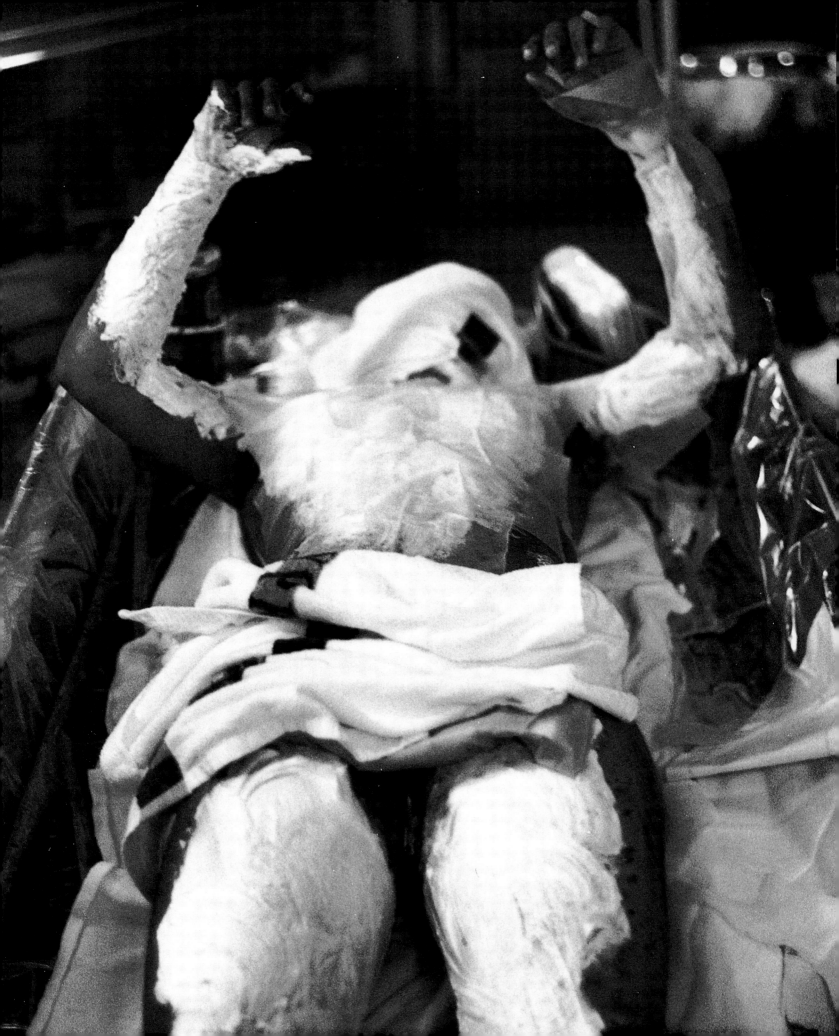

getting it all to work in a transplant. Pigs could thus be turned into factories churning out not only human-friendly organs but also cells and tissues.

In 1997 stunning successes in animal cloning brought that idea closer to reality. By creating a sheep named Dolly, Ian Wilmut, an embryologist at the Roslin Institute in Scotland, went where scientists had only lightly gone before. He removed a cell from an adult ewe's udder—a cell that carries all the genetic instructions to produce a duplicate individual—inactivated it by withholding proteins, then fused it into another sheep's egg cell from which the nucleus had been removed. A jolt of electricity reactivated the transplanted cell's "sleeping" genes, which tricked the egg cell into "believing" that it had been fertilized. Dividing repeatedly and forming an embryo, the egg cell was implanted into another female sheep, and, voilà, in five months Dolly was born—bearing the identical genetic makeup of the donor ewe. Although other scientists had created genetically identical animals by dividing embryos soon after they were formed by the union of eggs and sperm, Dolly was the world's first animal cloned from an adult.

A few months later Wilmut's team, using a similar technique, produced a lamb with a human gene in every one of its cells. They added new genes, including human ones, to skin cells from fetal sheep. Next, they replaced the genetic material in a sheep egg with that of one of the skin cells, which then directed the formation of a baby lamb—called Polly—with human genes in each of its cells. Before this, scientists who wanted to improve livestock were able only to inject genes into a fertilized egg, a hit-or-miss method of genetic engineering. Removing and replacing genes à la Polly is a far more precise method, one essential in mass-producing animals that would provide organs for transplant, better meat and milk, or even pharmacologically useful proteins.

While cloned sheep have begun to revolutionize bioengineering—and, indeed, have raised the highly controversial possibility of cloning humans one day—pigs still occupy the minds of many researchers. Transgenic, or genetically altered, pigs can now be bred to produce human shield proteins, and the next step has also been taken—transplanting organs from such made-to-order pigs into primates, a prerequisite to grafting those organs into

humans. Early results indicate that modified organs, hearts among them, survive far longer than organs from ordinary pigs. Whether animal organs will ever be routinely transplanted into humans is a question that scientists are yet unable to answer. More likely, animal organs will be used to keep patients alive while they wait for a suitable human donor. In 1995 the U.S. Food and Drug Administration permitted the Duke University Medical Center to test transgenic pig livers in a few patients with end-stage liver disease. Because it is yet uncertain whether a pig liver can function as well as a human liver for a long period of time, one way to take advantage of the animal organ's functions, without threat of rejection, is to use it as an outside source. The pig liver is kept alive outside the patient, whose blood is drawn through a catheter, shunted through an oxygenator and then into the liver, and finally fed back into the patient. Several patients have been maintained this way as researchers plan their next move, transplantation of a transgenic liver into a human patient.

Perhaps a more appealing variation on xenografting is the use of animal cells and tissues rather than organs. One reason for this is that cells and tissues lack the blood vessels and membrane linings that are attacked during a bout of hyperacute rejection, and thus are less apt to touch off a vigorous immune response. There is ample precedent for such transplants in the use of human cells for transplant. Bone-marrow transplants provide the stem cells that produce the human body's 30 trillion red blood cells, infection-fighting white cells, and the platelets important to clotting. Cells removed from aborted fetuses—a beneficial but controversial procedure—are administered, with some success, to treat autoimmune diseases, Parkinson's, and some forms of diabetes. Doctors have removed islets of Langerhans—the clusters of pancreatic cells that produce insulin—from cadavers and human fetuses, and have transplanted them into diabetics, again with varied success. An alternative might be pig islets, which are readily available and have also been tested in humans, and which produce insulin that is nearly identical to the human form. Other disorders, including Parkinson's and Huntington's diseases, may one day be corrected by the use of pig cell grafts. Whether from animals or humans, cell transplantation provides

Iron in hand, teenager Dima Sitnik stands steady on high-tech legs of polyethylene, titanium alloy, and composited carbon fibers as he provides a living testimonial to the grit of many amputees—and to the wonders of modern prosthetic design.

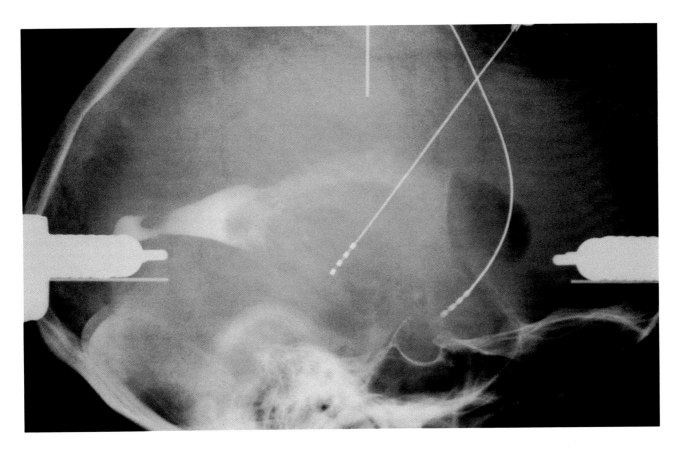

almost unlimited opportunities—muscle cell grafts to treat muscular dystrophy, nerve cells for spinal cord injuries, and liver cells to cure life-threatening metabolic disorders such as Tay-Sachs disease, an enzyme deficiency.

Probably the ultimate, most fundamental type of transplant is gene therapy, a futuristic treatment for the more than 5,000 inborn diseases in which one or two genes are determinative, and—since virtually all diseases have some genetic component—even such afflictions as heart disease and cancer. Gene therapy is a catchall term for a variety of approaches, the aim of which is to transfer healthy, corrective genes into a patient's cells to cure diseases caused by other, faulty, genes. Since many human genetic abnormalities involve a protein deficiency, a tailored gene that could produce the necessary protein would be a revolutionary therapeutic weapon.

Gene therapy is a logical extension of the Human Genome Project, the ambitious effort to decipher our hereditary instructions. It will be DNA's Rosetta stone,

ushering in an exhilarating new era of molecular medicine. From conception to death, life's many mysteries will, when all the genes are analyzed in detail, be illuminated, and so, too, will the full role of genes in diseases both mental and physical.

CASE REPORT. March 2, 2050

The patient, an 18-year-old woman, awaits word of the three-second genetic test that was part of her annual physical. A technician had simply rolled the tip of a cotton swab against the inside of her cheek and collected a tiny sample of mucus. It was inserted into an analyzer that matched the DNA in the mucosal cells against a computer database. The doctor enters the room. "Your genetic profile is as I expected," he says, "given your family history, but no great cause for concern. We can take care of it fairly easily, as an outpatient." The doctor explains that the woman's overall genetic inheritance is satisfactory, with one exception: a mutation in a cancer-

Chaos-control: Electrodes implanted deep within a brain afflicted with Parkinson's disease (above) can stop the trembling of this devastating neurological disorder. Its indiscriminate effects are evident in an athlete never before known for shuffling gait and stooped posture, Muhammad Ali (opposite).

causing gene that is linked to familial colorectal cancer. Three close members of the woman's family have had colon polyps—precursors of cancer—removed, and one distant relative had colon cancer. "Now," the doctor says, "the mutation by itself isn't at all life threatening. But, simply put, it can cause some misspelling, let's say, in those DNA paragraphs of yours, and that means some possible other mutations down the road. That might direct cells to make proteins that don't work right, and cancer results. In our experience the mutation means a 20 to 30 percent lifetime risk of developing colon cancer. Better to be safe than sorry, so let's take care of it now. Okay?"

If the patient had colon cancer, it could be treated with monoclonal antibodies, custom-made proteins injected alone or armed with suppressor drugs or radioisotopes that would home in on the cancer cells and neutralize them. "In this case, though," the doctor says, "we'll get right to where it can start. We'll stop that genetic mutation in its tracks, and get those genes to make the right proteins. One injection of our designer viruses will do the trick. Don't worry, they're harmless. As you probably know, viruses are DNA or RNA packets that invade a cell, insert their genetic material into the cell nucleus, and then make copies of themselves. Before you know it, you're sick. We take the viruses' reproductive genes away, add a batch of healthy replacement genes for those mutants, and shoot it all in with a painless lasodermic. The viruses commandeer the cell like they're supposed to and deposit the new genes. The genes travel into the cells' DNA, and they all start manufacturing the right protein, overwhelming the old. We can now correct a genetic defect that your family has had for generations."

GENE THERAPY

The routine discovery of "disease" genes—those with defects or mutations—not only has given scientists a better understanding of inherited diseases but also has spurred genetic testing for many of them. Thousands of fetuses, children, and adults are screened every year for such devastating disorders as Down's syndrome, cystic fibrosis, Tay-Sachs disease, thalassemia, Huntington's disease, some rare familial types of Alzheimer's, Duchenne muscular dystrophy, and breast cancer, two

genes of which were discovered in 1996. Scientists now estimate that one million people carry misspelled copies of two genes involved in a hereditary form of colon cancer. They have identified genes that cause some cases of Parkinson's disease and those that give rise to age-related macular degeneration. In a discovery with enormous implications for longevity, a gene that regulates glucose metabolism in a tiny worm and may enhance its longevity has been found to have a counterpart in humans. And, in 1997, scientists isolated the first gene to control social interactions in a mammal. That gene was discovered in mice; when it is lacking, the animals apparently do not mingle as easily as their fully endowed relatives.

Picking the gene responsible for a disease out of the human genome—which contains enough letters to fill a thousand one-thousand-page telephone books—and then analyzing each segment is a challenge and a half. Clues are sought—in an altered protein in diseased tissue, in a missing snippet of chromosome that may give some idea of the gene's location on a chromosome, in a marker of the disease that may reveal which chromosome carries the errant gene, or by comparing the DNA from a disease-carrier to that of relatives. Once found, the gene has to be analyzed, segment by segment. In the case of cystic fibrosis, a lethal hereditary disease of Caucasians, the most common error in its gene's sequence of bases is a deletion of but three bases out of a total of 250,000.

Finding the gene responsible for a disease is the easy part. Once it is isolated—not nearly enough to begin treatment of the disease it brings—the protein it makes and its role in the body must be identified. Essential to the structure of all organs and metabolic activities, each of the tens of thousands of proteins is made up of hundreds of thousands of amino acid molecules. Some proteins direct the structure of cellular walls, some determine that of various tissues. Others—enzymes—speed up the thousands of changes that take place regularly in our bodies. Still others—hormones—are chemical regulators that stimulate some physical or emotional reaction. The functions of proteins prompt scientists trying to devise a therapy to ask questions. Does an altered gene produce too little protein, or none at all? Is the protein itself flawed? How does the protein change cause the disease? Should treatment focus on the gene or the protein?

Our futuristic case report above is a bit fanciful, but it follows, with some license, the script researchers are already preparing for gene therapy. When that technique is perfected, many drugs now in use will, in comparison, have the cachet of an alchemist's potion. Radiation will seem crude; surgery will look like a brutal assault. Gene therapy is based on a fairly simple premise—if it's broken, fix it by replacing what's broken. To compensate for a genetic error, a physician might attempt to substitute a drug or normal protein for one that is missing or malfunctioning. The clinician might interfere with the genetic instructions that manufacture the defective protein, changing their blueprint for disaster to one of hope. There are many ways to accomplish such cures. Say the patient lacks a protein that particularly affects the brain. While an infant is in the womb, it receives from the protein eaten by the mother vital amino acids its brain cells need for growth. After birth, the infant gets its amino acids from mother's milk and other foods. If the supply is cut back, either while the mother is carrying the child or just after its birth, the infant's brain growth is stunted, learning slows, and the malnourished child may have a lower than normal IQ. Deliberately undernourished rats have been shown to have significantly smaller brains, but when they are fed an adequate, protein-rich diet, their brains grow to normal size in a few months. Studies with severely malnourished children suggest a similar pattern: The downswing in their intellectual function can be reversed if diet is improved.

Dietary management is a relatively simple—and morally acceptable—approach compared to what gene therapists have been trying to do for years. In 1971 an early laboratory exercise in what was termed viral gene therapy was reported by scientists at NIH. They focused on galactosemia, a disorder in which lack of an enzyme prevents the body from properly using the milk sugar galactose. Left untreated, the disease can cause retardation, cataracts, malnutrition, and death. The NIH gene scientists used a specially developed virus armed with the necessary code for galactose, then cultivated it with deficient human cells; some of the genetic message packed in the virus was transferred to the cells, the enzyme was produced, and the deficiency was corrected.

It was not until 20 years later that W. French Anderson, then at NIH, performed the first federally sanctioned gene therapy on a patient. He used a genetically engineered virus. The patient was a four-year-old girl with a serious immune-deficiency disease caused by a defective gene that could not produce an essential enzyme. Anderson's team removed some of the girl's white blood cells and exposed them to viruses that had been designed to carry a normal complement of the defective gene. The treated cells were then transfused back into the ailing child, where they manufactured the missing enzyme. The experiment was a success. But until numerous technological problems are overcome—among them getting the stripped-down viruses past the ever vigilant immune system, and once there, getting them to find the right cells and then express themselves—gene transplanters are progressing slowly. While research is generally confined to laboratory animals, some 200 human gene therapy trials are currently under way throughout the world. More than a dozen diseases have been targets—cancer, cystic fibrosis, and rheumatoid arthritis among them—but they are small-scale clinical experiments. Addressing the fledgling nature of gene therapy, a report prepared for NIH warned that expectations of gene therapy protocols had been oversold.

Still, the promise is great, and predictions are that various forms of gene therapy will play a major role in the treatment of most diseases by the date of our hypothetical patient scenarios. Viruses will undoubtedly still be used to distribute genes by "infecting" target cells. Some of those viruses may seem inappropriate. Scientists have found, for instance, that the AIDS virus could be a prime candidate for gene therapy since it is a master cell integrator. By disarming the virus, rendering it incapable of causing AIDS, but retaining its ability to infiltrate cells, researchers have created a viral vehicle that can deposit genes in rat neurons, where they remained stable for some six months. Another job for the virus is more direct. A disarmed AIDS virus combined with a killer-gene can be injected directly into a tumor to destroy it. Yet another promising delivery method is an artificial chromosome, a synthesis of selected portions of a human chromosome, which would overcome some of the drawbacks of viral carriers. Indeed, the creation of an artificial gene is not far-fetched at all. In 1970 Nobel laureate Har Gobind Khorana, then at the University of Wisconsin, constructed a yeast gene for the first time. Back then, little was known about the biochemical "start-stop" signals that enable a cell's machinery to utilize the coded information in various processes. In 1976 Khorana's team fabricated, from ordinary laboratory chemicals, the first synthetic gene that was fully functional in a living organism, including its start-stop mechanism. The scientists then introduced it into a bacterium with a mutant gene that prevented the normal production of a particular protein. The created gene canceled the mutant gene's stop signal, enabling the bacterium to make normal proteins. More recently, a team

at Case Western Reserve University made an artificial chromosome, which, when inserted into human cell cultures, was accepted naturally by the cells and then copied during cell division. The Case experiment was another major step toward finding a suitable way of delivering therapeutic genes to cells.

SPARE-PART MEDICINE: IMPLANTS

Leonardo da Vinci may have been correct when he observed that human subtlety, with all its variety, will never devise an invention more beautiful, more simple, or more direct than does nature. Nature can be commended for not leaving much out of its inventions—or incorporating much that is superfluous—but it did leave a few things for us to tinker with. Ever since ancient Egyptians first experimented with dental implants, scientists have sought—with the blessing of patients demanding improved health and longevity—artificial replacements for body parts.

The dialysis machine, or artificial kidney, is now a commonplace device that takes over when the kidneys fail, efficiently clearing waste products from the blood. Heart-lung machines maintain blood circulation and oxygenation during cardiac surgery. Robert Jarvik's famed artificial heart kept dentist Barney Clark alive for 112 days during 1982 and 1983, but its cumbersome external power supply and its tendency to trigger blood clots and strokes have since put it on hold. Researchers, however, continue to pursue the dream of an in-the-chest artificial heart, and at least two show promise. One, under development at Penn State, consists of a compact electric-motor-driven heart and an implanted control system that regulates the motor and blood flow according to the system's assessment of the body's needs. The patient would carry a battery pack to power the system, which could be plugged into an outside power source during sleep. Two plastic pumping chambers with polyurethane sacs replace the patient's natural left and right ventricles, while the upper portion of the patient's heart, containing both atria, remain intact. Implanted in calves, the device has kept them alive for at least a hundred days; researchers hope it will be widely available by the year 2005.

Progress also has been made in developing an artificial lung that can be implanted in the chest. One such device, under development at Northwestern University, preserved lung function in an animal model for 24 hours. Essentially, the artificial lung is a bundle of fibers with tiny holes that exchange oxygen for carbon dioxide. It is attached to the main pulmonary artery and returns oxygenated blood to the left atrium. Its principal use may eventually be as a bridge to lung transplant.

Not as awe-inspiring as artificial organs—but still a major player in the new millennium's medical road show— are medical implants. The first artificial heart valve was implanted in 1952; the first artificial hip replacement took place two years later. According to the National Center for Health Statistics, some 1.5 million artificial

Fluorescent mice, created by injecting mouse ova with DNA that, in jellyfish, encodes a light-producing protein (above), can transmit their luminescence to their offspring (opposite). More than mere lab magic, the achievement eventually may lead to human applications, such as targeting tumor cells by their glow.

joints and a quarter-million heart valves have been inserted into patients. Engineering materials, such as biodegradable polymers for site delivery of drugs, and improved implantation techniques have added to the explosive growth of spare-part medicine. The current scientific terminology for this is biomimetics—an emerging interdisciplinary field that combines the study of biological structures and their functions with physics, mathematics, chemistry, and engineering. Biomimetics has given us artificial heart valves and alloy-coated knee joints; it has given us shunts, contact and intraocular lenses, dental implants, fracture plates, breast-enlarging implants, vascular grafts, and under-the-skin polymer implants containing pain-killing narcotics.

There are pumps galore, both in service and in prototype form. An insulin pump that dangles from a belt around a patient's waist delivers insulin through a needle inserted in the abdomen. An improved experimental model is an implant the size of a contact lens which, ideally, will sense the patient's blood glucose level and regulate insulin doses accordingly. Another pump administers precisely controlled doses of morphine.

Yet another type of pump is swallowed instead of implanted. Invented by the late Takeru Higuchi of Kansas University, it is manufactured by coating a powdered drug with a water-absorbing membrane; when it reaches the digestive system, fluids seep through the coating and dissolve the powder, which leaks out through a tiny laser-drilled hole in the pill. Cochlear implants, multielectrode arrays, stimulate the auditory nerve, bypassing the missing or nonfunctioning cells in a cochlea that no longer transforms sound waves into the nerve impulses that transmit sound to the brain. Thousands of people with varying degrees of hearing impairment have attained some degree of sound perception with these implants, often dramatically so.

"Partial hearts," also known as left-ventricular assist devices (LVADs) increase the staying power of nature's master pump and help prevent formation of deadly clots; stainless steel stents prop open narrowed arteries to improve blood flow to heart muscle. And, in what has been termed "an emergency room implanted in the chest," cardiac defibrillators are inserted under the collarbone to control irregular heartbeats by sending electrical "jump-start" shocks directly to the heart when a built-in computer senses arrhythmia.

In 1997 the FDA approved a brain implant device to help control the hand and arm tremors of Parkinson's disease, thereby enabling many patients to resume the daily activities that most of us take for granted—eating, drinking, writing. Some 1.5 million Americans have the disorder, which can also cause mental deterioration. Called the Activa Tremor Control System, the new device consists of an electrode implanted in the thalamus—the brain's relay station through which sensory messages from all parts of the body are routed to appropriate areas of the cortex—and connected by a wire under the skin to a battery-powered pulse generator implanted in the chest. When activated, the device beams a steady stream of electrical pulses to the brain, blocking the tremors. To turn the stimulator on or off, a patient merely touches a handheld magnet over the pulse generator.

On another neurological level, implants have been newly pressed into service as scaffolds, or nerve conduits, on which severed spinal and facial nerves can be brought together. Once the cut ends are rejoined and growth factors are added, the nerves often manage to regenerate. Such scaffolding is attractive to neurosurgeons because one of the difficulties they face is coaxing the ends of cut nerves to line up and link up. Nerve fibers seem to have minds of their own, and are apt to sprout in unintended and unwanted directions, an attitude which, in facial surgery, might mean the difference between a smile and a frown. Harvard Medical School researchers have found a way to train these errant nerves.

First they prepare a mold with hundreds of fine, precisely positioned wires that mimic the size and direction of real nerve fibers. They then pour an inert and degradable polymer into the mold, and, after it solidifies, remove the wires, leaving behind plastic conduits less than a tenth of an inch in diameter and containing as many as 250 channels. Impregnated with slow-releasing, growth-enhancing substances, these conduits have been implanted in rats with damaged sciatic nerves. Results of such experiments, which will be applied to humans in the foreseeable future, have been heartening: Only weeks after manipulation, researchers applied heat to the rats' feet—and the animals responded, proof that their sciatic nerves had regenerated.

Implants and artificial organs, however, are not invulnerable to rejection. As we have seen, the body does not readily accept intruders. Sometimes the body walls off implants with scar tissue; some devices, such as glucose sensors or blood vessel replacements, are more affected than others. The use of degradable polymers in implants circumvents the nettling problem of rejection because, like surgical stitches made of the same materials, they eventually disappear. Another implant, a newly developed copolymer surgical screw that replaces the permanent titanium ones now used in delicate skull and facial surgery, dissolves within months of implantation.

Yet another way to circumvent the problem of rejection—knowing that many organs and other structures cannot be made from hardware due to their complexity—would be to custom grow body parts such as skin, cartilage, and even internal organs in the laboratory. A heart or a liver, for example, might one day be cultured and grown in a laboratory from a tiny tissue sample, then kept in storage, ready for implantation whenever its parent organ showed signs of wear and tear. Another solution might be to somehow, perhaps with engineered genes, endow organs and other structures with the power to regenerate themselves.

REGENERATING LOST PARTS

Regeneration is not a new notion, nor is it impossible. In the 1700s Lazzaro Spallanzani, an Italian experimenter who first observed isolated bacterial cells divide, asked: "If frogs are able to renew their legs when young, why should they not do the same when farther advanced? If [other] animals, either aquatic or amphibian, recover their legs when kept on dry ground, how comes it to pass that other land animals, such as are accounted perfect, and are better known to us, are not endowed with the same power?"

Some years ago, Harvard anatomist Marcus Singer addressed Spallanzani's questions. "It is probably safe to assume," he said, "that every organ has the power to regrow lying latent within it, needing only the appropriate 'useful dispositions' to bring it out." Given the steady stream of data pouring from the labs of the gene hunters, it seems more than likely that one day a cell's genetic program might indeed be reset to unleash that power, or to encourage a cell or group of cells to manufacture duplicates of organs and bone on a lab workbench. Regeneration and wound healing are regulated by highly specialized biochemical cellular processes. About all that humans seem able to renew with any great success on their own—aside from the natural healing of fractures and wounds—are hair, nails, skin, and portions of the liver. Yet frogs, salamanders, and crabs grow new appendages to replace lost ones. Chop a hydra—a simple, multicellular freshwater animal—into pieces, and each piece develops into a complete new organism; place small bits of chopped hydra next to one another, and they reform into another hydra. Cut a goldfish's optic nerve, and it grows a new one. It's also been known for years that when as much as 80 to 90 percent of the liver is surgically removed from an experimental animal, virtually the entire organ will regrow. The process occurs in humans as well, as long as the remaining liver tissue has an adequate blood supply.

Over time, researchers have managed to partially regenerate rats' severed limbs by implanting small electrical devices in the stumps. A similar technique also has been used to help heal human bone fractures. Animals treated with demineralized powdered bone and teeth have been able to differentiate connective tissue into cartilage, bone, and marrow.

More recently scientists have caused rat spinal cord nerves, once considered impossible to restore, to regrow by adding growth factors and building scaffolds—out of cells from peripheral nerves—for them to sprout on. A major breakthough was the discovery that genes responsible for the growth of nerve fibers in the brain shut down at a very early age, which could explain why nerves there do not regenerate in adults.

Geneticists have also identified a specific gene, bcl-2, that promotes the regrowth of severed axons. This discovery means that intrinsic genetic factors, not just the tissue environment, are crucial to the regeneration mystery. Genes such as bcl-2 may hold the key to inducing nerves to regenerate. In other research mesenchymal stem cells—embryonic-like cells that direct the formation of cartilage, bone, muscle, tendons, and ligaments—implanted in animals have regrown tendon that had been surgically removed.

One promising avenue for custom growth of real body parts in the laboratory is the creation of "organoids," implantable biomaterial lattices coated with genetically

FOLLOWING PAGES: Savings bank with human interest, a sperm bank stores its deposits in a super cooled, liquid-nitrogen environment, ready for future in vitro fertilization in cases of infertility and other reproductive disorders.

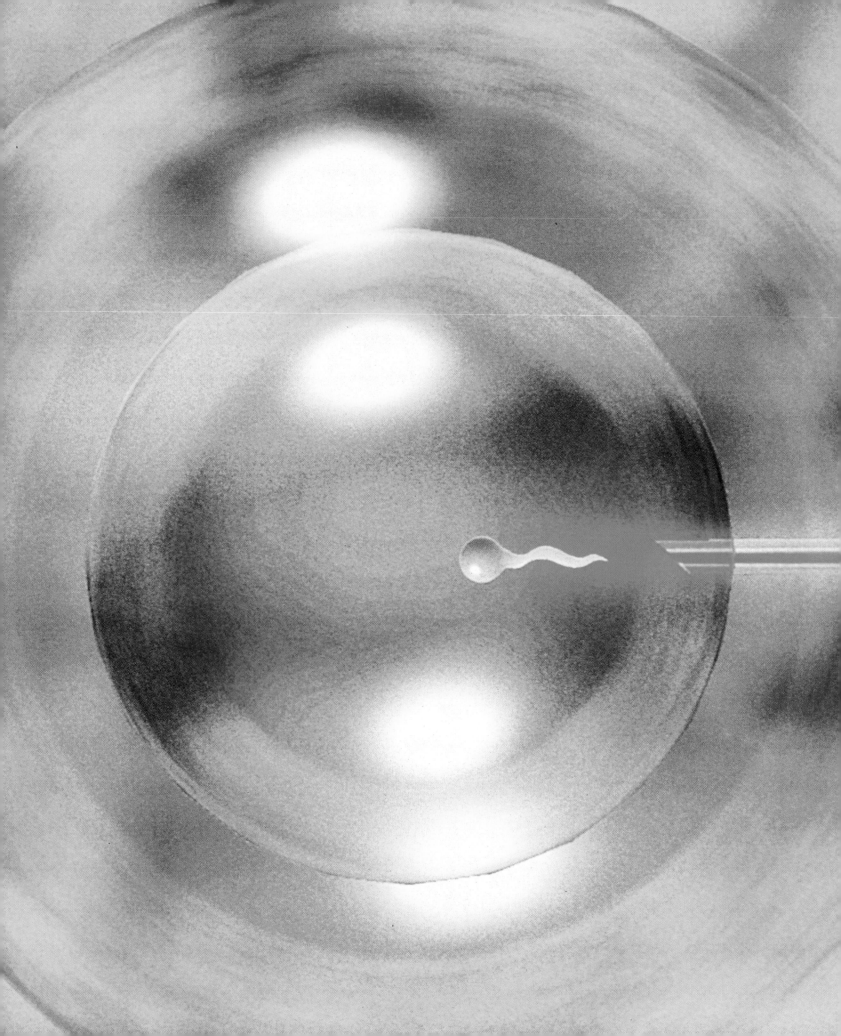

engineered cells that deliver beneficial proteins. Researchers have identified a gene, CBFA1, that encodes a protein which, in turn, can induce lab-grown, immature cells to start maturing into bone-producing cells. Manipulating genes that direct the formation of bone would mean that scientists could grow live human bone in the lab and implant it into patients who need it.

Using biomimetic strategies, researchers have imitated biological processes enough to grow a variety of viable tissues. Appropriate cells are grown in culture, then are seeded onto biodegradable mesh and other jungle-gym-style lattices that have been shaped to conform to whatever structure is to be created. Bathed with nutrients and tissue-inducing substances, they eventually grow into the needed structure. This technique has produced usable skin (although without sweat glands and pigment-producing cells), tissue-engineered cartilage in specific shapes for reconstructive surgery, blood vessels, and working heart valves.

Theoretically, large organs and smaller organelles could be synthesized in this manner. Indeed, researchers have had some experimental success with seeded liver cells. As with the creation of artificial organs, however, the complexity of some structures, such as the liver, with its multitude of diverse functions and high metabolic activity, is a barrier that even tissue engineering may have difficulty overcoming soon. Nonetheless, the reality is that there are not enough organs for transplant to go around, that there is considerable human aversion to xenografting, and that some implants still give the impression they have no business inside a human body. By all that is medical and logical, if a donor and recipient are the same person—someone for example, who receives a heart valve tissue engineered from his or her own cells—serious obstacles vanish, and nature's plan has not been too seriously violated.

The day is sure to come when tissue-engineered implants will be—as conventional spare-part medicine has already become—not just a therapeutic approach for end-stage disease but, in the words of an NIH consensus statement, "a method of elective restoration of chronically damaged structures [that] may someday be considered for preventive maintenance in early-stage disease."

DRUGS AND DOSAGES

No discussion of medicine can escape allusions to drugs, those concoctions, decoctions, quintessences, infusions, injections, tinctures, pastes, potions, and pills that have sustained humankind at least since the Bible book of Genesis revealed that the mandrake root acts as an aphrodisiac. An ancient Egyptian medical text, the *Ebers Papyrus*, lists nearly a thousand medicines, castor oil and the forerunners of opium among them. Inca healers prescribed cinchona bark, a source of quinine, for malaria, and cocaine-laden coca leaf as a calmative and a stimulant. "For every ill a pill," wrote Malcolm Muggeridge. "Tranquilizers to overcome angst, pep pills to wake us up, life pills to ensure blissful sterility." With an 80-billion-dollar U.S. pharmaceutical market and some two billion prescriptions written every year in the United States alone, those words are no exaggeration.

Drugs that are in use today will probably be with us for many years, but they, and a host of others that perform wondrous feats, will be designed by a far more diverse group of scientists—toxicologists, pharmacologists, organic chemists, physiologists, molecular biologists, and statisticians—and will be delivered more efficiently. The new preparations will know far more intimately the enemy agents that bear disease, and will have the capacity to attack their vulnerable points with an exquisite precision only dreamed of in years past. Compared to what is on the horizon, most of today's pharmaceuticals—in their manufacture, administration, and efficacy—will be as archaic as the mixture of saffron, myrrh, and vinegar that made up an allegedly vivifying elixir in the 15th century. As more and more is learned about the human genome and the nature of the proteins that disease-genes control, pharmacology in the new millennium is certain to produce drugs effective against virtually every human ill, from cancer to memory loss.

Drugs work in a panoply of ways. Whether they act as replacements for a missing hormone or some other natural substance or just kill bugs, however, they interfere with and affect biological systems. Often, in their zeal to wipe out an invading virus or a cancer cell,

"Movement is the cause of all life," observed Leonardo da Vinci. Rushing headlong out of a micro-needle to penetrate an egg in this computer artwork of in vitro fertilization, a solitary sperm delivers what is hoped to be the right stuff.

drugs level shotgun blasts at healthy cells, causing devastating side effects. Dosages, generally based on age and weight, are not always right for those who metabolize some drugs faster or slower than other individuals. Another difficulty with drugs is that the usual ways of delivering them—by mouth or by injection—are not always efficient. Swallowed, some may not get through the intestine's tough lining and into the bloodstream; others, proteins and enzymes, may be digested too quickly to have any effect. Injections, too, have drawbacks, not the least of which is the average person's aversion to being punctured. Both methods also cause drug levels in the blood to fluctuate, a potentially hazardous situation when powerful drugs are administered.

New drug delivery methods in use or on the drawing boards can surmount some obstacles. Drug-impregnated skin patches bypass the digestive system entirely, administering drugs such as scopalamine for motion sickness and nitroglycerin for angina. Since the nose's mucous membrane is more permeable than skin, nasal sprays—which can be used to deliver insulin as well as allergy and migraine medications— can deliver larger molecules that patches cannot handle, such as proteins, enzymes, and hormones. Undergoing testing are biodegradable implantable drugs that work like solid air fresheners, continuously releasing medica-

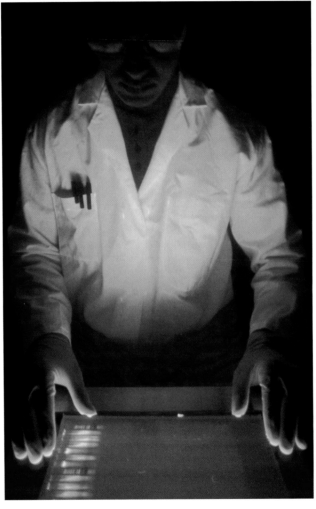

tion as they dissolve, a distinct advantage when steady flow of a drug is required, as for contraception. One promising method of achieving proper dosages uses drug-filled, bioerodible plastic beads, or microspheres. Once swallowed, they stick to intestinal tissues and slowly erode, steadily delivering drugs into the bloodstream. Under development at Brown University, this technique is aimed at inflammatory bowel disorders, but since it is capable of delivering small protein molecules and large DNA molecules to a target area and gluing them there, it may have future use in administering DNA-based vaccines, cancer chemotherapy, and gene therapy.

As promising as the new transport systems are, however, they are nothing without cargo, preferably the kind that can be produced in sufficient quantity and will work in a specific way with few side effects. As far back as 1982 genetic engineers solved the production end for one drug—insulin—by splicing a human insulin gene into the DNA of the common intestinal bacterium, *E. coli*. As enormous quantities of the bacteria grew in tanks, they turned out large amounts of human insulin along with their own protein products, and a vast industry was born. Today, many drugs and vaccines are prepared via biotechnology, essentially the same process, aside from the gene-splicing, that spurs yeast to make bread.

Geneticists' Fort Knox: Gingerly, a gloved genetic engineer (above) organizes a gene bank, deep-freezing a packet of tubes that collectively contain all the DNA in a human cell. Sample tube (opposite) stores a particular region of DNA found on one of the 46 chromosomes that normally occur in human cells.

But designing drugs from scratch is not simply a matter of brewing or baking. Consider, first, the huge cost: an average of $360 million to develop a new drug out of the thousands of chemical compounds that must be made and tested. Modern medicine and pharmacology, now more than ever, must go hand in hand with molecular and cellular biology.

Any compound that can adhere to and disrupt a disease-causing substance may have value, but medicinal chemists often search for a molecule's smallest functional unit because smaller molecules are easier to deliver as drugs. Proteins are the workhorse molecules of all biological systems, so drug-designers often focus on them, particularly on their structure since this is quite often associated with function. Scientists can determine the order of amino acids in a protein fairly easily. But coming up with a three-dimensional image of that protein can be a gigantic task—even with the help of computers— since it's not always clear just how proteins fold into their final, active structure. Yet knowing the folded shape can be crucial to understanding a protein.

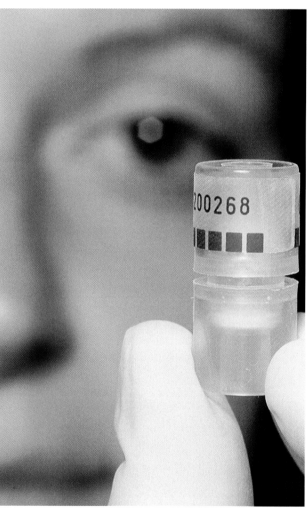

magnetic resonance spectroscopy, protein shape can be deduced from how atoms in a molecule interact. Once protein shape is determined and rendered into a computerized image, models of potentially useful drug molecules are directed toward it by a pharmacologist's joystick. If the pieces of the model drug fit the receptor, a drug is designed. It may be a synthetic protein to replace a malfunctioning one or, since drugs act on specific enzyme conformations, one that interferes with an enzyme that forces blood pressure to rise or that causes arthritis.

Monoclonal antibodies are one variation of a host of immune system proteins that protein engineers are trying to make more active and more selective in their docking sites. For example, a tumor-fighting white blood cell called a tumor infiltrating lymphocyte (TIL) can be made more potent by giving it a gene that codes for another tumor-destroying substance, tumor necrosis factor (TNF). One day genetically engineered TIL cells might be returned to a patient, where they would invade the tumor and produce TNF to destroy it. It may be impossible to predict whether techniques now being developed will mean cures for cancer and AIDS, but researchers know that the answers lie at the molecular level, and that future physicians will possess an unprecedented arsenal of weapons with which to launch an effective strike.

Jelling the bright vision of 21st-century health care will require more ingredients than designer drugs, alternative therapies, transplants, implants, and bloodless surgery. Realizing that vision will take more, too, than data from the Human Genome Project, as awesome as its

DESIGNER DRUGS

A target molecule may be a protein receptor on a virus or a cellular membrane. It may be a disease-linked enzyme; sometimes its shape can be inferred from the shape of a drug known to interact with it. Using nuclear

impact will be, and more than a better understanding of the molecular basis of disease and of the complexities of the human body. A large gap remains between today's and tomorrow's medical science, one that can be bridged only by commitment and education. The intrusions of technology and market forces into medicine have widened distances between health-care professionals and patients. Physicians now have less time to spend with their charges. Throughout the world, more than 12 million children under age five die every year, 3 million of them before they are a week old. Two million of those deaths could be prevented with vaccines, but too many children still slip through the net. Cloned sheep and genes may provide us with better things for better living, but such feats raise serious moral, ethical, social, and legal questions if they are extended to the cavalier manufacture of new human lives. Xenotransplants may well become the grafts of choice, but the possibility, even remote, of the emergence of new viruses that could be transmitted from animals to humans cannot be taken lightly, as many researchers appear to do. The absence of women in biomedical studies of disease and the short shrift given to how some diseases affect women differently than men are not the results of chivalry; they are realities because women have traditionally meant only fertility and reproduction to generations of investigators.

Medical advances promise a bright future. Hopefully, sensitivity and social awareness will accompany those advances. British writer Aldous Huxley's 1946 novel *Brave New World* describes a future without them. Created by excess of scientific and social development, it is a stark, sterile, and emotionless existence in which each person's every move is programmed.

There are, fortunately, encouraging signs of at least heightened awareness, and of some improvements. A number of medical schools, aware that technology cannot replace a skilled and compassionate physician, require that students have direct contact with patients in their first year, not waiting until the more traditional third and fourth. Increasingly, science is being blended with clinical issues, and the ethical ramifications of a doctor's and a researcher's work are part of the core curricula. Recent surveys of medical students exposed to such teaching methods indicate that they relate better to patients, collect more personal information, and generally act in a more empathetic manner.

Launched in 1990, the Children's Vaccine Initiative—brainchild of WHO, UNICEF, UNDP, the World Bank, and the Rockefeller Foundation—now pursues a radical agenda to improve the global supply and quality of vaccines, spur research into new ones, and develop strategies to ensure that immunization is affordable for use in developing countries. The Human Genome Project devotes 5 percent of its budget to research geared toward anticipating and resolving ethical, legal, and social issues likely to arise from genetic research. Commendably, this is one of the few times scientists have paid attention to the potential consequences of their research before a crisis arises. To address the fear that high-tech medical devices are often too complicated for the doctors, nurses, and technicians who use them—which sometimes creates new opportunities for error—the American Medical Association, assisted by concerned physicians and engineers, has created a National Patient Safety Foundation to improve interaction between new medical systems and their human operators.

Then there is the Women's Health Initiative, a 688-million-dollar, 15-year project sponsored by NIH, which is attempting to redress many of the inequities in women's health research. Involving 164,500 women between the ages of 50 and 79, it is one of the most definitive clinical trials of women's health ever undertaken in the U.S. When it is concluded, an enormous amount of practical information will be available about hormone replacement therapy, calcium and vitamin D supplements, and dietary patterns and their effects on the prevention of heart disease, cancer, and osteoporosis.

Certainly, the Women's Health Care Initiative, medical school curriculum changes, and an aroused social consciousness among researchers were long in coming. But they are here now. While their importance to health care in the new millennium may not be as compelling or as fascinating as the intricate inner workings of the human body, they will ensure that the grim vision of *Brave New World* remains, as Huxley intended it, a satire rather than a reality.

....

Genetic makeover: Animal cells are reprogrammed with microinjections of foreign genes, a step toward understanding gene expression and control. Ultimately, gene transfer will help correct human genetic diseases. FOLLOWING PAGES: Dolly, the world's first mammal successfully cloned from a single cell taken from an adult, looks out on a brave new world.

....

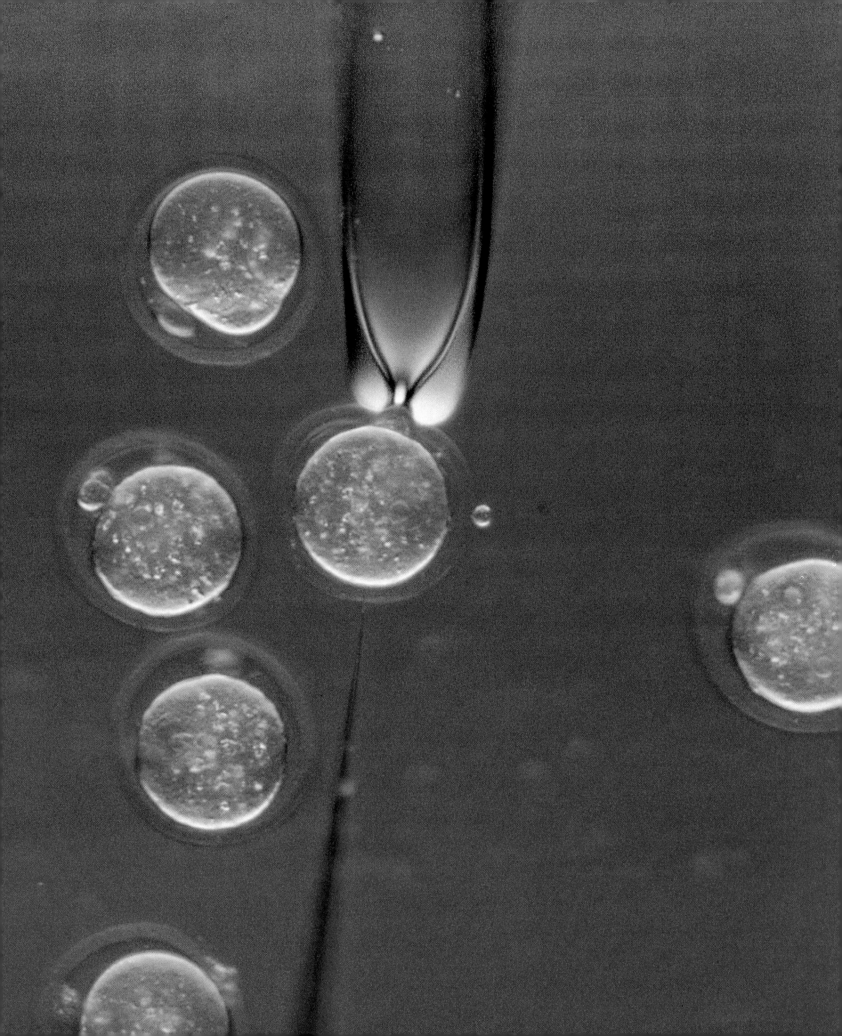

AUTHORS' NOTES

Sherwin B. Nuland (FOREWORD) is Clinical Professor of Surgery at the Yale School of Medicine. He is the author of *How We Die: Reflections on Life's Final Chapter*, which won the National Book Award in 1995, and, in 1997, *The Wisdom of the Body*, in addition to numerous articles and essays for general and medical readers. His column, "The Uncertain Art," appears regularly in the *American Scholar*.

Boyce Rensberger (ORIGINS) is a longtime science writer who has worked for the *New York Times* and the *Washington Post*. He has written four books, one of which—*Life Itself: Exploring the Realm of the Living Cell*—deals with many of the same topics as the "Origins" chapter. Rensberger currently directs the Knight Science Journalism Fellowship program at M.I.T.

Carole Howard (SYSTEMS) is the author of *Dolphin Chronicles*, a book about her research with dolphins, and numerous magazine articles. **Joel Shurkin,** a former science editor of the *Philadelphia Inquirer* and Reuters, and a science writer at Stanford University, has authored nine books. The two live in northern California with their daughter.

Lisa Marguerite Krieger (BODILY DEFENSES) is an award-winning journalist in the San Francisco Bay area, where for 14 years she has covered medical news for the *San Francisco Examiner* and the *San Jose Mercury News*. She lives with her daughter in Los Gatos, California.

Roberta Conlan (BRAIN) is an editor and writer whose work has appeared in publications for National Geographic Books, the National Academy of Sciences, and Sunset Books. She is a former senior editor at Time-Life Books, where she directed various series on neuroscience, computers, and astronomy.

Steven N. Austad (AGING) is a prize-winning research gerontologist and professor at the University of Idaho. Author of more than 50 scientific papers, he also writes about science for the general public in journals such as *Natural History, International Wildlife, Scientific American,* and others. His recent book, *Why We Age,* will soon be available in five languages, in addition to English. He is also a science advisor for National Public Radio.

Science journalist and author **John Langone** (NEW DIRECTIONS) is a former staff writer for *Discover* and *Time* magazines. He was a Kennedy Fellow in medical ethics at Harvard University and a Fellow at the Center for Advanced Study in the Behavioral Sciences at Stanford University. As a Fulbright Fellow at the University of Tokyo, he examined the social impacts of technology.

ACKNOWLEDGMENTS

The Book Division wishes to thank the many individuals, groups, and organizations mentioned or quoted in this publication for their help and guidance. In addition, we are especially grateful to the overall consultants, Sherwin B. Nuland and William Knaus, and to the following: The National Institutes of Health; Charlotte Catz, National Institute of Child Health; Rex L. Chisholm, Northwestern University Medical School; Charles Delgado; Sharon Durham, National Human Genome Research Institute; Caleb Finch, University of Southern California; David Harrison, Jackson Labs; Jonni Henofer, University of Virginia; Brenda R. Kiessling; Amy Koozer, University of Virginia; Ed Lakatta, National Institute of Aging; Donna M. Lindemuth; Sue Moenter; Robert L. Nicholls; Janice A. O'Brien; Richard Tasca, National Institutes of Health; Dan Theodorescu, University of Virginia; U.S. Centers for Disease Control and Prevention; Mary Lee Vance, University of Virginia.

ADDITIONAL READING

The reader may wish to consult the National Geographic Index for related articles and books, including *The Incredible Machine.* The following sources may also be of interest:
ORIGINS : Boyce Rensberger, *Life Itself: Exploring the Realm of the Living Cell*; Bruce Alberts, et al., *Molecular Biology of the Cell*; Keith L. Moore and T.V.N. Persaud, *The Developing Human: Clinically Oriented Embryology*; James Darnell, Harvey Lodish, and David Baltimore, *Molecular Cell Biology*; Ernst Mayr, *The Growth of Biological Thought: Diversity, Evolution, and Inheritance*; Clifford Grobstein, *Science and the Unborn*.
SYSTEMS : Curt Suplee, *Everyday Science Explained*; Alvin Silverstein, *Human Anatomy and Physiology*; James Bevan, *The Simon and Schuster Handbook of Anatomy and Physiology*; Marvin R. Barnum, *Human Form and Function, A Basic Approach*; Charles Clayman, *The Human Body, An Illustrated Guide to Its Structure, Function, and Disorders*.
BODILY DEFENSES : William R. Clark, *The New Healers*; Bernard Dixon, *Power Unseen*; Jeffrey A. Fisher, *The Plague Makers*; Arlo Karlen, *Man and Microbes*; Peter Radetsky, *The Invisible Invaders*; Barry E. Zimmerman and David J. Zimmerman, *Nature's Curiosity Shop*.
BRAIN : Roberta Conlan, ed., *Journey Through the Mind and Body*; Ronald Kotulak, *Inside the Brain: Revolutionary Discoveries of How the Mind Works*; Joseph LeDoux, *The Emotional Brain: The Mysterious Underpinnings of Emotional Life*; Daniel L. Schacter, *Searching for Memory: The Brain, the Mind, and the Past*; J. Allen Hobson, *The Chemistry of Conscious States: How the Brain Changes Its Mind*.
AGING : Steven Austad, *Why We Age*; Robert Arking, *Biology of Aging*; Caleb E. Finch, *Longevity, Senescence, and the Genome*; W. R. Hazzard, et al., ed., *Principles of Geriatric Medicine and Gerontology*; Edward L. Schneider, ed., *Handbook of the Biology of Aging*.
NEW DIRECTIONS : Anthony DiGioia, et al., *Robotics and Computer Assisted Medical Interventions*; R. A. Robb and B. Cameron, *Interactive Technology and the New Paradigm for Health Care*; Russell Taylor, et al., eds., *Computer Integrated Surgery*; Workshop on Alternative Medicine, *Alternative Medicine: Expanding Medical Horizons*.

ILLUSTRATIONS CREDITS

FRONT MATTER : 1 Nancy Kedersha/SPL/Photo Researchers; 2-3 Prof. P. Motta/SPL/Custom Medical Stock Photo; 4-5 Kenneth Jarecke/ Contact Press Images; 6 Peter Menzel; 10-11 Scott Camazine/Brian Camazine

ORIGINS : 12-13 Tim Bieber/The Image Bank; 14 Lennart Nilsson; 16 (upper) SPL/Photo Researchers; 16 (center) Dr. Dennis Kunkel/Phototake; 16 (lower) Alfred Pasieka/SPL/Photo Researchers; 17 (upper left) Jason Burns/Ace/Phototake; 17 (upper right) Lennart Nilsson; 17 (center) Michael Abbey/Photo Researchers; 17 (lower left) CNRI/SPL/Photo Researchers; 17 (lower right) Prof. P. Motta/Photo Researchers; 18 Lennart Nilsson; 19 Driscoll, Youngquist and Baldeschwieler/Caltech/SPL/Custom Medical Stock Photo; 23 Superstock; 24-25 David L. Arnold; 26 Hattie Young/SPL/ Photo Researchers; 28 (all), 29 Conly L. Rieder/Tony Stone Images; 30 Superstock; 31 Alison Wright/Photo Researchers; 33 (left) David Parker/ SPL/Photo Researchers; 33 (right) Michael Siegel/Phototake; 34 (all) Lennart Nilsson; 35 Superstock; 36 (all) C. Edelmann/La Villette/Photo Researchers; 37 Lennart Nilsson; 38 J.L. Carson/Custom Medical Stock Photo; 39 Lennart Nilsson; 40 Henry Sathananthan/Tony Stone Images; 41 Lennart Nilsson; 42, 43 (upper left and upper right) Dr. Nikas/Jason Burns/Phototake; 43 (center) Lennart Nilsson; 43 (lower) Dr. Nikas/Jason Burns/Phototake; 44,

45, 46 (all), 47, 49, 50, 51 Lennart Nilsson; 55, 56-57 Neil Bromhall/Oxford Scientific Films; 58 Matt Meadows/SPL/Photo Researchers; 60 Superstock; 62-63 Custom Medical Stock

SYSTEMS: 64-65 Robb Kendrick; 66 Centre Jean-Perrin Clermont-Ferrand/CNRI/Phototake; 68 Scott Camazine/Sue Trainor; 69 Prof. P. Motta/SPL/Photo Researchers; 70 Graham French/Masterfile; 71 Professors P.M. Motta, P.M. Andrews, K.R. Porter and J. Vial/SPL/Photo Researchers; 73 (left) Quest/SPL/Photo Researchers; 73 (right) G.W. Willis/Tony Stone Images; 74, 75 SPL/Photo Researchers; 76 Scott Camazine/Sue Trainor; 77 (upper) Darryl Torckler/Tony Stone Images; 77 (lower left) Gould/DeAnza/Peter Arnold, Inc.; 77 (lower right) Darryl Torckler/Tony Stone Images; 78-79 CNRI/Phototake; 80 (left) Prof. Castano/ Phototake; 80 (center) David Phillips/SS/Photo Researchers; 80 (right) Quest/SPL/Photo Researchers; 81 Omikron/SS/Photo Researchers; 82-83 David Burnett/Contact Press Images; 84 (upper) Martin Dohrn/Royal College of Surgeons/SPL/Photo Researchers; 84 (lower), 85 Prof. P. Motta/SPL/Photo Researchers; 86, Cary Wolinsky; 87, Alfred Pasieka/SPL /Photo Researchers; 89 Lennart Nilsson; 90-91 Robert W. Madden/NGS; 92 Jerome Tisne/Tony Stone Images; 94 Scott Camazine/Sue Trainor; 95 Alexander Tsiaras/SPL/Photo Researchers; 96-97, 97 (upper) Lennart Nilsson; 97 (lower) Alfred Pasieka/SPL/Photo Researchers; 98 (both) Prof. P. Motta/SPL/Photo Researchers; 99 Jan Lindberg; 102 Dennis Kunkel/CNRI/Phototake; 102-103 Lennart Nilsson; 104 CNRI/SPL/Photo Researchers; 105 (left) Professors P.M. Motta and T. Fujita/SPL/Photo Researchers; 105 (right) Lennart Nilsson; 106-107 Michael S. Yamashita; 109 CNRI/SPL/Photo Researchers; 110-111 Cary Wolinsky; 112 Biophoto Associates/Photo Researchers; 113 Richard Wehr/Custom Medical Stock Photo; 115 Sheila Terry/SPL/Photo Researchers

BODILY DEFENSES: 116-117 Thomas B. England/Photo Researchers; 118 Superstock; 120 Gallerie Dell'Accadamia Venice; 121 Nicole Duplaix; 122-123 Museo Del Prado, Madrid; 124 Biology Media/SPL/Photo Researchers; 127 Philadelphia Museum of Art, Smithkline Beckman Corporation Fund; 128 (upper) G.I. Bernard/Oxford Scientific Films; 128 (lower) Dr. Gopal Murti/SPL/Photo Researchers; 129 London School of Hygiene/SPL/Photo Researchers; 130 Dr. Gary Settles/SPL/Photo Researchers; 131 Graham French/Masterfile; 132-133 Prof. P. Motta/Dept. of Anatomy/University "La Sapienza." Rome/SPL/Photo Researchers; 133 Lennart Nilsson; 135 (upper left) 135 (upper right) Erich Schrempp/Photo Researchers; 135 (lower) Quest/SPL/Photo Researchers; 136-137 CNRI/SPL/Photo Researchers; 138, 139 Lennart Nilsson ©Boehringer-Ingelheim International GmbH; 141 Microworks/CNRI/Phototake; 142 (left) Lennart Nilsson ©Boehringer-Ingelheim International GmbH; 142 (right) Professors S.H.E. Kaufmann and R.J.R. Golecki/SPL/Photo Researchers; 143 SIU Medical Communications/Photo Researchers; 144 Andrew Syred/Tony Stone Images; 145 Tom McHugh/Photo Researchers; 146-147 Martha Cooper and Joan Nowicke/Smithsonian Institution; 148 Jeff Cadge/The Image Bank; 152-153 Karen Kasmauski; 154 (all) Petit Format/C. Dauguet/C. Edelmann/Photo Researchers; 156-157 Jonathan Nourok/Tony Stone Images; 160 Donato Pineider/Biblioteca Medicea Laurenziana, Italy; 161 Karen Kasmauski; 162-163 Stuart Franklin; 164-165 Karen Kasmauski; 166 Michael S. Yamashita; 169 Superstock; 171 CNRI/SPL/Photo Researchers; 172 (upper left) A.B. Dowsett/SPL/Photo Researchers; 172 (upper right) Andrew Syred/Tony Stone Images; 172 (center upper left) CNRI/SPL/Photo Researchers; 172 (center upper right) M. Abbey/Photo Researchers; 172 (center lower left) Petit Format/J.P. Mornon/C. Edelmann/Photo Researchers; 172 (center lower center) NIBSC/SPL/Photo Researchers; 172 (center lower right) Hans Gelderblom/Tony Stone Images; 172 (lower left) CNRI/SPL/Photo Researchers; 172 (lower right) C.V.L. Weybridge/SPL/Photo Researchers; 173 (upper left), 173 (upper right) A.B. Dowsett/SPL/Photo Researchers; 173 (center upper left) Superstock; 173 (center upper right) Oliver Meckes/E.O.S./Gelderblom/Photo Researchers; 173 (center lower left) Barry Dowsett/SPL/Photo Researchers; 173 (center lower right) CNRI/SPL/Photo Researchers; 173 (lower left) NIBSC/SPL/Photo Researchers; 173 (lower right) Lennart Nilsson ©Boehringer-Ingelheim International GmbH; 175 London Scientific Films/Oxford Scientific Films

BRAIN: 176-177 V.I. Lab E.R.I.C./FPG; 178 Joe McNally; 180 Alex Grey/Peter Arnold, Inc.; 181 Roy Ooms/Masterfile; 182-183 Manfred Kage/Peter Arnold, Inc.; 183 Don Fawcett/Science Source/Photo Researchers; 184-185 Grafton M. Smith/The Stock Market; 186 Scott Camazine/Sue Trainor; 187 (upper left) Ed Reschke/Peter Arnold, Inc.; 187 (upper center) J&L Weber/Peter Arnold, Inc.; 187 (upper right) Dr. Dennis Kunkel/Phototake; 187 (center left) Scott Camazine/Sue Trainor; 187 (center upper right) Manfred Kage/Peter Arnold, Inc.; 187 (center lower right) V.I. Lab E.R.I.C./FPG; 187 (lower) Don Fawcett/Science Source/Photo Researchers; 190-191 Quest/SPL/Photo Researchers; 192 Bruce Rowell/Masterfile; 193 Quest/SPL/Photo Researchers; 194 Louie Psihoyos; 195 Professors P. Motta and T. Nagura/SPL/Photo Researchers; 196-197 Karen Kasmauski; 198 (left) Dave Roberts/SPL/Photo Researchers; 198 (center) Dr. G. Oran Bredberg/SPL/Photo Researchers; 198 (right) Professors P.M. Motta and A. Caggiati/SPL/Photo Researchers; 199 Tony Stone Images; 202 Bill Aron/Tony Stone Images; 203 Richard H. Smith/FPG; 204 (left) Bruce Ayres/Tony Stone Images; 204 right, 205 Telegraph Colour Library/FPG; 208-209 Alfred Pasieka/SPL/Photo Researchers; 209 M.W. Davidson/Photo Researchers; 210 Faye Norman/SPL/Photo Researchers; 212-213 CNRI/SPL/Photo Researchers; 214, 215 Scott Camazine/Sue Trainor; 216 Tim Beddow/SPL/Photo Researchers; 216-217, 218-219 Joe McNally; 221 Sinclair Stammers/SPL/Photo Researchers; 222 (upper) Dr. Lewis Baxter/Peter Arnold, Inc.; 222 (lower) Wellcome Dept. of Cognitive Neurology/SPL/Photo Researchers; 224-225 Joe McNally; 227 Michael Williamson ©1995 the Washington Post, Reprinted with permission; 228 (upper left), 228 (upper right), 229 (upper left), 229 (upper right) Ted Spagna/SPL/Photo Researchers; 228 (lower left), 228 (lower center) Hank Morgan/Photo Researchers; 228 (lower right) Hank Morgan/Mt. Sinai School of Medical Research/Rainbow; 229 James Holmes/SPL/Photo Researchers; 230-231 Dr. Ray Clark and Mervyn Goff/SPL/Photo Researchers; 234-235, 236 Steve McCurry; 237 Joe McNally

AGING: 238-239 Joel Sartore; 241 John Launois/Black Star; 243 Peter Lansdorp/Terry Fox Laboratory, BC Cancer Research Center, Vancouver, BC; 244 D.R. & T.L. Schrichte/Tony Stone Images; 246-247 Karen Kasmauski; 248 Dr. Gopal/SPL/Photo Researchers; 250-251 Frank Siteman/Tony Stone Images; 252 (both) Dan McCoy/Rainbow; 254 Frank Fournier/Contact Press Images; 256 upper) Superstock; 256 (lower) Scott Camazine/Sue Trainor; 257 (both) Superstock; 258 David Alan Harvey; 260 Denise Hesse/Phototake; 261 (all) National Eye Institute/National Institutes of Health; 262 AlfredPasieka/SPL/Photo Researchers; 264-265 Karen Kasmauski; 266 CNRI/SPL/Photo Researchers; 267 Karen Kasmauski; 268-269 (all) Lennart Nilsson; 272-273 Rob Crandall/Rainbow; 274-275 CNRI/SPL/Photo Researchers; 275 Lunagrafix/Photo Researchers; 277 Kevin Horan; 280-281 Karen Kasmauski; 285 E. Robert/Sygma

NEW DIRECTIONS: 286-287 George Steinmetz; 288 NASA; 289 Ken Eward/Biografx/Photo Researchers; 290 Thierry Dosogne/The Image Bank; 293 Scott Camazine/Sue Trainor; 294-295 Robert Chase/Photo Researchers; 298 Superstock; 299 Joe McNally/Life Magazine, ©Time Inc.; 302-303, 304-305 Joe McNally; 307 Max Aguilera-Hellweg, Institute for the Advancement of Bloodless Medicine and Surgery, Englewood Hospital and Medical Center; 308-309 Roger Ressmeyer; 313 P. Plailly/Eurelios/Phototake; 314-315 Gerd Ludwig; 316 Hank Morgan/Rainbow; 320 Organogenesis, Inc.; 321 Max Aguilera-Hellweg; 322-323 Enrico Ferorelli; 325 NovaCare Sabolich; 326 Kenneth Jarecke/Contact Press Images; 327 C. Pouedras/Eurelios/SPL/Photo Researchers; 330, 331 M. Iwafuji/Eurelios/Phototake; 334-335 Phillip Hayson/Science Source/Photo Researchers; 336 Mehau Kulyk/SPL/Photo Researchers; 338 Superstock; 339 Klaus Guldbrandsen/SPL/Photo Researchers; 341 M. Baret/Science Source/Photo Researchers; 342-343 Remi Benali and Stephen Ferry/Life Magazine/Gamma Liaison

DUST JACKET: Front: Dan McCoy/Rainbow. Back: (upper left) Centre Jean-Perrin Clermont-Ferrand/CNRI/Phototake; (upper right) Neil Bromhall/Oxford Scientific Films; (right center) Phillip Hayson/SS/Photo Researchers; (lower left) Superstock; (lower right) Joe McNally/Life Magazine, © Time Inc.

Library of Congress Cataloging-in-Publication Data

Incredible voyage : exploring the human body / prepared by the Book Division, National Geographic Society.
    p.   cm.
    Includes index.
    ISBN 0-7922-7148-3 (reg). —ISBN 0-7922-7149-1 (dlx)
    1. Human physiology—Popular works. 2. Human anatomy—Popular works. I. National Geographic Society (U.S.). Book Division.
QP38.I536 1998
612—dc21

98-4635
CIP

CREDITS

Published by
The National Geographic Society

John M. Fahey, Jr.
President and Chief Executive Officer

Gilbert M. Grosvenor
Chairman of the Board

Nina D. Hoffman
Senior Vice President

Prepared by
The Book Division

William R. Gray
Vice President and Director

Charles Kogod
Assistant Director

Barbara A. Payne
Editorial Director and Managing Editor

David Griffin
Design Director

Staff for this book

Tom Melham
Managing Editor

Toni Eugene
Text Editor

Greta Arnold
Illustrations Editor

Cinda Rose
Art Director

Rebecca Beall Barns
Sallie M. Greenwood
Kimberley A. Kostyal
Joyce Marshall
Researchers

Steven N. Austad
Catherine Howell
Lisa M. Krieger
John M. Langone
Tom Melham
Boyce Rensberger
Picture Legends

Mimi Harrison
Picture Researcher

Suez B. Kehl
Assistant Designer

Julie Mao
Research Assistant

Peyton H. Moss, Jr.
Dust Jacket Copy

R. Gary Colbert
Production Director

Lewis R. Bassford
Production Project Manager

Richard S. Wain
Production

Janet A. Dustin
Illustrations Assistant

Peggy J. Candore
Dale-Marie Herring
Staff Assistants

Anne Marie Houppert
Indexer

Manufacturing and Quality Management

George V. White
Director

John T. Dunn
Associate Director

Polly P. Tompkins
Executive Assistant

Printed and bound by R. R. Donnelley & Sons, Willard, Ohio.
Color separations by CMI Color Graphix, Inc., Huntingdon Valley, Pennsylvania.
Dust jacket printed by Miken Companies, Inc., Cheektowaga, New York.
Visit the Society's Web site at www.nationalgeographic.com